APPLIED STATISTICS
AND THE SAS®
PROGRAMMING LANGUAGE

APPLIED STATISTICS AND THE SAS® PROGRAMMING LANGUAGE

Fifth Edition

Ronald P. Cody
Robert Wood Johnson Medical School

Jeffrey K. Smith
Rutgers University

PEARSON

Prentice Hall

Upper Saddle River, New Jersey 07458

Library of Congress Cataloging-in-Publication Data

Cody, Ronald P.
  Applied statistics and the SAS programming language / Ronald P. Cody, Jeffrey K.
Smith — 5th ed.
    p. cm.
  Includes bibliographical references and index.
  ISBN 0-13-146532-5
    1. SAS (Computer file) 2. Mathematical statistics — Data processing. I. Smith, Jeffrey K.
  II. Title.
QA276.4.C53 2006
519.5′ 0285′ 536 — dc22   2004060178

Executive Acquisitions Editor: *Petra Recter*
Editor-in-Chief: *Sally Yagan*
Project manager: *Jacqueline Riotto Zupic*
Full-Service Project Manager: *Heather Meledin/Progressive Publishing Alternatives*
Total Concept Coordinator: *Lynda Castillo*
Assistant Manufacturing Manager/Buyer: *Michael Bell*
Marketing Assistant: *Rebecca Alimena*
Editorial Assistant: *Joanne Wendelken*
Art Director: *Jayne Conte*
Director of Creative Services: *Paul Belfanti*
Cover Designer: *Bruce Kenselaar*
Cover Image: *Getty Images*

 © 2006, 1997, 1991 Pearson Education, Inc..
Pearson Prentice Hall
Pearson Education, Inc.
Upper Saddle River, NJ 07458

Published by Elsevier Science Publishing Co., Inc. 1985, and 1987.

Pearson Prentice Hall® is a trademark of Pearson Education, Inc.

Printed in the United States of America

10 9 8 7 6 5 4 3

ISBN: 0-13-146532-5

Pearson Education LTD., *London*
Pearson Education Australia PTY, Limited, *Sydney*
Pearson Education Singapore, Pte. Ltd
Pearson Education North Asia Ltd, *Hong Kong*
Pearson Education Canada, Ltd., *Toronto*
Pearson Educación de Mexico, S.A. de C.V
Pearson Education—Japan, *Tokyo*
Pearson Education Malaysia, Pte. Ltd

Contents

Chapter 20   Syntax Examples   495

Solutions to Odd-Numbered Problems

Index

Preface to the Fifth Edition

As we went about creating this fifth edition, several facts became clear: First, SAS® continues to evolve and improve. Second, our programming techniques have also improved. Third, we felt it was time to replace all the text-based graphics with SAS Graph™ output.

We have met many readers of earlier editions at meetings and conferences and were delighted to hear good things and constructive criticisms of the book. These we have taken to heart and attempted to improve old material and add relevant new topics. This fifth edition is the result of such reader reaction.

Most researchers are inherently more interested in the substance of their research endeavors than in statistical analyses or computer programming. Yet, conducting such analyses is an integral link in the research chain: all too frequently, the weak link. This condition is particularly true when there is no resource for the applied researcher to refer to for assistance in running computer programs for statistical analyses. *Applied Statistics and the SAS Programming Language* is intended to provide the applied researcher with the capacity to perform statistical analyses with SAS without wading through pages of technical documentation.

The reader is provided with the necessary SAS statements to run programs for most of the commonly used statistics, explanations of the computer output, interpretations of results, and examples of how to construct tables and write up results for reports and journal articles. Examples have been selected from business, medicine, education, psychology, and other disciplines.

We should mention that you could run some statistical analyses with SAS and even manipulate your data using the Enterprise Guide™ that is an interactive "front-end" to SAS. If you are using the SAS Learning Edition™, (Version 9 or above), or you have a separate license to the Learning Edition, you may want to give it a try. There are those of us who have a bias against interactive front-ends like the Enterprise Guide and prefer to do our own programming. However, the Enterprise Guide is quite powerful and you can even see the SAS code it generates. You may find it particularly useful for straight-forward statistical analyses and an easy way to get started with SAS Graph™.

We like to describe SAS as a combination of a statistical package, a data base management system, and a high-level programming language. Like SPSS, BMDP, Systat, and other statistical packages, SAS can be used to describe a collection of data and produce a variety of statistical analyses. However, SAS is much more than just a statistical package.

Many companies and educational institutions use SAS as a high-level data-management system and programming language. It can be used to organize and transform data and to create reports of all kinds. Also, depending on which portions of the SAS system you have installed on your computer (and depending on what type of computer system you are running), you may be using the SAS system for interactive data entry or an on-line system for order entry or retrieval.

The SAS system is a collection of products, available from the SAS Institute in Cary, North Carolina. The major products available from the SAS Institute are:

Base SAS® The main SAS module that provides some data manipulation and programming capability and some elementary descriptive statistics

SAS/Stat® The SAS product that includes all the statistical programs except the elementary ones supplied with the base package

SAS/Graph® A package that provides high quality graphs and maps. Note that "line graphics" (the graphs and charts that are produced by normal character plots, are available in the base and SAS/STAT packages. SAS/GRAPH adds the ability to produce high quality camera-ready graphs, maps and charts.

SAS/FSP® The initials stand for the Full Screen Product. This package allows you to search, modify, or delete records directly from a SAS data file. It also provides for data entry with sophisticated data checking capabilities. Procedures available with FSP are FSBROWSE, FSEDSIT, FSPRINT, FSLIST, and FSLETTER.

SAS/AF® AF stands for the SAS Applications Facility. Data processing professionals use this product to create "turn-key" or menu systems for their users. It is also used to create instructional modules relating to the SAS system.

SAS/ETS® The Econometric and Time Series package. This package contains specialized programs for the analysis of time-series and econometric data.

SAS/OR® A series of operations research programs.

SAS/QC® A series of programs for quality control.

SAS/IML® The Interactive Matrix Language module. The facilities of IML used to be included in PROC MATRIX in the version 5 releases. This very specialized package allows for convenient matrix manipulation for the advanced statistician.

SAS, SAS/STAT, SAS/GRAPH, SAS/FSP, SAS/AF, SAS/ETS, SAS/OR, SAS/QC, and SAS/IML are registered trademarks or trademarks of SAS Institute Inc. in the USA and other countries.

Learning to program is a skill that is difficult to teach well. While you will find our examples easy to read and their logic easy to follow, learning to write your own programs is another matter. Only through lots of practice will your programming skills develop. So, when you finish a chapter, please spend the time doing as many problems as you can. We wish you happy programming.

RON CODY
JEFFREY SMITH

You can download all the programs and data files used in this book (as well as data for the problems at the end of the chapter) by visiting the website: **www.prenhall.com/cody**

Acknowledgments

We owe special thanks to Mike Zdeb and Paul Grant who thoroughly reviewed the fifth edition. It was hard work and we really appreciate it.

It also takes a professional and dedicated team to bring a book from concept to the store shelf. The members of the Prentice Hall team that we would like to thank are: Acquisition Editor—Petra Recter; Total Concept Coordinator—Lynda Castillo; Editorial Assistant—Joanne Wendelken; Project Managers—Jacquelyn Riotto Zupic and Michael Bell. Heather Meledin of Progressive Publishing Alternatives managed the book through all production stages.

APPLIED STATISTICS
AND THE SAS®
PROGRAMMING LANGUAGE

A SAS Tutorial

A.   INTRODUCTION

For the novice, engaging in statistical analysis of data can seem as appealing as going to the dentist. If that pretty much describes your situation, perhaps you can take comfort in the fact that this is the fifth edition of this book—meaning that the first four editions sold pretty well, and this time we may get it right. Our purpose for this tutorial is to get you started using SAS. The key objective is to get one program to run successfully. If you can do that, you can branch out a little bit at a time. Your expertise will grow.

SAS is a combination of programs originally designed to perform statistical analysis of data. Other programs you may have heard of are SPSS, BMDP, or SYSTAT. If you look at personal computer magazines, you might run across other programs primarily designed to run on personal computers. Since its inception, SAS has grown to where it can perform a fairly impressive array of nonstatistical functions. We'll get into a little of that in later chapters. For now, we want to learn the most basic rudiments of the SAS system. If you skipped over it, the Preface to the fifth edition contains some history of SAS software development and a more complete overview of the capabilities of SAS software.

To begin, SAS runs on a wide variety of computers and operating systems (computer people call these platforms), and we don't know which one you have. You may have a personal computer running a Windows operating system (such as Windows XP or Windows 2002). You may be running UNIX or LINUX, or you may be connected to a network or a mainframe computer by way of a modem or network connection. You may only have a sophisticated VCR, which you think is a computer. If you are unsure of what platform you are using or what version of SAS you are using, ask someone. As a matter of fact, right now would be a good time to invite the best computer person you know to lunch. Have that person arrive at your office about an hour before lunch so you can go over some basic elements of your system. You need to find out what is necessary on your computer to get SAS running. What we can teach you here is how to use SAS and how to adapt to your computer system.

If you are running on a mainframe, you may well be submitting what are called "batch" jobs. When you run batch jobs, you send your program (across a phone line or a network from your personal computer or terminal) to the computer. The computer runs your program and holds it until you ask for it, or it prints out the results on a high-speed printer. You may even need to learn some Job Control Language (which you have to get from your local computer folks), and then you can proceed.

If you are running on a personal computer, or running in what is called interactive mode on a minicomputer or mainframe, then you need to learn how to use the SAS Display Manager or the Enterprise Guide (a "front end" for SAS on Windows platforms that allows you many point-and-click operations as well as the ability to write SAS programs). The look and feel of SAS, once you are in the Display Manager, is pretty much the same, whatever platform you are using. If you are undaunted, take a deep breath and plunge into the real content in the next section. If you are already daunted, take a minute and get that lunch scheduled, then come back to this.

You can download all the programs and data files used in this book, as well as the data for the problems at the end of the chapters, by visiting the web site: **www.prenhall. com/cody**

B.   COMPUTING WITH SAS: AN ILLUSTRATIVE EXAMPLE

SAS programs communicate with the computer by means of SAS "statements." There are several kinds of SAS statements, but they share a common feature—they end in a semicolon. A semicolon in a SAS program is like a period in English. Probably the most common error found in SAS programs is the omission of the semicolon. This omission causes the computer to read two statements as a run-on statement and invariably fouls things up.

SAS programs are comprised of SAS statements. Some of these statements provide information to the system, such as how many lines to print on a page and what title to print at the top of the page. Other statements act together to create SAS data sets, while other SAS statements act together to run predefined statistical or other routines. Groups of SAS statements that define your data and create a SAS data set are called a DATA step; SAS statements that request predefined routines are called a PROC (pronounced "prock") step. DATA steps tell SAS programs about your data. They are used to indicate where the variables are on data lines, what you want to call the variables, how to create new variables from existing variables, and several other functions we mention later. PROC (short for PROCEDURE) steps indicate what kind of statistical analyses to perform and provide specifications for those analyses. Let's look at an example. Consider this simple data set:

SUBJECT NUMBER	GENDER (M or F)	EXAM 1	EXAM 2	HOMEWORK GRADE
10	M	80	84	A
7	M	85	89	A
4	F	90	86	B
20	M	82	85	B
25	F	94	94	A
14	F	88	84	C

We have five variables (SUBJECT NUMBER, GENDER, EXAM 1, EXAM 2, and HOMEWORK GRADE) collected for each of six subjects. The unit of analysis (the thing being studied—people in this example) is referred to as an "observation" in SAS terminology. If you are familiar with SQL (Structured Query Language), you would refer to observations as "rows" (of a table). SAS uses the term "variable" to represent each piece of information we collect for each observation. In SQL terms, you would call variables "columns" or "fields." Before we can write our SAS program, we need to assign a variable name to each variable. We do this so that we can distinguish one variable from another when doing computations or when requesting statistics. SAS variable names must conform to a few simple rules: They must start with a letter or the underscore character (_) and must not be more than 32 characters (letters, underscores, or digits) in length. Using blanks or special characters, such as commas, semicolons, etc., are not allowed. The underscore character (_) is particularly useful as part of SAS variable names since it can be used to make variable names more readable. Therefore, our column headings of "SUBJECT NUMBER" or "EXAM 1" are not valid SAS variable names. Logical SAS variable names for this collection of data would be:

```
SUBJECT   GENDER   EXAM1   EXAM2   HW_GRADE
```

It is prudent to pick variable names that help you remember which name goes with which variable. We could have named our five variables VAR1, VAR2, VAR3, VAR4, and VAR5, but we would then have to remember that VAR1 stands for "SUBJECT NUMBER" and so forth.

To begin, let's say we are interested only in getting the class means for the two exams. In reality it's hardly worth using a computer to add up six numbers, but it does provide a nice example. In order to do this, we could write the following SAS program:

```
DATA TEST;  ①
    INPUT SUBJECT 1-2 GENDER $ 4 EXAM1 6-8 EXAM2 10-12  ②
        HW_GRADE $ 14;
DATALINES;  ③
```
                                                        [Continued]

```
10 M   80   84 A
7  M   85   89 A
4  F   90   86 B
20 M   82   85 B
25 F   94   94 A
14 F   88   84 C
;
PROC  MEANS  DATA=TEST;  ④
RUN;  ⑤
```

The first four lines make up the DATA step. In this example, the DATA step begins with the word DATA and ends with the word DATALINES. Older versions of SAS software used the term CARDS instead of DATALINES. Either term is still valid. (If you don't know what a computer card is, ask an old person.) Line ① tells the program that we want to create a SAS data set called TEST. The next two lines ② show an INPUT statement that gives the program two pieces of information: what to call the variables and where to find them on the data line. Notice that this single SAS statement occupies two lines. The SAS system understands this is a single SAS statement because there is a single semicolon at the end of it. The first variable is SUBJECT and can be found in columns 1 and 2 of the data line. The second variable is GENDER and can be found in column 4. The dollar sign after GENDER means that GENDER is a character (alphanumeric) variable, that is, a variable that can have letters or numbers as data values. (More on this later.) EXAM1 is in columns 6–8, etc. The DATALINES statement ③ says that the DATA statements are done and the next thing the program should look for are the data themselves. The next six lines are the actual data. In this example, we are including the data lines directly in the program. Later on in this book, we will show you how to read data from external files. You may also be importing your data from other programs such as Microsoft Excel™ or Access™.

Great latitude is possible when putting together the data lines. Using a few rules will make life much simpler for you. These are not laws; they are just suggestions. First, put each new observation on a new line. Having more than one line per observation is sometimes necessary (and no problem), but don't put two observations on one line (at least for now). Second, line up your variables. Don't put EXAM1 in columns 6–8 on one line and in columns 9–11 on the next. SAS software can actually handle some degree of sloppiness here, but sooner or later it'll cost you. Third, right-justify your numeric data values. If you have AGE as a variable, record the data as follows:

Correct	Problematic
87	87
42	42
9	9
26	26
4	4
Right-justified	Left-justified

Again, SAS software doesn't care whether you right-justify numeric data or not, but other statistical programs will, and right-justification is standard. Fourth, think carefully if you want a variable to be stored as a numeric or character value. Take HW_GRADE, for example. We have HW_GRADE recorded as a character value. But we could have recorded it as 0–4 (0 = F, 1 = D, etc.). As it stands, we cannot compute a mean grade. Had we coded HW_GRADE numerically, we could get an average grade. Enough on how to code data for now.

Back to our example. A SAS program knows that the data lines are completed when it finds a SAS statement or a single semicolon. When you include your data lines in the program, as in this example, we recommend placing a single semicolon on the line directly below your last line of data. The next SAS statement ④ is a PROC statement. PROC says "Run a procedure" to the program. We specify which procedure right after the word PROC. Here we are running a procedure called MEANS. Following the procedure name (MEANS), we place the option DATA= and specify that this procedure should compute statistics on the data set called TEST. In this example, we could omit the option DATA=TEST, and the procedure would operate on the most recently created SAS data set, in this case, TEST. We recommend that you always include the DATA= option on every procedure since, in more advanced SAS programs, you can have procedures that create data sets as well as have many data sets "floating around." By including the DATA= option, you can always be sure your procedure is operating on the correct data set.

The MEANS procedure calculates the mean for any variables you specify. The RUN statement ⑤ is necessary only when SAS programs are run under the Display Manager. The RUN statement tells SAS that there are no more statements for the preceding procedure and to go ahead and do the calculations. If we have several PROCs in a row, we need only a single RUN statement at the end of the program. However, as a stylistic standard, we prefer to end every procedure with a RUN statement.

When this program is executed, it produces something called the SAS LOG and the SAS OUTPUT. The SAS LOG is an annotated copy of your original program (without the data listed). It's a lot like a phone book: Usually it's pretty boring, but sometimes you need it. Any SAS error messages will be found there, along with information about the data set that was created. The SAS LOG for this program is shown here:

```
NOTE: AUTOEXEC processing completed.

1      DATA TEST;
2          INPUT SUBJECT 1-2 GENDER $ 4 EXAM1 6-8 EXAM2 10-12
3                HW_GRADE $ 14;
4      DATALINES;

NOTE: The data set WORK.TEST has 6 observations and 5 variables.
NOTE: DATA statement used (Total process time):
          real time            0.03 seconds
          cpu time             0.00 seconds
11     ;

                                                      [Continued]
```

```
12    PROC MEANS DATA=TEST;
13    RUN;


NOTE: There were 6 observations read from the data set WORK.TEST.
NOTE: PROCEDURE MEANS used (Total process time):
      real time           0.00 seconds
      cpu time            0.01 seconds
```

The more important part of the output is found in the OUTPUT window if you are using the Display Manager. It contains the results of the computations and procedures requested by our PROC statements. This portion of the output from the previous program is shown next:

```
The MEANS Procedure

Variable    N          Mean       Std Dev        Minimum        Maximum
          ─────────────────────────────────────────────────────────────
SUBJECT     6     13.3333333     7.9916623      4.0000000     25.0000000
EXAM1       6     86.5000000     5.2057660     80.0000000     94.0000000
EXAM2       6     87.0000000     3.8987177     84.0000000     94.0000000
          ─────────────────────────────────────────────────────────────
```

If you don't specify which variables you want, SAS software will calculate the mean and several other statistics for every numeric variable in the data set. Our program calculated means for SUBJECT, EXAM1, and EXAM2. Since SUBJECT is just an arbitrary ID number assigned to each student, we aren't really interested in its mean. We can avoid getting it (and using up extra CPU cycles) by adding a new statement under PROC MEANS:

```
PROC MEANS DATA=TEST;
   VAR EXAM1 EXAM2;    ⑥
RUN;
```

The indentation used is only a visual aid. The VAR statement ⑥ specifies on which variables to run PROC MEANS. PROC MEANS not only gives you means, it also gives you the number of observations used to compute the mean, the standard deviation, the minimum score found, and the maximum score found. PROC MEANS can compute many other statistics such as variance and standard error. You can specify just which pieces you want in the PROC MEANS statement. For example:

```
PROC MEANS DATA=TEST N MEAN STD STDERR MAXDEC=1;
   VAR EXAM1 EXAM2;
RUN;
```

will get you only the number of nonmissing values for each variable (N), mean (MEAN), standard deviation (STD), and standard error (STDERR) for the variables EXAM1 and EXAM2. In addition, the statistics will be rounded to one decimal place (because of the MAXDEC=1 option). Chapter 2 describes most of the commonly requested options used with PROC MEANS.

C.   ENHANCING THE PROGRAM

The program as it is currently written provides some useful information, but with a little more work, we can put some "bells and whistles" on it. The bells-and-whistles version shown here adds the following features: It computes a final grade, which we will let be the average of the two exam scores; it assigns a letter grade based on that final score; it lists the students in student number order, showing their exam scores, their final grades, and their homework grades; it computes the class average for the exams and final grade and a frequency count for gender and homework grade; finally, it gets you a cup of coffee and tells you that you are a fine individual.

```
DATA EXAMPLE;   ①
   INPUT SUBJECT GENDER $ EXAM1 EXAM2   ②
         HW_GRADE $;
   FINAL = (EXAM1 + EXAM2) / 2;   ③
   IF FINAL GE 0 AND FINAL LT 65 THEN GRADE='F';   ④
   ELSE IF FINAL GE 65 AND FINAL LT 75 THEN GRADE='C';   ⑤
   ELSE IF FINAL GE 75 AND FINAL LT 85 THEN GRADE='B';
   ELSE IF FINAL GE 85 THEN GRADE='A';
DATALINES;   ⑥
10   M   80   84   A
 7   M   85   89   A
 4   F   90   86   B
20   M   82   85   B
25   F   94   94   A
14   F   88   84   C
;
PROC SORT DATA=EXAMPLE;   ⑦
   BY SUBJECT;   ⑧
RUN;   ⑨
```

[Continued]

```
PROC PRINT DATA=EXAMPLE;   ⑩
   TITLE "Roster in Student Number Order";
   ID SUBJECT;
   VAR EXAM1 EXAM2 FINAL HW_GRADE GRADE;
RUN;

PROC MEANS DATA=EXAMPLE N MEAN STD STDERR MAXDEC=1;   ⑪
   TITLE "Descriptive Statistics";
   VAR EXAM1 EXAM2 FINAL;
RUN;

PROC FREQ DATA=EXAMPLE;   ⑫
   TABLES GENDER HW_GRADE GRADE;
RUN;
```

As before, the statements from DATA EXAMPLE to DATALINES constitute our DATA step. Statement ① is an instruction for the program to create a data set whose data set name is "EXAMPLE." (Data set names follow the same conventions as variable names.) Statement ② is an INPUT statement that is different from the one in the previous example. We could have used the same INPUT statement as in the previous example but wanted the opportunity to show you another way that SAS programs can read data. Notice that there are no column numbers following the variable names.

This form of an INPUT statement is called list input. To use this form of INPUT, the data values must be separated by one or more blanks (or other separators, which computer people call delimiters). If you use one of the other possible delimiters, you need to modify the program accordingly (see Chapter 12, Section C). The order of the variable names in the list corresponds to the order of the values in the line of data. In this example, the INPUT statement tells the program that the first variable in each line of data represents SUBJECT values, the next variable is GENDER, the third EXAM1, and so forth. If your data conform to this "space-between-each-variable" format, then you don't have to specify column numbers for each variable listed in the INPUT statement. You may want to anyway, but it isn't necessary. (You still have to follow character variable names with a dollar sign.) If you are going to use "list input," then every variable on your data lines must be listed. Also, since the order of the data values is used to associate values with variables, we have to make special provisions for missing values. Suppose that subject number 10 (the first subject in our example) did not take the first exam. If we listed the data like this:

```
10   M   84   A
```

with the EXAM1 score missing, the 84 would be read as the EXAM1 score, the program would read the letter "A" as a value for EXAM2 (which would cause an error since the program was expecting a number), and, worst of all, the program would look on the next line for a value of homework grade. You would get an error message in the SAS LOG telling you that you had an invalid value for EXAM2 and that SAS went to a new line when the INPUT statement reached past the end of a line. You wouldn't understand these error messages and might kick your dog.

To hold the place of a missing value when using a list INPUT statement, use a period to represent the missing value. The period will be interpreted as a missing value by the program and will keep the order of the data values intact. When we specify columns as in the first example, we can use blanks as missing values. Using periods as missing values when we have specified columns in our INPUT statement is also OK but not recommended. The correct way to represent this line of data, with the EXAM1 score missing, is:

```
10 M . 84 A
```

Since list input requires one or more blanks between data values, we need at least one blank before and after the period. We may choose to add extra spaces in our data to allow the data values to line up in columns.

Line ③ is a statement assigning the average of EXAM1 and EXAM2 to a variable called FINAL. The variable name "FINAL" must conform to the same naming conventions as the other variable names in the INPUT statement. In this example, FINAL is calculated by adding together the two exam scores and dividing by 2. Notice that we indicate addition with a + sign and division by a / sign. We need the parentheses because, just the same as in handwritten algebraic expressions, SAS computations are performed according to a hierarchy. Multiplication and division are performed before addition and subtraction. Thus, had we written:

```
FINAL = EXAM1 + EXAM2 / 2;
```

the FINAL grade would have been the sum of the EXAM1 score and half of EXAM2. The use of parentheses tells the program to add the two exam scores first, and then divide by 2. To indicate multiplication, we use an asterisk (*); to indicate subtraction, we use a − sign. Exponentiation, which is performed before multiplication or division, is indicated by two asterisks. As an example, to compute A times the square of B, we write:

```
X = A * B**2;
```

The variable FINAL was computed from the values of EXAM1 and EXAM2. That does not, in any way, make it different from the variables whose values were read in from the raw data. When the DATA step is finished, the SAS procedures that follow will not treat variables such as FINAL any differently from variables such as EXAM1 and EXAM2.

The IF statement ④ and the ELSE IF statements ⑤ are logical statements that are used to compute a letter grade. They are fairly easy to understand. When the condition specified by the IF statement is true, the instruction following the word THEN is executed. The logical comparison operators used in this example are GE (greater than or equal to) and LT (less than). So, if a FINAL score is greater than or equal to 0 and less than 65, a letter grade of 'F' is assigned. The ELSE statements are only executed if a previous IF statement is not true. For example, if a FINAL grade is 73, the first IF statement ④ is not true, so the ELSE IF statement ⑤ is tested. Since this statement is true, a GRADE of 'C' is assigned, and all the following ELSE IF statements are skipped.

Other logical operators and their equivalent symbolic form are shown in the following table:

Expression	Symbol	Meaning
EQ	=	Equal
LT	<	Less than
LE	<=	Less than or equal
GT	>	Greater than
GE	>=	Greater than or equal
NE	∧=	Not equal
NOT	∧	Negation

NOTE: The symbols for NOT and NE may vary, depending on your system.

The "DATALINES" statement ⑥ indicates that the DATA step is complete and that the following lines contain data.

Notice that each SAS statement ends with a semicolon. As mentioned before, the semicolon is the logical end of a SAS statement. We could have written the first four lines like this:

```
DATA EXAMPLE; INPUT SUBJECT GENDER $
EXAM1 EXAM2 HW_GRADE $; FINAL =
(EXAM1 + EXAM2) / 2;
```

The program would still run correctly. Using a semicolon as a statement delimiter is convenient since we can write long SAS statements on several lines and simply put a semicolon at the end of the statement. However, if you omit a semicolon at the end of a SAS statement, the program will attempt to read the next statement as part of previous statement, causing an error. This may not only cause your program to die, it may also result in a bizarre error message emanating from the SAS system. Notice also that the data lines, since they are not SAS statements, do not end with semicolons.

Following the DATALINES statement are our data lines. Remember that if you have data that have been placed in preassigned columns with no spaces between the data values, you must use either the form of the INPUT shown earlier, with column specifications after each variable name, or formatted input (see Chapter 12). We have used a RUN statement to end every procedure. Each RUN statement tells the system that we are finished with a section of the program and to do the computations just concluded. Remember, when using the Display Manager, only the last RUN statement is absolutely necessary; the others are really only a matter of programming style.

D.   SAS PROCEDURES

Immediately following the data is a series of PROCs. They perform various functions and computations on SAS data sets. Since we want a list of subjects and scores in subject order, we first include a SORT PROCEDURE ⑦, ⑧, and ⑨. Line ⑦ indicates that we plan to sort our data set; line ⑧ indicates that the sorting will be by SUBJECT number. Sorting can be multilevel if desired. For example, if we want separate lists of male and female students in subject number order, we write:

```
PROC SORT DATA=EXAMPLE;
   BY GENDER SUBJECT;
RUN;
```

This multilevel sort indicates that we should first sort by GENDER (Fs followed by Ms—character variables are sorted alphabetically), then in SUBJECT order within GENDER.

The PRINT procedure ⑩ requests a listing of our data (which is now in SUBJECT order). The PRINT procedure is used to list the data values in a SAS data set. We have followed our PROC PRINT statement with three statements that supply information to the procedure. These are the TITLE, ID, and VAR statements. As with many SAS procedures, the supplementary statements following a PROC can be placed in any order. Thus:

```
PROC PRINT DATA=EXAMPLE;
   ID SUBJECT;
   TITLE "Roster in Student Number Order";
   VAR EXAM1 EXAM2 FINAL HW_GRADE GRADE;
RUN;
```

is equivalent to

```
PROC PRINT DATA=EXAMPLE;
   TITLE "Roster in Student Number Order";
   ID SUBJECT;
   VAR EXAM1 EXAM2 FINAL HW_GRADE GRADE;
RUN;
```

SAS programs recognize the keywords TITLE, ID, and VAR and interpret what follows in the proper context. Notice that each statement ends with its own semicolon.

The words following TITLE are placed in single or double quotes and will be printed across the top of each of the SAS output pages. (Note: You can leave off the quotes on a TITLE statement if you wish, but we recommend their use as a matter of style.) The ID variable, SUBJECT in this case, will cause the program to print the variable SUBJECT in the first column of the report, omitting the column labeled OBS (observation number), which the program will print when an ID variable is absent. The variables following the keyword VAR indicate which variables, besides the ID variable, we want in our report. The order of these variables in the list also controls the order in which they appear in the report.

The MEANS procedure ⑪ is the same as the one we used previously. Finally, the FREQ procedure ⑫ (you're right: pronounced "PROC FREAK") requests a frequency count for the variables GENDER, HW_GRADE, and GRADE. That is, what is the number of Males and Females, the number of As, Bs, etc., as well as the percentages of each category. PROC FREQ will compute frequencies for the variables listed on the TABLES statement. The reason that SAS uses the keyword TABLES instead of VAR for this list of variables is that PROC FREQ can also produce n-way tables (such as 2 × 3 tables).

Output from the complete program is shown here:

```
Roster in Student Number Order     14:13 Wednesday, February 18, 2004   1

SUBJECT    EXAM1    EXAM2     FINAL    HW_GRADE    GRADE

   4         90       86      88.0        B          A
   7         85       89      87.0        A          A
  10         80       84      82.0        A          B
  14         88       84      86.0        C          A
  20         82       85      83.5        B          B
  25         94       94      94.0        A          A

Descriptive Statistics             14:13 Wednesday, February 18, 2004   2

The MEANS Procedure

Variable      N      Mean     Std Dev      Std Error
───────────────────────────────────────────────────────
EXAM1         6      86.5       5.2           2.1
EXAM2         6      87.0       3.9           1.6
FINAL         6      86.8       4.2           1.7
───────────────────────────────────────────────────────

Descriptive Statistics             14:13 Wednesday, February 18, 2004   3

The FREQ Procedure

                                   Cumulative      Cumulative
GENDER      Frequency      Percent   Frequency       Percent
──────────────────────────────────────────────────────────────
F               3          50.00         3            50.00
M               3          50.00         6           100.00
```

HW_GRADE	Frequency	Percent	Cumulative Frequency	Cumulative Percent
A	3	50.00	3	50.00
B	2	33.33	5	83.33
C	1	16.67	6	100.00

GRADE	Frequency	Percent	Cumulative Frequency	Cumulative Percent
A	4	66.67	4	66.67
B	2	33.33	6	100.00

The first part of the output (the page number is shown at the extreme right of each page) is the result of the PROC PRINT on the sorted data set. Each column is labeled with the variable name. Because we used an ID statement with SUBJECT as the ID variable, the left-most column shows the SUBJECT number instead of the default OBS column, which would have been printed if we did not have an ID variable.

Page 2 of the output lists each of the variables listed on the VAR statement and produces the requested statistics (N, mean, standard deviation, and standard error) all to the tenths place (because of the MAXDEC=1 option).

Page 3 is the result of the PROC FREQ request. Notice that the title "Descriptive Statistics" is still printed at the top of each page. Titles remain in effect until the end of the current SAS session or if you change it to another title line. This portion of the listing gives you frequencies (the number of observations with particular values) as well as percentages. The two columns labeled "Cumulative Frequency" and "Cumulative Percent" are not really useful in this example. In other cases, where a variable represents an ordinal quantity, the cumulative statistics may be more useful.

E.   OVERVIEW OF THE SAS DATA STEP

Let's spend a moment examining what happens when we execute a SAS program. This discussion is a bit technical and can be skipped, but an understanding of how SAS software works will help when you are doing more advanced programming. When the DATA statement is executed, SAS software allocates a portion of a disk and names the data set "EXAMPLE," our choice for a data set name. Before the INPUT statement is executed, each of the character and numeric variables is assigned a missing value. Next, the INPUT statement reads the first line of data and substitutes the actual data values for the missing values. These data values are not yet written to our SAS data set EX-AMPLE but to a place called the Program Data Vector (PDV). This is simply a "holding" area where data values are stored before they are transferred to the SAS data set. The computation of the final grade comes next ③ and the result of this computation is added to the PDV. Depending on the value of the final grade, a letter grade is assigned by the series of IF and ELSE IF statements. The DATALINES line triggers the end of the DATA step, and the values in the PDV are transferred to the SAS data set. The program then returns control back to the top of the DATA step, and the INPUT statement is executed again to read the next line of data, compute a final grade, and write the next observation to the SAS data set. This reading, processing, and writing cycle continues until no more observations remain. Wasn't that interesting?

F.   SYNTAX OF SAS PROCEDURES

As we have seen, SAS procedures can have options. Also, procedures often have statements, like the previous VAR statement, which supply information to the procedure. Finally, statements can also have options. We will show you the general syntax of SAS procedures and then illustrate it with some examples. The syntax for all SAS procedures is:

```
PROC PROCNAME options;
STATEMENTS / statement options;
     .
     .
     .
STATEMENTS / statement options;
RUN;
```

First, all procedures start with the word PROC followed by the procedure name. If there are any procedure options, they are placed, in any order, between the procedure name and the semicolon, separated by spaces. If we refer to a SAS manual, under PROC MEANS we will see a list of options to be used with the procedure. As mentioned, N, MEAN, STD, STDERR, and MAXDEC= are some of the available options. A valid PROC MEANS request for statistics from a data set called EXAMPLE, with options for N, MEAN, and MAXDEC would be:

```
PROC MEANS DATA=EXAMPLE N MEAN MAXDEC=1;
RUN;
```

Next, most procedures need statements to supply more information about which type of analysis to perform. An example would be the VAR statement used with PROC MEANS. Statements follow the procedure, in any order (except for certain statistical procedures such as ANOVA or GLM). They each end with a semicolon. So, to run the previous PROC MEANS procedure on the variables EXAM1 and EXAM2 and to supply a title, we would enter:

```
PROC MEANS DATA=EXAMPLE N MEAN STD MAXDEC=1;
   TITLE "Descriptive Statistics on Exam Scores";
   VAR EXAM1 EXAM2;
RUN;
```

The order of the TITLE and VAR statements can be interchanged with no change in the results. Finally, some procedure statements also have options. Statement options are placed between the statement keyword and the semicolon and are separated from the statement by a slash. To illustrate, we need to choose a procedure other than PROC MEANS. Let's use PROC FREQ as an example. As we saw, PROC FREQ will usually have one or more TABLES statements following it. There are TABLES options that control which statistics can be placed in a table. For example, if we do not want the cumulative statistics printed, the statement option NOCUM is used. Since this is a statement option, it is placed between the TABLES statement and the semicolon, separated by a slash. The PROC FREQ request in the earlier example, modified to remove the cumulative statistics, would be:

```
PROC FREQ DATA=EXAMPLE;
   TABLES GENDER HW_GRADE GRADE/ NOCUM;
RUN;
```

To demonstrate a procedure with procedure options and statement options, we use the ORDER= option with PROC FREQ. This useful option controls the order in which the values can be arranged in our frequency table. One option is ORDER=FREQ, which enables the frequency table to be arranged in frequency order, from the highest frequency to the lowest. So, to request frequencies in descending order of frequency and to omit cumulative statistics from the output, we write our PROC FREQ statements as follows:

```
PROC FREQ DATA=EXAMPLE ORDER=FREQ;
   TABLES GENDER HW_GRADE GRADE/ NOCUM;
RUN;
```

G.   COMMENT STATEMENTS

Before concluding this chapter, we introduce one of the most important SAS statements—the comment statement. (No, we're not kidding!) A properly commented program indicates that a true professional is at work. A comment, inserted in a program, is one or more lines of text that are ignored by the program—they are there only to help the programmer or researcher when he or she reads the program at a later date.

There are two ways to insert comments into a SAS program. One is to write a comment statement. Begin it with an asterisk (*) and end it with a semicolon. There are many possible styles of comments using this method. For example:

```
*Program to Compute Reliability Coefficients
Ron Cody
September 18, 2004
Program Name: FRED stored in directory C:\MYDATA
Contact Fred Cohen at 555-4567;
```

Notice the convenience of this method. Just enter the * and type as many lines as necessary, ending with the semicolon. Just make sure the comment statement doesn't contain a semicolon. Some programmers get fancy and make pretty boxes for their comments, like this:

```
*-----------------------------------------------------*
| Program Name: FRED stored in C:\MYDATA              |
| Purpose: To compute reliability coefficients        |
| Contact: Fred Cohen at 555-4567                     |
| Date: September 18, 2004                            |
| Programmer: Ron Cody                                |
*-----------------------------------------------------*;
```

Notice that the entire box is a SAS comment statement since it begins with an asterisk and ends in a semicolon. Notice also that the box cries out, "I need a life!"

You may also choose to comment individual lines by resorting to one of the following three ways:

```
QUES = 6 - QUES; *Transform QUES VAR;
X = LOG(X);        *LOG Transform of X;
```

or

```
*Transform the QUES Variable;
QUES = 6 - QUES;
*Take the LOG of X;
X = LOG(X);
*True professonal at work;
```

or

```
*
*Transform the QUES Variable
*;
QUES = 6 - QUES;
*
*Take the LOG of X
*;
X = LOG(X);
*
*True professional at work
*;
```

For visual effect, the last example uses more than one asterisk to set off the comment. Note, however, that each group of three lines is a single comment statement since it begins with an asterisk and ends with a semicolon.

An alternative commenting method begins a comment with a /* and ends with a */. This form of comment can be embedded within a SAS statement and can include semicolons within the comment itself. It can occur any place a blank can occur. A few examples:

```
/* This is a comment line */
```

or

```
/*-----------------------------------------------*
| This is a pretty comment box using the slash star |
| method of commenting. Notice that it begins with  |
| a slash star and ends with a star slash.          |
*-----------------------------------------------*/
```

or

```
DATA EXAMPLE; /* The data statement */
    INPUT SUBJECT /* Note SUBJECT is numeric */ GENDER $
    EXAM1 /* EXAM1 is the first exam score */
    EXAM2 /* EXAM2 is the second exam score */
    HW_GRADE $;
    FINAL = (EXAM1 + EXAM2)/2; /* Compute a composite grade */
DATALINES;
```

Let us show you one final, very useful trick using a comment statement before concluding this chapter. Suppose you have written a program and have run several procedures. Now, you return to the program and wish to run additional procedures. You could edit the program, remove the old procedures, and add the new ones (if you are using the Display Manager, you could mark the block you want to submit—SAS will only submit the marked block if this is done). Or, you could "comment them out" by preceding the section with a /* and ending with a */, making the entire section a comment. As an example, our commented program could look like this:

```
DATA MYPROG;
   INPUT X Y Z;
DATALINES;
1 2 3
3 4 5
;
/*********************
PROC PRINT DATA=MYPROG;
   TITLE "MY TITLE";
   VAR Y Z;
RUN;
*********************/
PROC CORR DATA=MYPROG;
   VAR X Y Z;
RUN;
```

The print procedure is not executed since it is treated as a comment; the correlation procedure will be run.

One final point: when running a batch program on IBM mainframes under MVS, a /* in columns 1 and 2 causes the program to terminate, so don't use column 1 if you are using JCL (Job Control Language).

A few extra minutes are needed to comment a SAS program, but that time is well spent. You will be thankful that you added comments to a program when it comes time to modify the program or if you expect other people to understand how your program works.

H.   REFERENCES

One of the advantages of using SAS software is the variety of procedures that can be performed. We cannot describe them all here nor can we explain every option of the procedures we do describe. You may, therefore, want to obtain one or more of the following manuals available from the SAS Institute Inc., Book Sales Department, SAS Campus Drive, Cary, NC 27513-2414 (Web site www.sas.com/pubs). The SAS Institute also takes phone orders. Call (919)677-8000 or (800)727-3228.

You can also obtain the Online Doc™ free from the following SAS Web site: http://support.sas.com/documentation/onlinedoc/index.html

Recommended books and manuals include (all available from the SAS Institute at www.sas.com/pubs):

SAS 9.1 Language Reference: Concepts. Order Number: 58940, ISBN: 1-59047-198-9

SAS OnlineDoc 9. Order Number: 58277, ISBN: 1-59047-056-7

SAS Procedures Guide, Version 8, Volumes 1 and 2. Order Number: 57238, ISBN: 1-58025-482-9

SAS 9.1 Language Reference: Dictionary, Volumes 1, 2, and 3. Order Number: 58941, ISBN: 1-59047-199-7

SAS/STAT 9.1 User's Guide, Volumes 1–7. Order Number: 59051, ISBN: 1-59047-243-8

SAS Functions by Example (Cody). Order Number: 59343, ISBN: 1-59047-378-7

Longitudinal Data and SAS: A Programmer's Guide (Cody). Order Number: 58176, ISBN: 1-58025-924-3

Cody's Data Cleaning Techniques Using SAS Software (Cody). Order Number: 57198, ISBN: 1-58025-600-7

SAS Programming by Example (Cody and Pass). Order Number: 55126, ISBN: 1-55544-681-7

The SAS Workbook and Solutions Set (Cody). Order Number: 55594, ISBN: 1-55544-762-7

SAS/GRAPH 9.1 Reference, Volumes 1 and 2. Order Number: 58932, ISBN: 1-59047-195-4

Here are some statistics books that we recommend:

Gravetter, F. J., and L. B. Wallnau. 2004. *Statistics for the Behavioral Sciences,* 6th ed. Belmont, CA: Wadsworth/Thomson Learning.

Kirk, R. E. 1995. *Experimental Design: Procedures for the Behavioral Sciences,* 3rd ed. Pacific Grove, CA: Brooks/Cole.

Pedhazur, E. J. 1997. *Multiple Regression in Behavioral Research,* 3rd ed. Orlando, FL: Harcourt College Publishers.

Stevens, J. P. 2001. *Applied Multivariate Statistics for the Social Sciences,* 4th ed. Mahwah, NJ: Lawrence Erlbaum Associates.

Tabachnick, B. G., and L. S. Fidell. 2001. *Using Multivariate Statistics,* 4th ed. Boston, MA: Allyn & Bacon.

Winer, B. J. 1991. *Statistical Principles in Experimental Design,* New York: Mc-Graw-Hill.

PROBLEMS

(Answers to the odd-numbered problems can be found at the end of this book; answers to even-numbered problems are available from Prentice Hall for instructors.)

Remember, you can download all the data sets and programs for these problems from the Web site: www.prenhall.com/cody

1.1 We have collected the following data on five subjects:

ID	AGE	GENDER	Grade Point Average (GPA)	College Entrance Exam Score (CSCORE)
1	18	M	3.7	650
2	18	F	2.0	490
3	19	F	3.3	580
4	23	M	2.8	530
5	21	M	3.5	640

(a) Write the SAS statements necessary to create a SAS data set.

(b) Add the statement(s) necessary to compute the mean grade point average and mean college entrance exam score.

(c) We want to compute an index for each subject, as follows:

```
INDEX = GPA + 3 × CSCORE/500
```

Modify your program to compute this INDEX for each student and to print a list of students in order of increasing INDEX. Include in your listing the student ID, GPA, CSCORE, and INDEX.

1.2 You have collected the following information on some participants in a diet program:

Variable Name	Description	Type	Starting Column	Ending Column
SUBJ	Subject number	Character	1	3
HEIGHT	Height in inches	Numeric	4	5
WT_INIT	Initial weight (in pounds)	Numeric	6	8
WT_FINAL	Final weight	Numeric	9	11

(a) Reading the following data lines, create a SAS data set using column input. Include in this data set three computed variables called BMI_INIT (BMI stands for body mass index), BMI_FINAL, and BMI_DIFF. Computed BMI is a person's weight (in kilograms) divided by their height (in meters) squared. To convert pounds to kilograms, divide by 2.2; to convert inches to meters, multiply by .0254. BMI_DIFF is BMI_FINAL minus BMI_INIT. Note: Large values of BMI indicate more overweight individuals. Therefore, if the diet program is working, BMI_DIFF should come out negative.

(b) Print out a listing of this data set in subject number order. Include in the listing *only* the variables SUBJ, HEIGHT, BMI_INIT, BMI_FINAL, and BMI_DIFF.

Here is some sample data for you to use:

```
00768155150
00272250240
00563240200
00170345298
```

1.3 Given the following set of data:

Social Security Number	Annual Salary	Age	Race
123874414	28,000	35	W
646239182	29,500	37	B
012437652	35,100	40	W
018451357	26,500	31	W

(a) Create a SAS data set using the previous data. Compute the average annual salary and average age. NOTE: Since you don't know yet how to read numeric values containing commas, you may enter the values without commas.

(b) If all subjects were in a 30% tax bracket, compute their taxes (based on annual salary) and print out a list, in Social Security number order, showing the annual salary and the tax.

1.4 You inherited the following SAS program. Add the necessary statements to:

(a) Create two new variables. Call one OVERALL, computed as the average of the IQ score, the MATH score, and the SCIENCE score divided by 500. Call the other GROUP, defined as 1 for IQ scores between 0 and 100 (inclusive), 2 for IQ scores between 101 and 140 (inclusive), and 3 for IQ scores greater than 140.

(b) Provide a listing of this data set in IQ order.

(c) Compute the frequencies for GROUP.

```
***Program to read in IQ and other test scores;
DATA IQ_AND_TEST_SCORES;
   INPUT ID      1-3
         IQ      4-6
         MATH    7-9
         SCIENCE 10-12;
DATALINES;
001128550590
002102490501
003140670690
004115510510;
```

1.5 What's wrong with the following program?

```
DATA MISTAKE;
INPUT ID 1-3 TOWN 4-6 REGION 7-9 YEAR 11-12 BUDGET 13-14
VOTER-TURNOUT 16-20
DATALINES;
00104005  0422 12345
  (more data lines)
;
PROC MEANS DATA=MISTAKE;
VAR ID REGION VOTER-TURNOUT;
N,STD,MEAN;
RUN;
```

1.6 You conducted a survey and the responses to the five questions were coded as A, B, C, D, or E. Using the following eight lines of data, write a SAS DATA step to read these data and create five variables (QUES1, QUES2, ..., QUES5). Remember that these must all be character variables (since the values are letters) and there are no spaces between the values so you must use column input. Next, generate frequency distributions for each of the five questions, omitting cumulative statistics from the listing.

The data:

```
ABCDE
AACCE
BBBBB
CABDA
DDAAC
CABBB
EEBBB
ACACA
```

*1.7 A large corporation is interested in who is buying their product. What the CEO wants is a profile of the "typical buyer." The variables collected on a sample of buyers are: age, gender, race, income, marital status, and homeowner/renter. Set up a layout for the observations, and write a SAS program to obtain a profile of the "typical buyer." NOTE: Slightly more difficult problems are preceded by an asterisk (*). Yes, this is a loosely defined problem and you can make up your own data to try out your program.

HINTS AND COMMENTS:

1. For variables such as "homeowner," it is easier to remember what you have if you let negative responses be 0 and positive responses be 1.
2. When grouping numerical variables into categories, make sure your grouping suits your needs and your data. For example, if your product is denture cream, a grouping of age, such as:

$$1 = {<}21 \quad 2 = 21{-}35 \quad 3 = 36{-}50 \quad 4 = {>}50$$

is virtually useless. You know such people are mostly over 50. You might use groupings such as:

$$1 = {<}50 \quad 2 = 50{-}59 \quad 3 = 60{-}69 \quad 4 = {>}69.$$

1.8 Employee salaries and job class are listed here. Write a program to create a SAS data set (call it anything you wish) containing the employee ID (EMPID), the salary (SALARY), the job class (JCLASS), and a new variable called BONUS, computed as follows: BONUS is 10% of salary for job class 1, 15% of salary for job class 2, and 20% of salary for job class 3 (the CEO and president of the company). Once you have the bonus amount, add it to salary to create one more variable called NEW_SALARY. Use PROC PRINT to list the contents of the data set. Here is the data you can use:

```
137   28000  1
214   98000     3
199  150000  3
355   57000     2
```

Notice that the columns don't line up. You may make JCLASS a character or a numeric variable. Make sure the SAS statements you write using JCLASS are compatible with which type you chose. Look in the SAS LOG to make sure there are no "character to numeric" or "numeric to character" conversion messages.

1.9 Given the data set:

ID	RACE	SBP	DBP	HR
001	W	130	80	60
002	B	140	90	70
003	W	120	70	64
004	W	150	90	76
005	B	124	86	72

(NOTE: SBP is systolic blood pressure, DBP is diastolic blood pressure, and HR is heart rate.)

Write the SAS statements to produce a report as follows:

Race and Hemodynamic Variables			
ID	RACE	SBP	DBP
003	W	120	70
005	B	124	86
001	W	130	80
002	B	140	90
004	W	150	90

NOTE: 1. To omit the default "OBS" column, use the PROC PRINT option NOOBS.
2. Data is in increasing order of SBP.
3. The variable HR is not included in the report.
4. The report has a title.

1.10 The amount of rain for three months (June, July, and August) was recorded for each of three cities. Using the data in the following table, create a SAS data set (RAIN) containing the variables CITY, RAIN_JUNE, RAIN_JULY, and RAIN_AUGUST. Create four additional variables in this data set: AVERAGE, PERCENT_JUNE, PERCENT_JULY, and PERCENT_AUGUST. AVERAGE is the average of the rain for the three months. Each of the PERCENT variables is computed as rainfall for the month as a percent of the AVERAGE. For example, PERCENT_JUNE would be $100 \times 23/26$. Provide a listing of the data set in alphabetical order of CITY, with CITY as the ID variable. Then compute the mean, standard deviation, and the 95% confidence interval (CI) for the three RAIN variables (with all statistics printed with two decimal places). Here are the data:

CITY	RAIN_JUNE	RAIN_JULY	RAIN_AUGUST
Trenton	23	25	30
Newark	18	27	22
Albany	22	21	27

CHAPTER 2

Describing Data

A. INTRODUCTION

Did you work the problems at the end of Chapter 1 as we asked you? We didn't think so. So, go back and do them now, and then send in a check to your local Public Broadcasting station.

One of the first steps for any data analysis project is to generate simple descriptive statistics for all the continuous variables. Besides the traditional mean and standard deviation, you may want to use histograms, stem-and-leaf plots, test for normality of distributions, and a variety of other descriptive measures.

B. DESCRIBING DATA

Even for complex statistical analysis, you must be able to describe the data in a straightforward, easy-to-comprehend fashion. This is typically accomplished in one of several ways. The first way is through descriptive summary statistics. Probably the most popular way to describe a sample of scores is by reporting: (1) the number of people in the sample (called the sample size and referred to by "n" in statistics books and SAS printouts); (2) the mean (arithmetic average) of the scores; and (3) the standard deviation of the scores. The standard deviation is a measure of how widely spread the scores are. Roughly speaking, if the scores form a "bell-shaped" (normal) distribution, 68% of the scores will fall within 1 standard deviation of the mean (plus or minus), and 95% of the scores will fall within 2 standard deviations.

Let's create a SAS data set to introduce some concepts related to descriptive statistics. Suppose we conducted a survey (albeit a very small one) where we record the gender, height, and weight of seven subjects. We collect the following data:

GENDER	HEIGHT	WEIGHT
M	68.5	155
F	61.2	99
F	63.0	115
M	70.0	205
M	68.6	170
F	65.1	125
M	72.4	220
M		188

NOTE: HEIGHT is missing for the last person.

There may be several questions we want to ask about these data. Perhaps we want to count how many males and females are in our sample. We might also want means and standard deviations for the variables HEIGHT and WEIGHT. Finally, we may want to see a histogram of our continuous variables and, perhaps, determine if the data can be considered to have come from a normal distribution. These are fairly simple tasks if only eight people are involved. However, we are rarely interested in such small samples. Once we begin to consider as many as 20–30 people, statistical analysis by hand becomes quite tedious. A SAS program to read these data and to compute some descriptive statistics is shown next:

```
***Program to create the HTWT data set
   and to compute descriptive statistics;
DATA HTWT; ①
   INPUT SUBJECT
         GENDER $
         HEIGHT
         WEIGHT; ②
DATALINES; ③
1 M 68.5 155
2 F 61.2 99
3 F 63.0 115
4 M 70.0 205
5 M 68.6 170
6 F 65.1 125
7 M 72.4 220
8 M   .  188
;
PROC MEANS DATA=HTWT; ④
   TITLE "Simple Descriptive Statistics"; ⑤
RUN; ⑥
```

In this example, our data are placed "in-stream" directly into our program. For small data sets, this is an appropriate method. For larger data sets, we usually place our data in a separate file and instruct our SAS program where to look to find the data (see Chapter 13, Section C). In this program, we have chosen to use the list form of input where each data value is separated from the next by one or more spaces. The first three statements define our DATA step. In ① we indicate that we are creating a SAS data set called HTWT. As mentioned in the tutorial, SAS data set names as well as SAS variable names are from 1 to 32 characters in length. They must start with a letter or an underscore (_). The other characters in SAS data set names or SAS variable names can, in addition, include numerals. Thus, our name HTWT meets these criteria and is a valid SAS data set name. Statement ② is the INPUT statement. This statement gives names to the variables we are going to read. In this form of INPUT, the order of the variable names corresponds to the order of the data values. Since our data values are arranged in SUBJECT, GENDER, HEIGHT, and WEIGHT order, our INPUT statement lists variable names in the same order. Notice that subject 8 does not have a value for HEIGHT. Since we are using list input, we need a "place holder" to indicate that we do not have a value for HEIGHT. We use a period (.) separated by one or more spaces to do this. Without the period, the value of WEIGHT (188) would be read as a HEIGHT value (a VERY tall person indeed). Again, in review, the following the variable name GENDER indicates that we are using character values for GENDER ('M' and 'F'). Statement ③ DATALINES, indicates that the in-stream lines of data will follow and that the DATA step is finished. Following the data is a lone semicolon (;), which is a convenient way to indicate there are no more data lines. Statement ④ is our request for descriptive statistics. Since we did not tell it to do otherwise, PROC MEANS will give us the number of observations used to calculate the descriptive statistics for each of our numeric variables, the mean, standard deviation, minimum, and maximum. In a moment, we will show you how to request only those statistics that you want rather than accepting the defaults. Let's look at the results of running this program:

```
Simple Descriptive Statistics

The MEANS Procedure

Variable    N          Mean        Std Dev       Minimum        Maximum
─────────────────────────────────────────────────────────────────────────
SUBJECT     8     4.5000000      2.4494897     1.0000000      8.0000000
HEIGHT      7    66.9714286      4.0044618    61.2000000     72.4000000
WEIGHT      8   159.6250000     43.9218055    99.0000000    220.0000000
─────────────────────────────────────────────────────────────────────────
```

For the variable of HEIGHT in our sample, we see that there are seven values used to compute statistics (one missing value). For SUBJECT and WEIGHT, eight values were used; the shortest person is 61.2 inches tall and the tallest is 72.4 inches (from the "minimum value" and "maximum value" columns); their mean height is 66.971 (rounded off); and the standard deviation is 4.004. Notice that we also obtain statistics on the variable called SUBJECT. The mean subject number is probably not going to be of interest to us. We will show you, shortly, how to compute statistics only for those variables of interest.

You can specify which statistics you want to compute by specifying options for PROC MEANS. Most SAS procedures have options that are placed between the procedure name and the semicolon. Many of these options are listed in this text; a complete list of options for all SAS procedures can be found in the manuals available from the SAS Institute. As mentioned in Chapter 1, the option MAXDEC=n controls the number of decimal places for the printed statistics, N prints the number of (nonmissing) observations, and MEAN produces the MEAN. So, if you want only the N and MEAN and you want three places to the right of the decimal, you would write:

```
PROC MEANS DATA=HTWT N MEAN MAXDEC=3;
```

You may also wish to specify for which numeric variables in your data set you want to compute descriptive statistics (so we can avoid getting descriptive statistics on numeric variables such as SUBJECT). We make this specification with a VAR statement. VAR (short for VARIABLES) is a statement that gives additional information to PROC MEANS (and many other procedures as well). The syntax is the word VAR followed by a list of variable names. So, if we want descriptive statistics only on HEIGHT and we want the sample size, the mean, and three decimal places, we would write:

```
PROC MEANS DATA=HTWT N MEAN MAXDEC=3;
   TITLE "Simple Descriptive Statistics";
   VAR HEIGHT;
RUN;
```

The order of the options is irrelevant. A list of the commonly requested options for PROC MEANS is as follows:

Option	Description
N	Number of observations for which the statistic was computed
NMISS	Number of observations with missing values for the variable of interest
MEAN	Arithmetic mean
MEDIAN	Median
STD	Sample standard deviation
STDERR	Standard error
CLM	Lower and upper two-sided 95% confidence interval (CI) for the mean
LCLM	Lower one-sided 95% CI for the mean. If both LCLM and UCLM are requested, a two-sided CI is computed; otherwise, this option gives you a one-sided interval.
UCLM	Upper one-sided 95% CI for the mean. If both LCLM and UCLM are requested, a two-sided CI is computed; otherwise, this option gives you a one-sided interval.

(Continued)

Option	Description
MIN	Minimum: lowest value for the data
MAX	Maximum: highest value for the data
SUM	Sum
VAR	Variance
Q1	First quartile (25th percentile)
Q3	Third quartile (75th percentile)
QRANGE	Interquartile range
CV	Coefficient of variation
SKEWNESS	Skewness
KURTOSIS	Kurtosis
T	Student's t-test, testing the null hypothesis that the population mean is zero.
PRT	Probability of obtaining a larger absolute value of t under the null hypothesis.
MAXDEC=n	Where n specifies the number of decimal places for printed statistics

One further example. Suppose we want the sample size, mean, standard deviation, standard error, and a 95% confidence interval about the sample mean. In addition, we want the statistics rounded to two decimal places. We would write:

```
PROC MEANS DATA=HTWT MAXDEC=2 N MEAN STD STDERR CLM;
```

Output from this request would look as follows:

```
Simple Descriptive Statistics

The MEANS Procedure

Variable     N       Mean      Std Dev      Std Error

HEIGHT       7      66.97         4.00           1.51
WEIGHT       8     159.63        43.92          15.53


Variable     Lower 95% CL for Mean      Upper 95% CL for Mean

HEIGHT                       63.27                      70.67
WEIGHT                      122.91                     196.34
```

The standard error of the mean is used to indicate a "confidence interval" about the mean, which is useful when our scores represent a sample of scores from some population. For example, if our eight subjects were a random sample of high school juniors in New Jersey, we could use the sample mean (66.97) as an estimate of the average height of all New Jersey high school juniors. The standard error of the mean tells us how far off this estimate might be. If our population is roughly normally distributed, the sample estimate of the mean (based on a random sample) will fall within one standard error (1.51) of the actual, or "true," mean roughly 68% of the time and within two standard errors (3.02) of the mean roughly 95% of the time. By specifying the CLM option, PROC MEANS will print a 95% confidence interval about our sample mean. Looking at our listing, we are 95% "confident" that the interval from 63.27 to 70.67 contains the true population mean for height. For weight, this interval is from 122.91 to 196.34. If you were to compute a 95% confidence interval by hand, you would need to add and subtract a t-value (based on the degrees of freedom) times the standard error of the mean.

As mentioned earlier in this chapter, PROC MEANS produces, by default, the N, mean, minimum, maximum, and standard deviation. Suppose you want standard error added to that list. If you request any statistic (MAXDEC= is not a statistic), PROC MEANS will print only that statistic. Therefore, if you decide to override the system defaults and request an additional statistic, you must specify them all. As an example, to add standard error to the default list, we would write:

```
PROC MEANS DATA=HTWT N MEAN STD MIN MAX STDERR;
```

C.   MORE DESCRIPTIVE STATISTICS

If you would like a more extensive list of statistics, including tests of normality, stem-and-leaf plots, and boxplots, **PROC UNIVARIATE** is the way to go. This extremely useful procedure can compute, among other things:

1. The number of observations (nonmissing)
2. Mean
3. Standard deviation
4. Variance
5. Skewness
6. Kurtosis
7. Uncorrected and corrected sum of squares
8. Coefficient of variation
9. Standard error of the mean
10. A t-test comparing the variable's value against zero
11. Maximum (largest value)
12. Minimum (smallest value)
13. Range
14. Median, 3rd, and 2nd quartiles
15. Interquartile range
16. Mode
17. 1st, 5th, 10th, 90th, 95th, and 99th percentiles

18. The five highest and five lowest values (useful for data checking)
19. W or D statistic to test whether data are normally distributed
20. Stem-and-leaf plot
21. Boxplot
22. Normal probability plot, comparing your cumulative frequency distribution to a normal distribution

To run PROC UNIVARIATE for our variables HEIGHT and WEIGHT, we would write:

```
PROC UNIVARIATE DATA=HTWT;
   TITLE "More Descriptive Statistics";
   VAR HEIGHT WEIGHT;
RUN;
```

By default, we get the first 18 items of the previous list of statistics. To request, additionally, a test of normality, a stem-and-leaf plot, and a boxplot, we would add the options NORMAL and PLOT as follows:

```
PROC UNIVARIATE DATA=HTWT NORMAL PLOT;
   TITLE "More Descriptive Statistics";
   VAR HEIGHT WEIGHT;
RUN;
```

A portion of the output from the previous request is shown next:

```
More Descriptive Statistics

The UNIVARIATE Procedure
Variable: HEIGHT

                              Moments

N                          7      Sum Weights                 7
Mean               66.9714286    Sum Observations        468.8
Std Deviation       4.0044618    Variance           16.0357143
Skewness           -0.2390545    Kurtosis           -1.1613152
Uncorrected SS       31492.42    Corrected SS       96.2142857
Coeff Variation    5.97935849    Std Error Mean      1.51354429
```

Basic Statistical Measures

Location		Variability	
Mean	66.97143	Std Deviation	4.00446
Median	68.50000	Variance	16.03571
Mode	.	Range	11.20000
		Interquartile Range	7.00000

Tests for Location: Mu0=0

Test		-Statistic-		-----p Value------	
Student's t	t	44.24808	Pr > |t|	<.0001
Sign	M	3.5	Pr >= |M|	0.0156
Signed Rank	S	14	Pr >= |S|	0.0156

Tests for Normality

Test		--Statistic---		-----p Value------	
Shapiro-Wilk	W	0.956642	Pr < W	0.7895
Kolmogorov-Smirnov	D	0.220093	Pr > D	>0.1500
Cramer-von Mises	W-Sq	0.038859	Pr > W-Sq	>0.2500
Anderson-Darling	A-Sq	0.224606	Pr > A-Sq	>0.2500

Quantiles (Definition 5)

Quantile	Estimate
100% Max	72.4
99%	72.4
95%	72.4
90%	72.4
75% Q3	70.0
50% Median	68.5
25% Q1	63.0
10%	61.2
5%	61.2
1%	61.2
0% Min	61.2

[Continued]

```
                     Extreme Observations

      ----Lowest----        ----Highest---

      Value     Obs     Value     Obs

      61.2       2      65.1       6
      63.0       3      68.5       1
      65.1       6      68.6       5
      68.5       1      70.0       4
      68.6       5      72.4       7


                       Missing Values

                        -----Percent Of-----
      Missing                        Missing
        Value    Count     All Obs       Obs

           .         1       12.50     100.00


      Stem Leaf              #     Boxplot
       72 4                  1        |
       70 0                  1     +-----+
       68 56                 2     *-----*
       66                          |  +  |
       64 1                  1     |     |
       62 0                  1     +-----+
       60 2                  1        |
          ----+----+----+----+
```

More Descriptive Statistics

The UNIVARIATE Procedure
Variable: HEIGHT

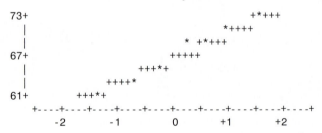

```
                   Normal Probability Plot
        73+                              +*+++
          |                             *++++
          |                         *  +*+++
        67+                         +++++
          |                     +++*+
          |                 ++++*
        61+             +++*+
          +----+----+----+----+----+----+----+----+----+----+
             -2       -1        0       +1       +2
```

More Descriptive Statistics

The UNIVARIATE Procedure
Variable: WEIGHT

```
                              Moments

N                          8        Sum Weights                    8
Mean                 159.625        Sum Observations            1277
Std Deviation     43.9218055        Variance                1929.125
Skewness          -0.0356013        Kurtosis               -1.482243
Uncorrected SS        217345        Corrected SS            13503.875
Coeff Variation   27.5156182        Std Error Mean         15.5287033


                    Basic Statistical Measures

          Location                          Variability

Mean         159.6250        Std Deviation            43.92181
Median       162.5000        Variance                     1929
Mode                .        Range                 121.00000
                             Interquartile Range    76.50000


                 Tests for Location: Mu0=0

Test                 -Statistic        -----p Value------

Student's t      t      10.27935       Pr > |t|      <.0001
Sign             M             4       Pr >= |M|     0.0078
Signed Rank      S            18       Pr >= |S|     0.0078


                    Tests for Normality

Test                   --Statistic---        -----p Value------

Shapiro-Wilk       W      0.954956      Pr < W         0.7609
Kolmogorov-Smirnov D      0.159749      Pr > D        >0.1500
Cramer-von Mises   W-Sq   0.027456      Pr > W-Sq     >0.2500
Anderson-Darling   A-Sq   0.19299       Pr > A-Sq     >0.2500


Quantiles (Definition 5)


Quantile              Estimate

100% Max              220.0
99%                   220.0
95%                   220.0
90%                   220.0
75% Q3                196.5
50% Median            162.5
25% Q1                120.0
10%                    99.0
5%                     99.0
1%                     99.0
0% Min                 99.0
```

[Continued]

```
                        Extreme Observations

       ----Lowest----          ----Highest---

      Value      Obs          Value      Obs

         99       2            155        1
        115       3            170        5
        125       6            188        8
        155       1            205        4
        170       5            220        7


     Stem Leaf                   #     Boxplot

       22 0                      1        |
       20 5                      1        |
       18 8                      1     +-----+
       16 0                      1     *--+--*
       14 5                      1     |     |
       12 5                      1     +-----+
       10 5                      1        |
        8 9                      1        |
          ----+----+----+----+
     Multiply Stem.Leaf by 10**+1


                  Normal Probability Plot
       230+                            *++++
          |                         *++++
          |                       *++++
          |                     +*++
          |                  ++*++
          |               ++++*
          |            ++++*
        90+       ++++*
          +----+----+----+----+----+----+----+----+----+
             -2        -1        0        +1        +2
```

There is a lot of information in a PROC UNIVARIATE output. Under the title "Moments" you will see a number of statistics. Most of them are self-explanatory. Here is a list with explanations:

N	Number of nonmissing observations (observations that have actual values)
Sum Wgts	Sum of weights (if WEIGHT statement used)

Mean	Arithmetic mean
Sum	Sum of the scores
Std Dev	Standard deviation
Variance	Variance
Skewness	Skewness (measure of the symmetry or asymmetry of the distribution)
Kurtosis	Kurtosis (measure of the flatness or the distribution)
USS	Uncorrected sum of squares (the sum of the scores squared— each score is squared and the squares are added together)
CSS	Corrected sum of squares (sum of squares about the mean, usually more useful than USS)
CV	Coefficient of variation (the standard deviation divided by the mean times 100%)
Std Mean	Standard error of the mean (the standard deviation divided by the square root of n)
T:Mean=0	Student's t-test for testing the hypothesis that the population mean is zero
Prob>|T|	The p-value for the t-statistic (two-tailed)
Sgn Rank	The Wilcoxon signed rank sum (usually used for difference scores)
Prob>|S|	The p-value for the Sign Rank test
Num $^\wedge$=0	Number of nonzero observations
W:Normal	Shapiro-Wilk statistic for a test of normality (SAS
(D:Normal)	will produce the Kolomogorov D:Normal test when n is larger than 2000)
Prob<W (Prob>D)	P-value testing the null hypothesis that the population is normally distributed (when the D:Normal test is done, the statistic is Prob>D)

Looking further at the output, we find under the heading "Quantiles(Definition =5)," two sets of useful information. The left-hand set lists quantiles, showing the highest score (100%), the score at the third quartile (75%), the median (50%), the score at the first quartile, and the lowest score (0%). The "Definition=5" in the heading indicates that SAS is using definition 5 listed in the SAS procedures manual. This definition is described as an "empirical distribution function with averaging." We refer those interested in the subtle, yet fascinating, differences of the five available definitions, to consult the SAS procedures manual under PROC UNIVARIATE. For the other 99.6% of our readers, we recommend the default.

The right-hand column lists other percentiles that are often of interest (99%, 95%, etc.). Below these two columns you will find the range, the interquartile range (Q3–Q1), and the mode.

The list of Extremes, which comes next, is extremely useful. This list also comes without our specifically asking for it. It lists the five lowest and five highest values in the data set. We find this useful for data checking. Obviously incorrect values can be spotted easily. Next to each extreme value is the corresponding observation number. This can be made more useful if an ID statement is used with PROC UNIVARIATE. The ID variable (usually a subject number) is printed next to each extreme value, in addition to the observation number. That way, when a data error is spotted, it is easier to locate the incorrect value in the data by referring to the ID variable. We can use our variable SUBJECT as an ID variable to demonstrate this. Our complete PROC UNIVARIATE request would be:

```
PROC UNIVARIATE DATA=HTWT NORMAL PLOT;
   TITLE "More Descriptive Statistics";
   VAR HEIGHT WEIGHT;
   ID SUBJECT;
RUN;
```

Before we leave the topic of extreme observations, there is a useful procedure option (NEXTROBS=n) that allows you to specify how many extreme high and low observations you want in this portion of the output. For example, the statement:

```
PROC UNIVARIATE DATA=HTWT NEXTROBS=10;
```

will provide a list of the 10 highest and lowest observations for the variables listed in the VAR list.

The next portion of the output from PROC UNIVARIATE is a result of the PLOT option. The left side of the page is a Tukey-style stem-and-leaf plot. This can be thought of as a sideways histogram. However, instead of using Xs to represent the bars, the next digit of the number after the "stem" is used. For example, the largest height is 72.4. Next to the stem value of 72, we see a '4,' which tells us not only is there a value between 72 and 74 but that the actual value is 72.4. Look at the stem labeled "68." Notice that there were two scores between 68 and 70. The "leaf" values of '5' and '6' indicate that there is one height equal to 68.5 and another height equal to 68.6. This style of "histogram" supplies us with additional information concerning the structure of values within a bar. In this small example, the stem-and-leaf plot is not too impressive, but with samples from about 20 to 200, it is very useful.

Another Tukey invention, the boxplot, is displayed on the right side of the page. The SAS conventions are as follows: The bottom and top of the box represent the sample 25th and 75th percentiles. The median is represented by the dashed line inside the box, and the sample mean is shown as a + sign. The vertical lines at the top and bottom of the box are called whiskers and extend as far as the data to a maximum distance of 1.5 times the interquartile range. Data values beyond the whiskers, but within three interquartile ranges of the box boundaries, are shown with 0s. Still more extreme values are shown with an asterisk (*). The purpose of the boxplot is to present a picture of the data, and it is useful for comparing several samples side by side. In the next section, we show you how to produce descriptive statistics broken down by one or more variables. When you do this with PROC UNIVARIATE (using a BY statement), you will obtain side-by-side boxplots automatically for each value of the BY variable or for each combination of the BY variables if you have more than one.

To make this clear, a boxplot of test scores, using 50 data points, was generated and is shown here. Notice on this plot, the 0s represent moderate outliers (more than 1.5 interquartile ranges above the third quartile or 1.5 interquartile ranges below the first quartile). The asterisk (at 140) represents more extreme outliers—those that are more than 3 interquartile ranges above or below the 3rd or 1st quartiles, respectively.

```
Variable:   score

    Stem Leaf                     #        Boxplot

    14 0                          1          *
    13
    12
    11
    10 7                          1          0
     9 58                         2          |
     8                                       |
     7 137                        3          |
     6 0111339                    7        +-----+
     5 11122355778899            14        *--+--*
     4 1123347                    7        |     |
     3 3345678                    7        +-----+
     2 046789                     6          |
     1 2                          1          |
     0 9                          1          |
       ----+----+----+----+
Multiply Stem.Leaf by 10**+1
```

The Normal Probability Plot, shown last (also a result of the PLOT option), represents a plot of the data compared to a normal distribution. The y-axis displays our data values, and the x-axis is related to the inverse of the standard normal function. The asterisks (*) mark the actual data values, and the plus signs (+) provide a reference straight line based on the sample mean and standard deviation. Basically, as the sample distribution deviates from the normal, the more the asterisks deviate from the plus signs. You will, no doubt, find the stem-and-leaf plot and boxplots a more intuitive way to inspect the shape of the distribution.

D. HISTOGRAMS, QQ PLOTS, AND PROBABILITY PLOTS

You can ask PROC UNIVARIATE to produce histograms (with a superimposed normal plot), QQ (Quantile-Quantile) plots, and probability plots by adding appropriate statements. To produce a histogram, use the HISTOGRAM statement. By adding a NORMAL option, you can superimpose a normal curve on the histogram. This will allow you to visually compare your distribution to a normal distribution. PROC UNIVARIATE will automatically create a histogram for all the variables in your VAR list. If you wish, you may select variables on the HISTOGRAM statement. For example, to produce a histogram for HEIGHT, with a superimposed normal curve and specifying midpoints for the histogram, submit the following SAS statements (Note: You must have SAS Graph™ registered to run the following example):

```
GOPTIONS RESET=ALL
         FTEXT='Arial/bo'
         CBACK=WHITE
         COLORS=(BLACK)
         GUNIT=PCT
         HTEXT=2
         HPOS=15;
PROC UNIVARIATE DATA=HTWT;
   TITLE "More Descriptive Statistics";
   VAR HEIGHT WEIGHT;
   HISTOGRAM HEIGHT / MIDPOINTS=60 TO 75 BY 5  NORMAL;
   INSET MEAN='Mean' (5.2)
         STD='Standard Deviation' (6.3)/ FONT='Arial'
                                         POS=NW
                                         HEIGHT=3;
   RUN;
   QUIT;
```

Please refer to SAS Online Doc™ or a manual on SAS Graph™ for more details of the graphics options. In this example, Arial bold is requested with text height set at two units (as a percent of the graph size). The INSET statement adds a box that can display selected statistics. We chose mean and standard deviation here. The (5.2) and (6.3) values in parentheses are formats for the mean and standard deviation, respectively. (Formats are specified by $w.d$, where w is the width [total number of spaces] for the number and d is the number of places to the right of the decimal point.) The position option tells SAS where to place the inset box. NW stands for the north-west corner of the graph. The requested histogram is displayed here:

More Descriptive Statistics

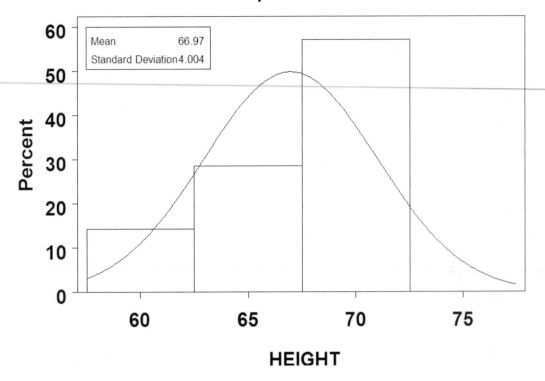

Mean   66.97
Standard Deviation 4.004

HEIGHT

To produce a Quantile-Quantile plot, use the QQ statement. As with the HISTOGRAM statement, you may select specific variables or, if the list of variables is omitted, produce QQ plots for all the variables listed in the VAR statement. The default for this statement is a normal distribution. The vertical axis displays your actual data values, and the horizontal axis shows the quantiles from the specified distribution for these values. If your variable is normally distributed, the QQ plot should be a straight line. You may also produce QQ plots for other distributions, such as lognormal (option LOGNORMAL), beta (BETA), or Weibull (WEIBULL). To produce a QQ plot for HEIGHT, submit the following statements:

```
PROC UNIVARIATE DATA=HTWT;
   TITLE "More Descriptive Statistics";
   VAR HEIGHT WEIGHT;
   QQPLOT HEIGHT;
RUN;
QUIT;
```

The resulting QQ plot is:

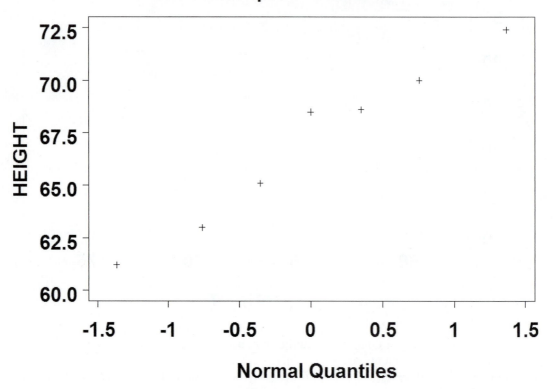

More Descriptive Statistics

Finally, to produce a probability plot, use the PROBPLOT statement. A normal probability plot is another graphical way of determining if your data values come from a distribution that is approximately normal. The vertical axis shows you the actual data values, while the horizontal axis shows you the expected percentiles from a standard normal distribution. If your data values in this plot fall along a straight line, you can assume that your data comes from a normal distribution. A probability plot for WEIGHT would be requested by the statement:

```
PROBPLOT WEIGHT;
```

This plot is shown here:

More Descriptive Statistics

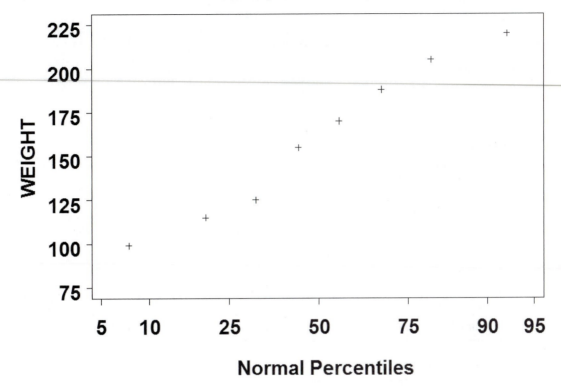

Normal Percentiles

E.   DESCRIPTIVE STATISTICS BROKEN DOWN BY SUBGROUPS

Instead of computing descriptive statistics for every subject in your data set, you may want these statistics for specific subgroups of your data. For example, in the data set of Section B, you might want the number, the mean, and the standard deviation of HEIGHT and WEIGHT separately for males and females. We can accomplish this in two ways. First, we can sort the data set by GENDER and then include a BY statement with PROC MEANS, like this (use of a BY statement requires that your data value be sorted by the same variable(s) or that they be indexed):

```
*Sort the data by GENDER;
PROC SORT DATA=HTWT;
   BY GENDER;
RUN;
                                                    [Continued]
```

```
*Run PROC MEANS for each value of GENDER;
PROC MEANS DATA=HTWT N MEAN STD MAXDEC=2;
   TITLE "The MEANS Procedure, Using a BY Statement";
   BY GENDER;   *This is the statement that gives the breakdown;
   VAR HEIGHT WEIGHT;
RUN;
```

The resulting output is shown here:

```
The MEANS Procedure, Using a BY Statement

GENDER=F

The MEANS Procedure

Variable      N       Mean     Std Dev
────────────────────────────────────────
HEIGHT        3      63.10        1.95
WEIGHT        3     113.00       13.11
────────────────────────────────────────


GENDER=M

Variable      N       Mean     Std Dev
────────────────────────────────────────
HEIGHT        4      69.88        1.82
WEIGHT        5     187.60       26.10
────────────────────────────────────────
```

Many SAS procedures allow you to include a BY statement that runs the procedure for each unique value of the BY variable. Remember to always sort the data set first by the same BY variable (or know in advance that the data set is already sorted). Sometimes in a long program, you ask for a data set to be sorted that you previously sorted (because you sorted it earlier and forgot that you did it). In SAS versions 6.08 and above, if you request a sort on a data set that has been previously sorted by SAS, the sort will not be performed, and you will be notified in the SAS LOG that the data set was already sorted in the order indicated.

An alternative to using a BY statement with PROC MEANS is the use of a CLASS statement. If you use a CLASS statement instead of a BY statement, the printed output will be similar, as you can see in the following output where a CLASS statement is used:

```
PROC MEANS DATA=HTWT N MEAN STD MAXDEC=2;
   TITLE "The MEANS Procedure, Using a CLASS Statement";
   CLASS GENDER; *You do NOT have to sort the data when
                  you use a CLASS statement;
   VAR HEIGHT WEIGHT;
RUN;
```

```
The MEANS Procedure, Using a CLASS Statement

The MEANS Procedure

GENDER     N Obs     Variable      N       Mean     Std Dev
─────────────────────────────────────────────────────────────
F             3      HEIGHT        3       63.10      1.95
                     WEIGHT        3      113.00     13.11

M             5      HEIGHT        4       69.88      1.82
                     WEIGHT        5      187.60     26.10
─────────────────────────────────────────────────────────────
```

The advantage of using a CLASS statement is that you do not have to sort the data set first. For large data sets, this can mean a large saving of processing time (and possibly money). On the downside, use of a CLASS statement requires considerably more memory than a BY statement, especially if there are several CLASS variables and many levels of each. In general, try using a CLASS statement first, and resort to using a BY statement only if memory limitations force you to (or the data set is already sorted in the appropriate order). We amplify the discussion about using a CLASS statement with PROC MEANS in Chapter 4, Section I.

F.   FREQUENCY DISTRIBUTIONS

Let's look at how to get SAS to count how many males and females there are in our sample. The following SAS program does this:

```
DATA HTWT;
    INPUT SUBJECT
          GENDER $
          HEIGHT
          WEIGHT;
DATALINES;
1 M 68.5 155
2 F 61.2 99
3 F 63.0 115
4 M 70.0 205
5 M 68.6 170
6 F 65.1 125
7 M 72.4 220
8 F  .   188
;
PROC FREQ DATA=HTWT;
    TITLE "Using PROC FREQ to Compute Frequencies";
    TABLES GENDER;
RUN;
```

This time, instead of PROC MEANS, we are using PROC FREQ. PROC FREQ is followed by a request for a table of frequencies for the variable GENDER. The word TABLES (or TABLE), used with PROC FREQ, is followed by a list of variables for which we want to count occurrences of particular values (e.g., how many males and how many females for the variable GENDER). Be sure to include a TABLES statement and to list only variables with a reasonable number of levels when you use PROC FREQ. That is, don't ask for frequencies on continuous variables such as AGE, since you will get the frequency for every distinct value. This may kill several trees.

Note that PROC FREQ does not use a VAR statement to specify on which variables to compute frequencies (as we did with PROC MEANS and PROC UNIVARIATE). This may seem inconsistent to you. You need to learn (or look up) the appropriate statements that can be used with each particular procedure. Chapter 20 includes some syntax examples that may help you.

Notice that the two statements, TITLE and TABLES, are indented several spaces to the right of the other lines (one author thinks three spaces are just right). The starting column of any SAS statement does not affect the program in any way. These lines are indented only to make it clear to the programmer that the TITLE and TABLES statements are part of PROC FREQ.

The next table shows the output from PROC FREQ. The column labeled "FREQUENCY" lists the number of people who are male or female; the column labeled "PERCENT" is the same information expressed as a percentage of the total number of people. The "CUMULATIVE FREQUENCY" and "CUMULATIVE PERCENT" columns give us the cumulative counts (the number and percentage, respectively) for each category of gender.

```
Using PROC FREQ to Compute Frequencies

The FREQ Procedure

                              Cumulative   Cumulative
GENDER   Frequency   Percent  Frequency    Percent
F               3     37.50          3       37.50
M               5     62.50          8      100.00
```

If you do not need (or want) the cumulative statistics that are automatically produced by a TABLES request, you may use the NOCUM statement option to omit the cumulative statistics. To review, statement options follow a slash (/) after the appropriate statement. Thus, to omit cumulative statistics, you would write:

```
TABLES GENDER / NOCUM;
```

To omit both cumulative statistics and percentages, you would include the NO-PERCENT TABLES option as well, like this:

```
TABLES GENDER / NOCUM NOPERCENT;
```

The order in which you place these options does not matter. The output from the previous request is shown next:

```
Using PROC FREQ to Compute Frequencies

The FREQ Procedure

GENDER      Frequency
─────────────────────
F                  3
M                  5
```

G.   BAR GRAPHS

We have seen the statistics that are produced by running PROC MEANS and PROC FREQ. It is an excellent way to get a summarization of our data. But then, a picture is worth a thousand words (p = 1000w), so let's move on to presenting pictures of our data. SAS software can generate a frequency bar chart showing the same information as PROC FREQ, using PROC CHART or PROC GCHART. The procedure without the "G" is the older procedure (part of base SAS) that produces rather ugly "printer-quality" charts. That is, it uses printing characters such as asterisks to create a bar. The GCHART procedure is part of SAS Graph™, and it produces very pretty graphics-quality bar charts. The syntax for the two procedures is similar. You can add graphics statements to the program to control aspects of the chart produced with PROC GCHART (such as the color or whether to fill in the box, give just an outline, or make a bar with crosshatches or diagonal lines). We will show the PROC GCHART program and output here. If you do not have SAS Graph™, just leave off the "G" and run PROC CHART instead.

The statements:

```
PROC GCHART DATA=HTWT;
   TITLE "Bar Chart from PROC GCHART";
   VBAR GENDER;
RUN;
```

were used to generate the frequency bar chart shown next:

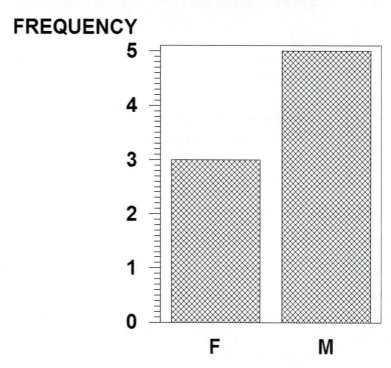

Bar Chart from PROC GCHART

FREQUENCY

GENDER

Note that the graphics options and pattern descriptions were set as follows for this graph:

```
GOPTIONS RESET=ALL
         FTEXT='Arial/bo'
         CBACK=WHITE
         CTEXT=BLACK
         HPOS=25
         GUNIT=PCT
         HTEXT=2;
PATTERN VALUE=X1 COLOR=BLACK;
```

Briefly, the font was set to Arial bold, the color of the text was set to black, and the background was set to white. The horizontal position was set to 25, and the text height was set to 2. Finally, the pattern chosen was X1 (crosshatch, other options are SOLID, EMPTY, L [left-slanting lines], and R [right-slanting lines]). You may follow the X (crosshatch), L, and R values with integers from 1 to 5 to specify the density of the lines (1 lightest, 5 heaviest). Remember that you can omit the GOPTIONS and PATTERN statements completely and still obtain a nice-looking graph using all the SAS default values.

The term HBAR (H for horizontal) in place of VBAR will generate a chart with horizontal bars instead of the vertical bars obtained from VBAR. When HBAR is used, frequency counts and percentages are also presented alongside each bar (see chart). (Note: you can omit the statistics to the right of the horizontal bar chart by using the HBAR option NOSTAT.) Also, the HBAR approach often allows for more groups to be presented as it takes up less space. The same graphics options (except for HPOS) as those in the vertical bar chart were used to produce the horizontal bar chart here:

Bar Chart from PROC GCHART

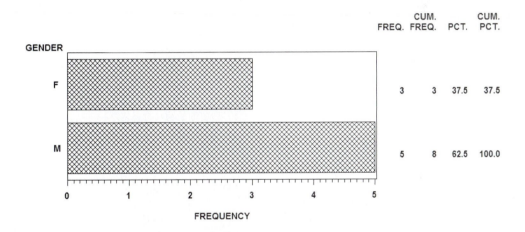

Now, what about the distribution of heights or weights? If we use PROC FREQ to calculate frequencies of heights, it will compute the number of subjects for every value of height (how many people are 60 inches tall, how many are 61 inches tall, etc.). If we use PROC GCHART instead, it will automatically place the subjects into height groups (unless we specify options to control how we wanted the data displayed). Since our sample is so small, a frequency distribution of heights or weights will look silly, but we'll show it to you anyway. The following SAS statements will generate a vertical bar chart:

```
PROC GCHART DATA=HTWT;
    TITLE "Distribution of Heights";
    VBAR HEIGHT / MIDPOINTS = 60 TO 74 BY 2;
RUN;
```

The option MIDPOINTS=60 to 74 by 2 is an instruction to group the heights so that the midpoints of the intervals will start from 60 and end at 74 and the intervals will be 2 units wide. The general form of the MIDPOINTS option is:

```
MIDPOINTS = start to end by interval;
```

If we leave out any options when we are charting a continuous variable, PROC GCHART will use its own grouping algorithm to select the number of levels and the midpoints for the plot. The output from the previous program is:

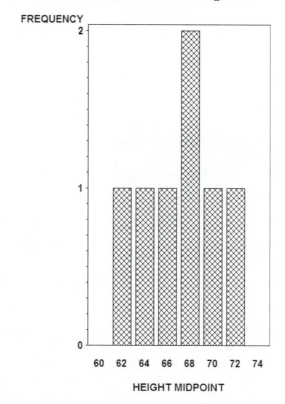

The VBAR and HBAR statements of PROC CHART or PROC GCHART have a variety of options. The general form of the VBAR and HBAR statements is:

```
VBAR or HBAR   list-of-variable(s) / list-of-options ;
```

An alternative to the MIDPOINTS= option is the option LEVELS=*n*.

An example would be:

```
VBAR HEIGHT / LEVELS=6;
```

There are times when we do not want PROC GCHART to divide our numerical variable into intervals. Look at the following chart, where frequencies of a variable called WEEK (day of the week) were displayed (using the statement VBAR WEEK;). Notice that SAS attempted to divide the x-axis into intervals (because it wasn't told that fractional values for WEEK were not appropriate).

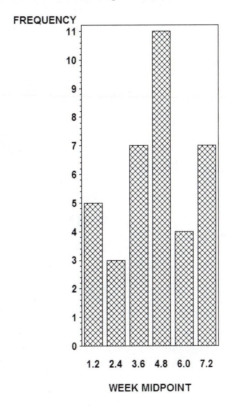

Chart of Week (values of 1 to 7)

To have SAS treat the WEEK values as discrete values (as if they were character values), use the DISCRETE option of the VBAR or HBAR statement.

```
VBAR WEEK / DISCRETE;
```

Remember that statement options are placed between the statement and the semi-colon, separated by a slash.

Notice the difference in the resulting chart.

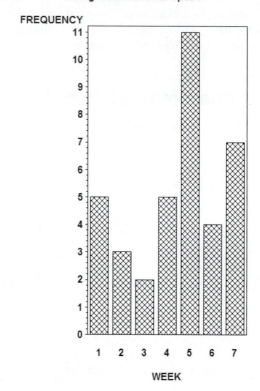

Chart of Week (values of 1 to 7)
Using the DISCRETE Option

Before leaving PROC GCHART, we demonstrate a few of the other options that are available. To do this, we have constructed another data set called STORE (see following program), which contains the variables YEAR, DEPT (department), and SALES.

```
DATA E_MART;
    INPUT YEAR
          DEPT $
          SALES;
DATALINES;
```

```
2001 TOYS 5000
2001 TOYS 4500
2001 TOYS 5500
2001 FOOD 4100
2001 FOOD 3300
2002 TOYS 6344
2002 TOYS 4567
2002 TOYS 4567
2002 TOYS 4567
2002 TOYS 4300
2002 FOOD 3700
2002 FOOD 3900
2003 TOYS 7000
2003 TOYS 7200
2003 TOYS 6000
2003 TOYS 7900
2003 FOOD 4000
2003 FOOD 5800
2003 FOOD 5600
;
```

The statements:

```
PATTERN VALUE=EMPTY COLOR=BLACK;
AXIS1 LABEL=('Department');
PROC GCHART DATA=E_MART;
   TITLE "Simple Frequency Bar GCHART";
   VBAR DEPT / MAXIS=AXIS1;
RUN;
```

will produce a simple frequency bar graph as shown here (Note: Same graphics as the first GCHART. This time we are selecting EMPTY to obtain only the bar outline, which can save a lot of ink or toner in your printer. The AXIS1 statement supplies a label for the midpoints, and the VBAR option MAXIS says to use the AXIS1 definition for the midpoint axis. You can use a LABEL statement instead of the AXIS statement to obtain the same result. As we mentioned earlier, you can leave out all these GOPTION, PATTERN, and AXIS statements and still get a reasonable-looking graph):

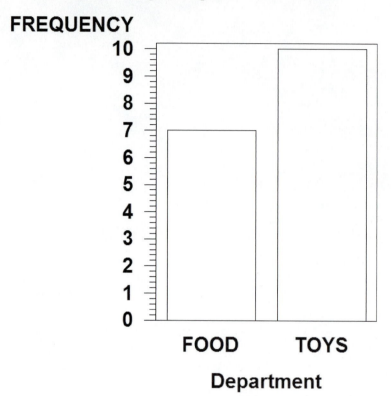

Simple Frequency Bar GCHART

FREQUENCY

The statements:

```
PROC GCHART DATA=E_MART;
    TITLE "Bar Chart on a Numerical Variable (SALES)";
    VBAR SALES;
RUN;
```

will produce a frequency distribution of SALES for all years and all departments.

Bar Chart on a Numerical Variable (SALES)

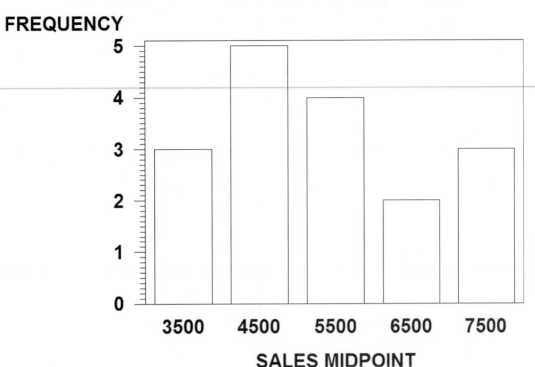

To see the sales distributions of each department side by side, we can use the GROUP option available with both VBAR and HBAR. The statement:

```
PATTERN VALUE=L2 COLOR=BLACK;
PROC GCHART DATA=E_MART;
   TITLE "Distribution of SALES by Department";
   VBAR SALES / GROUP=DEPT MIDPOINTS=4500 TO 5500 BY 1000;
   FORMAT SALES DOLLAR8.0;
RUN;
```

will produce the side-by-side graph like the one shown next (Note: This time we used the L2 style [left-slanting lines], specified midpoints for the variable SALES, and formatted SALES with a DOLLAR format so that the dollar signs and commas would be printed). Here it is:

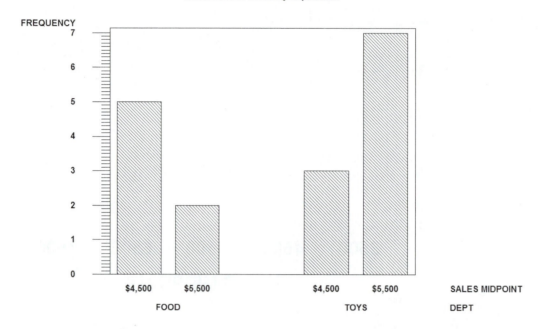

Another way to display these data is to have the y-axis represent a sales sum rather than a frequency or count. This is done by using a SUMVAR option with VBAR or HBAR. The keyword SUMVAR is followed by a variable whose sum we want displayed on the y-axis. We also use the SUMVAR option to display a mean value on the y-axis by adding the keyword TYPE=MEAN to the list of VBAR or HBAR options. We will show you a chart using the SUMVAR and TYPE options. Since we are displaying a sum of sales for each department, the TYPE= option is redundant but is included to remind you that it is available to display other statistics on the y-axis. Notice the DISCRETE option is needed here to treat the YEAR variable as discrete values and not to attempt to create midpoints for the YEAR variable. The statements:

```
PROC GCHART DATA=E_MART;
    TITLE "Sum of SALES by YEAR";
    VBAR YEAR / SUMVAR=SALES TYPE=SUM DISCRETE;
    FORMAT SALES DOLLAR8.;
RUN;
```

produces the following graph:

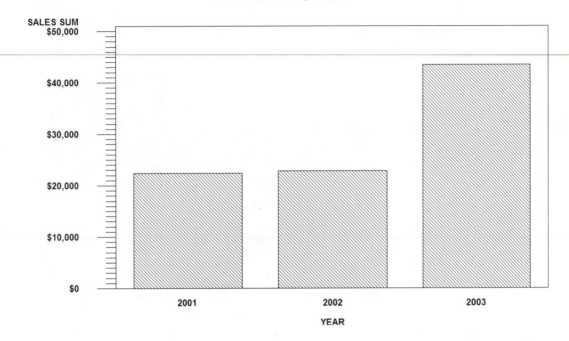

Sum of **SALES** by YEAR

Other valid values for the TYPE = option are:

Option	Result
TYPE=FREQ	Frequency counts
TYPE=PCT	Percentages
TYPE=CFREQ	Cumulative frequencies
TYPE=CPCT	Cumulative percentages
TYPE=SUM	Totals
TYPE=MEAN	Means

So, to see the mean sales by year, you would use:

```
PROC GCHART DATA=E_MART;
   TITLE "Mean Sales by YEAR";
   VBAR YEAR / SUMVAR=SALES TYPE=MEAN DISCRETE;
   FORMAT SALES DOLLAR8.;
RUN;
```

with the resulting chart:

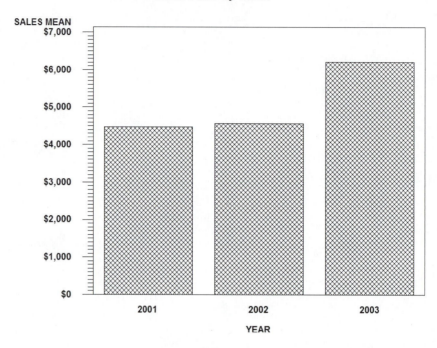

One final option used with VBAR and HBAR is SUBGROUP. The values of the SUB-GROUP variable are displayed within each bar. With PROC CHART, the first character of the value of the SUBGROUP variable is used to create the bar; in PROC GCHART, the bar is broken up into either different colors or crosshatch patterns. If we write:

```
PATTERN1 COLOR=BLACK VALUE=X1;
PATTERN2 COLOR=BLACK VALUE=L2;

PROC GCHART DATA=E_MART;
   TITLE "Demonstrating the SUBGROUP Option";
   VBAR YEAR / SUBGROUP=DEPT
              SUMVAR=SALES
              TYPE=SUM
              DISCRETE;
   FORMAT SALES DOLLAR8.;
```

```
RUN;
QUIT;
```

the department values will be shown within each bar (please see following). (Note: Here we used two PATTERN statements to define the patterns for each subgroup. Again, if you leave out all the PATTERN statements, you still obtain an acceptable-looking chart):

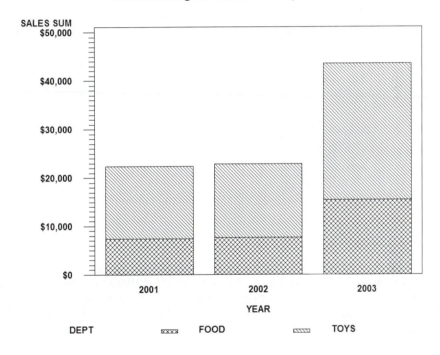

Demonstrating the SUBGROUP Option

Finally, if you would like to see the mean sales by YEAR and department (DEPT), you can include both a GROUP and SUMVAR variable as shown next:

```
PROC GCHART DATA=E_MART;
    TITLE "SALES Broken Down by YEAR and DEPT";
    VBAR YEAR / GROUP=DEPT SUMVAR=SALES TYPE=SUM DISCRETE;
    FORMAT SALES DOLLAR8.;
RUN;
```

The resulting chart is shown here:

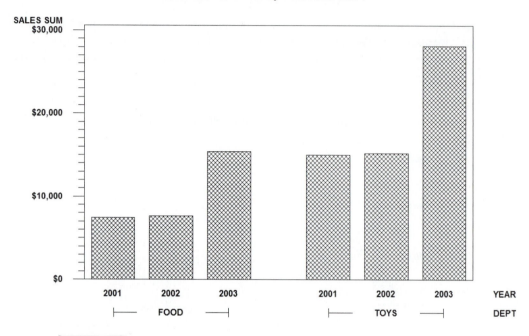

H. PLOTTING DATA

We now investigate the relationship between height and weight in the HTWT data set (earlier in the chapter). Our intuition tells us that these two variables are related: The taller a person is, the heavier (in general). The best way to display this relationship is to draw a graph of height versus weight. We can have our SAS program generate this graph by using PROC PLOT (or use the SAS Graph™ equivalent, PROC GPLOT). The statements:

```
PROC PLOT DATA=HTWT;
    TITLE "Scatter Plot of WEIGHT by HEIGHT";
    PLOT WEIGHT*HEIGHT;
RUN;
```

generate a scatter plot or WEIGHT (on the y-axis) versus HEIGHT (on the x-axis). In general, the variable before the asterisk in the PLOT statement will represent the variable on the y-axis, and the variable after the asterisk will represent the variable on the x-axis. (Note: This is a plot using the original data set of eight people, one with a missing value.)

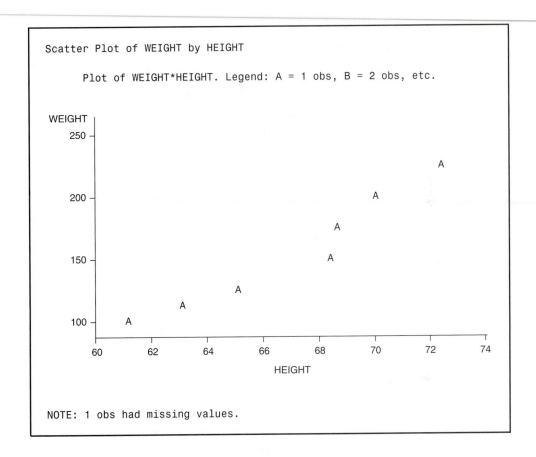

PROC PLOT uses the letters A, B, C, etc. as the default plotting symbols. Since this procedure was originally intended to produce output on a printing device such as a typewriter terminal, the plotting symbols needed to be in fixed rows and columns. Thus, two or more data points that were close in value would wind up in the same location on the plot. A 'B' would indicate that there were two points at that location, a 'C' three points, etc. Using PROC GPLOT eliminates this problem since output from GPLOT is a true graphics image. The GPLOT statements to produce a similar plot to the previous one are:

```
PROC GPLOT DATA=HTWT;
   TITLE "Scatter Plot of WEIGHT by HEIGHT";
   TITLE2 "Using all the Defaults";
   PLOT WEIGHT*HEIGHT;
RUN;
```

Notice the plus signs are the default plotting symbols.

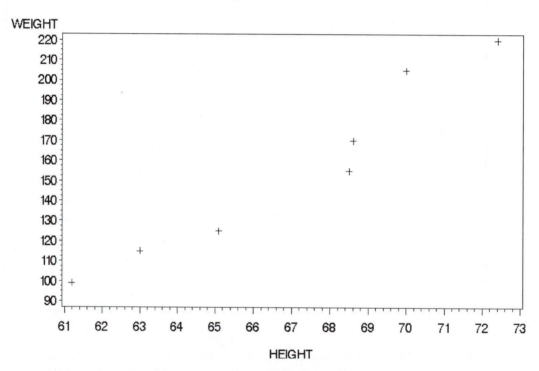

To change the plotting symbols using PROC GPLOT, use a SYMBOL statement. You can change many of the attributes of the plot, such as the plotting symbol, the color of the points, whether to join the points, etc., with this statement. As an example, let's produce the previous graph using dots as the plotting symbols and making sure the data points are printed in black. The statement:

```
SYMBOL VALUE=DOT COLOR=BLACK;
```

placed before PROC GPLOT will do just that. Some alternatives to DOT are CIRCLE, SQUARE, PLUS, and TRIANGLE. Please see the SAS Online Doc™ or a reference on SAS Graph™ for more choices of plotting symbols and for other plotting options. If you are producing only a single plot, the default color is black, however, adding a COLOR option will not do any harm. It is important to note that attributes changed in a SYMBOL statement remain in effect until you close your SAS session and that they are additive. That is, if you have selected a plotting symbol with one SYMBOL statement and you change the color in another SYMBOL statement, both attributes (the plotting symbol and color) will be changed. Please see the following graph, produced by the previous SYMBOL statement:

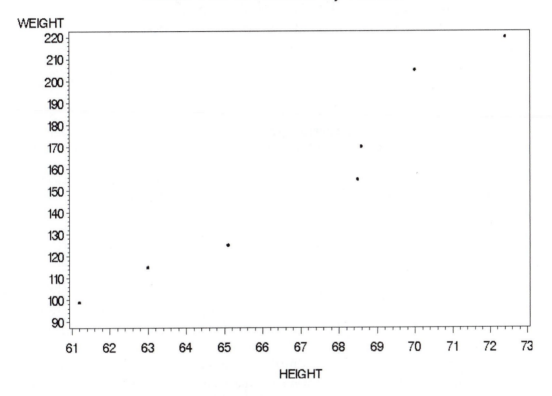

One form of the plot procedure is:

```
PROC PLOT DATA=data_set_name; /* Same syntax for GPLOT */
    PLOT Y-variable * X-variable;
RUN;
```

You might be thinking that plotting male and female height and weight on the same graph is misleading. Can we obtain a plot of height versus weight for males and one for females? The answer is 'yes,' and it is quite easy to do. Just as we used a BY variable with PROC MEANS earlier, we can use the same BY statement with PROC PLOT to create separate plots for males and females. First, we need to have the SAS program sort our data by GENDER. Once this is done, we can use PROC PLOT to produce the desired graphs.

Our program will look as follows:

```
PROC SORT DATA=HTWT;
   BY GENDER;
RUN;
PROC PLOT DATA=HTWT;
   TITLE "Separate Plots by GENDER";
   BY GENDER;
   PLOT WEIGHT*HEIGHT;
RUN;
```

The result of this program is a separate graph for males and another for females. If we omit a BY statement with PROC PLOT, the program ignores the fact that the data set is now sorted by GENDER.

We can generate another graph that displays the data for males and females on a single graph, but instead of the usual plotting symbols, we will use separate plotting symbols for the males and females. The statements:

```
PROC PLOT DATA=HTWT;
   TITLE "Using GENDER to Generate the Plotting Symbol";
   PLOT WEIGHT*HEIGHT=GENDER;
RUN;
```

accomplishes this. If you are using PROC PLOT, the first character of the value of GENDER is used as the plotting symbol; if you are using PROC GPLOT, you can use two SYMBOL statements (SYMBOL1 and SYMBOL2) to select a plotting symbol for males and females (and the procedure will provide a key at the bottom of the graph). It is important to use the COLOR=BLACK option when you have more than one SYMBOL statement. Without this, even with two different VALUE (V) options, SAS will use only one plotting symbol and will use different colors for the two sets of data points. The reason for this is that SAS will, by default, cycle through different colors of one SYMBOL statement before moving on to the next SYMBOL statement. Perhaps the best way of avoiding problems relating to the color of plotting symbols is to use the single statement GOPTIONS CSYMBOL=BLACK. Note that the data set does not have to be sorted to use this form of PROC PLOT or PROC GPLOT. Using the following PROC GPLOT statements

```
SYMBOL1 V=CIRCLE COLOR=BLACK;
SYMBOL2 V=SQUARE COLOR=BLACK;
***Note: SYMBOL and SYMBOL1 are equivalent;
PROC GPLOT DATA=HTWT;
   TITLE "Using GENDER to Generate the Plotting Symbol";
   PLOT WEIGHT*HEIGHT=GENDER;
RUN;
```

results in the following graph:

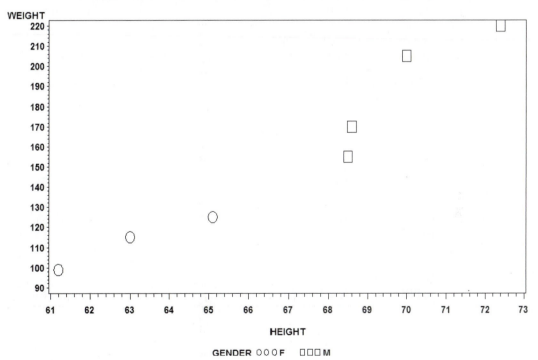

Using **GENDER** to Generate the Plotting Symbol

If you are using PROC PLOT and would like to choose a plotting symbol instead of the SAS default of A, B, C, etc., you may follow the PLOT request with an equal sign and a plotting symbol of your choice in quotes. If you wanted an asterisk as your plotting symbol, the plot request would read:

```
PLOT WEIGHT*HEIGHT='*';
```

If you are using PROC GPLOT, you could indicate a plotting symbol by referring to a particular SYMBOL statement. For example, the statement

```
PLOT WEIGHT*HEIGHT=2;
```

would result in squares being used as plotting symbols, given the two SYMBOL statements used earlier.

PROBLEMS

Remember, you can download all the data sets and programs for these problems from the web site: www.prenhall.com/cody

2.1 Add the necessary statements to compute the number of males and females in problem 1.1.

2.2 Using the following data, create a SAS data set called CLINIC. In the DATA step, include a statement to compute average blood pressure (call it AVE_BP) computed as the diastolic blood pressure (DBP) plus one-third the difference of the systolic blood pressure (SBP) and the DBP. (For those interested in this, since the heart spends more time in diastole (relaxed state) than it does in systole (when it contracts), a weighted average is used to represent the "average" blood pressure—one-third of the SBP and two-thirds of the DBP.) Use PROC MEANS to compute the number of nonmissing values, the mean, the standard deviation, the 95% CI on the mean, and the median value for SBP, DBP, and AVE_BP.

The data layout:

Variable	Description	Starting Column	Length	Type
ID	Subject ID	1	3	Char
GENDER	Gender	4	1	Char
RACE	Race of subject	5	1	Char
HR	Heart rate	6	3	Num
SBP	Systolic blood pressure	9	3	Num
DBP	Diastolic blood pressure	12	3	Num
N_PROC	Number of procedures	15	2	Num

Sample data:

```
001MW08013008010
002FW08811007205
003MB05018810002
004FB   10806801
005MW06812208204
006FB101   07404
007FW07810406603
008MW04811207006
009FB07719011009
010FB06616410610
```

2.3 Modify the program from problem 1.3 to create a new variable (AGE_GROUP) that has a value of 1 for ages between 0 and 35, and 2 for ages greater than 35. Compute the number of Whites (W) and Blacks (B) and the number in each age group. Use the appropriate option to omit cumulative statistics from the output.

2.4 Using the data set from problem 2.2, generate frequencies and bar charts for GENDER. Omit the cumulative statistics and percentages for your table.

2.5 Run the following program to create a SAS data set called PROB2_5, containing variables X, Y, Z, and GROUP:

```
DATA PROB2_5;
    LENGTH GROUP $ 1;
    INPUT X Y Z GROUP $;
DATALINES;
2   4   6   A
3   3   3   B
1   3   7   A
7   5   3   B
1   1   5   B
2   2   4   A
5   5   6   A
;
```

(a) Write the SAS statements to generate a frequency bar chart (histogram) for GROUP. You may use either the base SAS procedures or the SAS Graphic procedures to answer these questions.

(b) Write the SAS statements to generate a plot of Y versus X (with "Y" on the vertical axis and "X" on the horizontal).

(c) Write the SAS statements to generate a separate plot of Y versus X for each value of the GROUP variables.

2.6 Using the data set from problem 2.2, write the SAS statements necessary to produce a stem-and-leaf plot and a boxplot for the variables SBP and DBP.

2.7 We have recorded the following data from an experiment:

SUBJECT	DOSE	REACT	LIVER_WT	SPLEEN
1	1	5.4	10.2	8.9
2	1	5.9	9.8	7.3
3	1	4.8	12.2	9.1
4	1	6.9	11.8	8.8
5	1	15.8	10.9	9.0
6	2	4.9	13.8	6.6
7	2	5.0	12.0	7.9
8	2	6.7	10.5	8.0
9	2	18.2	11.9	6.9
10	2	5.5	9.9	9.1

Use PROC UNIVARIATE to produce histograms, normal probability plots, and boxplots, and test the distributions for normality. Do this for the variables REACT, LIVER_WT, and SPLEEN, first for all subjects and then separately for each of the two DOSES.

2.8 Using the data set from problem 2.2, make two charts (using either PROC CHART or PROC GCHART). One should represent the frequencies of each GENDER for each value of RACE, as shown here. The other should show a distribution of heart rate for each value of GENDER (also shown here). Note that the x-axis for the second chart uses midpoints of 50, 60, 70, etc.

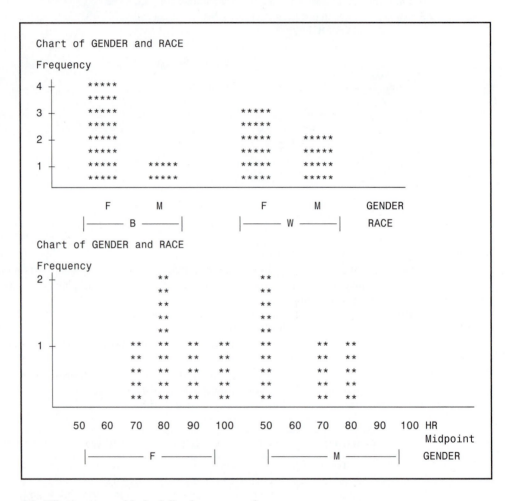

2.9 What's wrong with the following program?

```
DATA 123;
   INPUT AGE STATUS PROGNOSIS DOCTOR GENDER STATUS2
   STATUS3;
(data lines)
;
```

```
PROC CHART DATA=123 BY GENDER;
    VBAR STATUS
    VBAR PROGNOSIS;
RUN;
PROC PLOT DATA=123;
    DOCTOR BY PROGNOSIS;
RUN;
```

2.10 Using the data set from problem 2.2, generate several plots:

(a) Showing heart rate (HR) on the x-axis and systolic blood pressure (SBP) on the y-axis. Select a plotting symbol of your choice and use either PROC PLOT or PROC GPLOT.

(b) A plot of diastolic blood pressure on the x-axis and systolic blood pressure (SBP) on the y-axis. Use values of GENDER as plotting symbols (PROC PLOT) or symbols of your choice (PROC GPLOT).

(c) Repeat (b), except produce a separate graph for RACE=W and RACE=B.

2.11 Given the data set:

Salesperson	Target Company	Number of Visits	Number of Phone Calls	Units Sold
Brown	American	3	12	28,000
Johnson	VRW	6	14	33,000
Rivera	Texam	2	6	8,000
Brown	Standard	0	22	0
Brown	Knowles	2	19	12,000
Rivera	Metro	4	8	13,000
Rivera	Uniman	8	7	27,000
Johnson	Oldham	3	16	8,000
Johnson	Rondo	2	14	2,000

(a) Write a SAS program to compare the sales records of the company's three salespeople. (Compute the sum and mean for the number of visits, phone calls, and units sold for each salesman.) Note that the values for "Units sold" contain commas. Since you don't know how to read data values with commas (you use a COMMA informat), omit the commas when you enter your data values.

(b) Plot the number of visits against the number of phone calls. Use the value of Salesperson (the first character in the name) as the plotting symbol (instead of the usual A, B, C, etc.).

(c) Make a frequency bar chart for each salesperson for the sum of "units sold."

*2.12 Using the data set from problem 2.2, create a chart showing the sum of procedures (N_PROC) for each GENDER with the contribution of each value of RACE shown within each bar (please see following chart):

Number of Procedures by GENDER and RACE

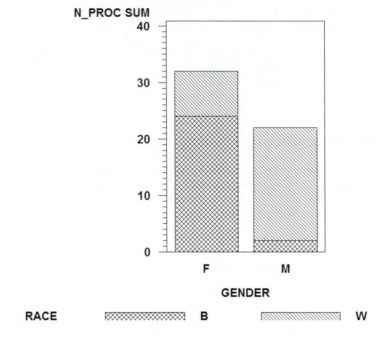

*2.13 You have completed an experiment and have recorded a subject ID and values for variables A, B, and C. You want to compute means for A, B, and C but, unfortunately, your lab technician, who didn't know SAS programming, arranged the data like this:

ID	TYPE	SCORE
1	A	44
1	B	9
1	C	203
2	A	50
2	B	7
2	C	188
3	A	39
3	B	9
3	C	234

Write a program to read these data and produce means. (HINT: A CLASS or BY statement might come in handy.)

CHAPTER 3

Analyzing Categorical Data

A.   INTRODUCTION

This chapter explains ways of analyzing categorical data as well as providing step-by-step instructions for analyzing questionnaires. Variables such as gender, sick or well, success or failure, and age group represent categories rather than numerical values. We use a number of statistical techniques such as tests of proportions and chi-square with variables of this type.

You may notice a substantial enlargement of this topic from our previous edition. This is the result of expanded techniques from the SAS procedures used on categorical data and some techniques such as meta-analysis that have become increasingly popular in many fields.

B.   QUESTIONNAIRE DESIGN AND ANALYSIS

A common way to collect certain types of data is by using a questionnaire. Although these can be designed in many ways, the accompanying Sample Questionnaire contains features that make it especially useful when the collected data are to be entered into a computer. We present an approach based on data being manually entered into the computer, which is still quite common. However, scanners are increasingly being used to enter data, and many researchers are collecting their data via computer or online, and the data are automatically entered into a data set. Researchers doing a lot of work with questionnaires should investigate that possibility.

For office use only

SAMPLE QUESTIONAIRE

ID ☐ ☐ ☐
☐ ☐

1. Age in years _____

(Questions 2–4: Please check the appropriate category.)

2. Gender _____ 1 = Male
_____ 2 = Female   ☐

3. Race _____ 1 = White
_____ 2 = African American
_____ 3 = Hispanic
_____ 4 = Other   ☐

4. Marital Status:
_____ 1 = Single
_____ 2 = Married
_____ 3 = Widowed
_____ 4 = Divorced   ☐

5. Education Level:
_____ 1 = High school or less
_____ 2 = Two-year college
_____ 3 = Four-year college (BA or BS)
_____ 4 = Post-graduate degree   ☐

For each of the following statements, please enter a NUMBER from the following list on the line to the LEFT of each question. Use the following codes:

1 = Strongly disagree   2 = Disagree   3 = No opinion
4 = Agree   5 = Strongly agree

_____ **6.** The president of the U.S. has been doing a good job.   ☐

_____ **7.** The arms budget should be increased.   ☐

_____ **8.** There should be more federal aid to big cities.   ☐

Designing a questionnaire presents several problems. First, you want to make the questions and directions as clear as possible. You want it to have a clean look and not be too long. Many multipage questionnaires wind up in the circular file. It is important to get as high a return rate as possible. If only highly conscientious individuals or those with an interest in the outcome of the survey complete a questionnaire, their answers are not likely to be representative of the population as a whole. They may be responding because they have "an axe to grind." This results in bias, a systematic skewing of the results.

The typical method of coding data from a questionnaire of this type is to enter the data into a computer by using a word processor, a specialized data-entry program, a spreadsheet (such as Lotus™ or Excel™), or by using a database management program (such as Paradox™, Access™, or DBase™). In the case of the word processor or data-entry program, we would probably set aside certain columns for each variable. Where a database management system is used, the data can either be written to a text file or automatically converted directly to a SAS data set by using the SAS Access Product™. For example, SAS Access™ for PC File Formats allows you to automatically convert such things as Excel™ spreadsheets and Access™ databases directly into SAS data sets. Another option is the use of an optical mark sense reader for large volume data-entry requirements. It is preferable to design the questionnaire so that the data can be entered directly from the questionnaire rather than having to be transcribed first to a coding form.

We might decide to enter our questionnaire data as follows:

Column	Description	Variable Name
1–3	Subject ID	ID
4–5	Age in years	AGE
6	Gender	GENDER
7	Race	RACE
8	Marital status	MARITAL
9	Education level	EDUCATION
10	President doing good job	PRESIDENT
11	Arms budget increased	ARMS
12	Federal aid to cities	CITIES

Typical lines of data would look like:

```
001091111232
002452222422
```

Notice that we have not left any spaces between the values for each variable. This is a common method of data entry since it saves space and extra typing. Therefore, we must specify the column location for each variable. Our INPUT statement for this questionnaire would be written:

```
INPUT ID         $ 1-3
      AGE          4-5
      GENDER     $   6
      RACE       $   7
      MARITAL    $   8
      EDUCATION  $   9
      PRESIDENT     10
      ARMS          11
      CITIES        12;
```

Remember, even though this INPUT statement takes several lines, it is a single SAS statement and ends with a semicolon. Each variable name is followed by its column designation. We chose to store data values for ID, GENDER, RACE, MARITAL, and EDUCATION as characters, even though we are coding the values as numbers (technically, character representations of numbers are called numerals). One reason for doing this is to save storage space on the computer. Another reason is that we would be unlikely to perform any type of arithmetic operation on a variable such as RACE. Storing these values as characters helps remind us that the numerical values are merely the names of categories.

A common occurrence with questionnaires is that some respondents do not answer all the questions. With a list INPUT statement (one in which we list only the variable names and not the column designations), we use a period to represent a missing value; with our column INPUT statement, we leave the column(s) blank. We can do this since it is the columns, not the order of the data, that determine which variable is being read.

A complete SAS program shown here, with some sample lines of data, calculates the mean age of the respondents and computes frequencies for all the other variables:

```
DATA QUEST;
   INPUT ID         $ 1-3
         AGE          4-5
         GENDER     $   6
         RACE       $   7
         MARITAL    $   8
         EDUCATION  $   9
         PRESIDENT     10
         ARMS          11
         CITIES        12;
DATALINES;
001091111232
002452222422
003351324442
004271111121
005682132333
006651243425
;
```

```
PROC MEANS DATA=QUEST MAXDEC=2 N MEAN STD CLM;
   TITLE "Questionnaire Analysis";
   VAR AGE;
RUN;


PROC FREQ DATA=QUEST;
   TITLE "Frequency Counts for Categorical Variables";
   TABLES GENDER RACE MARITAL EDUCATION
          PRESIDENT ARMS CITIES;
RUN;
```

We have chosen to supply PROC MEANS with options to print statistics to two decimal places and to compute N, the number of nonmissing observations, the mean, the standard deviation, and the 95% confidence interval for the mean.

We listed the variables GENDER, RACE, MARITAL, EDUCATION, PRESI-DENT, ARMS, and CITIES in our PROC FREQ TABLES statement. A shortcut method for specifying a list of variables in a SAS data or PROC step is the -- (two dashes together) notation. The convention

```
variable_name_1 -- variable_name_2
```

means to include all the variables from *variable_name_1* to *variable_name_2* in the order they exist in the SAS data set. A TABLES statement equivalent to the previous one using this shortcut method, is:

```
TABLES GENDER -- CITIES;
```

The statement "in the order they exist in the SAS data set" is particularly important when using this shortcut method of specifying variable lists. In this case, the order is the same as the order on the INPUT statement. As you will see later, such SAS statements as LENGTH, ARRAY, and RETAIN can affect this order.

While we are on the topic of variable list notation, *ROOTn - ROOTm* is used to refer to all variables with the same alphabetic root, from the *n*th to the *m*th numeric ending. For example, ABC1-ABC5 is equivalent to: ABC1 ABC2 ABC3 ABC4 ABC5. It is convenient to name certain variables using the same root with a number ending so we can use the single-dash notation any time we want to refer to part or all of the list. If we recorded the response to 50 multiple-choice questions in a test, convenient variable names would be QUES1, QUES2, ... up to QUES50. Then, if we wanted frequencies on all 50 variables, the tables request:

```
TABLES QUES1-QUES50;
```

would do the trick. It is not necessary for the variables to be in any particular order in the data set when this notation is used.

In this questionnaire, we have not requested any statistics for the variable ID. The ID number only serves as an identifier if we want to go back to the original questionnaire to check data values. This is a highly recommended procedure. Without an ID variable of some sort, when we discover an error in the data, it is difficult to find the original questionnaire to check on the correct value. Even if you did not include an ID on the questionnaire, number the questionnaires as they are returned, and enter this number in the computer along with the other responses.

A sample of the output from PROC FREQ is shown here:

```
Frequency Counts for Categorical Variables

The FREQ Procedure

                                   Cumulative    Cumulative
GENDER    Frequency     Percent    Frequency      Percent

1              4         66.67          4          66.67
2              2         33.33          6         100.00

                                   Cumulative    Cumulative
RACE    Frequency      Percent     Frequency      Percent

1            3          50.00           3          50.00
2            2          33.33           5          83.33
3            1          16.67           6         100.00

                                   Cumulative    Cumulative
MARITAL    Frequency     Percent    Frequency      Percent

1              2         33.33          2          33.33
2              2         33.33          4          66.67
3              1         16.67          5          83.33
4              1         16.67          6         100.00

                                   Cumulative    Cumulative
EDUCATION    Frequency    Percent   Frequency      Percent

1              2         33.33          2          33.33
2              2         33.33          4          66.67
3              1         16.67          5          83.33
4              1         16.67          6         100.00

                                   Cumulative    Cumulative
PRESIDENT    Frequency    Percent   Frequency      Percent

1              1         16.67          1          16.67
2              1         16.67          2          33.33
3              1         16.67          3          50.00
4              3         50.00          6         100.00

                                   Cumulative    Cumulative
ARMS    Frequency      Percent     Frequency      Percent

2            3          50.00           3          50.00
3            2          33.33           5          83.33
4            1          16.67           6         100.00
```

CITIES	Frequency	Percent	Cumulative Frequency	Cumulative Percent
1	1	16.67	1	16.67
2	3	50.00	4	66.67
3	1	16.67	5	83.33
5	1	16.67	6	100.00

C.   ADDING VARIABLE LABELS

We can improve considerably on this output. First, we have to refer back to our coding scheme to see the definition of each of the variable names. Some variable names like GENDER and RACE need no explanation; others like PRESIDENT and CITIES do. We can associate a variable label with each variable name by using a LABEL statement. These labels will be printed along with the variable name in certain procedures such as PROC FREQ and PROC MEANS. The general form of a LABEL statement is:

```
LABEL variable_name = "description"
              .
              .
              .
      variable_name = "description";
```

The "description" can contain up to 256 characters (each blank counts as a character) and must be enclosed in single or double quotes. If the label contains single quotation marks, use double quotes to start and end the label. The LABEL statement can be placed anywhere in the DATA step. Our program, rewritten to include variable labels, follows:

```
DATA QUEST;
    INPUT ID          $ 1-3
          AGE           4-5
          GENDER      $   6
          RACE        $   7
          MARITAL     $   8
          EDUCATION   $   9
          PRESIDENT      10
          ARMS           11
          CITIES         12;
                                            [Continued]
```

```
LABEL MARITAL    = "Marital Status"
      EDUCATION  = "Education Level"
      PRESIDENT  = "President Doing a Good Job"
      ARMS       = "Arms Budget Increase"
      CITIES     = "Federal Aid to Cities";
DATALINES;
001091111232
002452222422
003351324442
004271111121
005682132333
006651243425
;
PROC MEANS DATA=QUEST MAXDEC=2 N MEAN STD CLM;
   TITLE "Questionnaire Analysis";
   VAR AGE;
RUN;

PROC FREQ DATA=QUEST;
   TITLE "Frequency Counts for Categorical Variables";
   TABLES GENDER RACE MARITAL EDUCATION
          PRESIDENT ARMS CITIES;
RUN;
```

Notice that we did not supply variable labels for all our variables. The ones you choose are up to you. Now, when we run our program, the labels will be printed along with the variable names in our PROC FREQ output. A partial listing of the output from this program is shown here:

```
Frequency Counts for Categorical Variables

The FREQ Procedure

                    Education Level
```

EDUCATION	Frequency	Percent	Cumulative Frequency	Cumulative Percent
1	2	33.33	2	33.33
2	2	33.33	4	66.67
3	1	16.67	5	83.33
4	1	16.67	6	100.00

President Doing a Good Job

PRESIDENT	Frequency	Percent	Cumulative Frequency	Cumulative Percent
1	1	16.67	1	16.67
2	1	16.67	2	33.33
3	1	16.67	3	50.00
4	3	50.00	6	100.00

Notice that the variable name and the label are used by this procedure.

D.   ADDING "VALUE LABELS" (FORMATS)

We would like to improve the readability of the output one step further. The remaining problem is that the values for our variables (1 = male, 2 = female, etc.) are printed on the output, not the names that we have assigned to these values. We would like the output to show the number of males and females, for example, not the number of 1s and 2s for the variable GENDER. We can supply the "value labels" in two steps.

The first step is to define our code values for each variable. For example, 1 = Male, 2 = Female will be used for our variable GENDER. The codes 1 = str disagree, 2 = disagree, etc. will be used for three variables: PRESIDENT, ARMS, and CITIES. SAS calls these codes formats, and we define the formats in a procedure by that name.

The second step, shown later, will be to associate a FORMAT with one or more variable names. The following is an example of PROC FORMAT used for our questionnaire:

```
PROC FORMAT;
    VALUE $SEXFMT '1' = 'Male'
                  '2' = 'Female'
          OTHER     = 'Miscoded';
    VALUE $RACE   '1' = 'White'
                  '2' = 'African Am.'
                  '3' = 'Hispanic'
                  '4' = 'Other';
    VALUE $OSCAR  '1' = 'Single'
                  '2' = 'Married'
                  '3' = 'Widowed'
                  '4' = 'Divorced';
    VALUE $EDUC   '1' = 'High Sch or Less'
                  '2' = 'Two Yr. College'
                  '3' = 'Four Yr. College'
                  '4' = 'Graduate Degree';
                                        [Continued]
```

```
        VALUE LIKERT    1 = 'Str Disagree'
                        2 = 'Disagree'
                        3 = 'No opinion'
                        4 = 'Agree'
                        5 = 'Str Agree';
     RUN;
```

The names $SEXFMT, $RACE, $OSCAR, $EDUC, and LIKERT are format names. They are from 1 to 32 characters in length (in versions prior to 9, format names had a maximum length of 8 characters). If they are going to define a format for character variables, they must begin with a dollar sign. (The dollar sign counts as one of the characters.) (An aside: We chose the name LIKERT since scales such as 1 = strongly disagree, 2 = disagree, 3 = no opinion, etc., are called Likert scales by psychometricians.) You may choose any name (consistent with naming conventions, with the exception that they cannot end in a number or be identical to a SAS supplied format) for your formats. It is best to give names that help you remember what the format will be used for. As a matter of fact, you can use the same name for a format and for a variable without confusion. Notice the silly format name $OSCAR used to format values of the variable called MARITAL. We did this just to emphasize the fact that format names do not have to be related to the variables they will be used to format.

Following the format name is a single value, a range of values, a keyword, or a combination of all three. For example, in defining the ranges for the format $SEXFMT, values of '1' will be associated with the formatted value "Male" and a value of '2' will be associated with the formatted value "Female." Notice the keyword OTHER in the $SEXFMT format. If this format is associated with the variable GENDER, then any value other than a '1' or '2' for GENDER will result in the label "Miscoded."

Formatted values, placed in quotes to the right of the = sign, can be up to 32,767 characters long, but only 8 or 16 of these characters will be used by certain procedures. When you use certain procedures, such as PROC PRINT, the formatted value will be printed in place of the internal, coded value. For example, if you associate the format $SEXFMT with the variable GENDER, instead of printing a 1, the formatted value "Male" will be printed instead.

Notice the format names $SEXFMT, $RACE, $OSCAR, and $EDUC. These formats will all be used to define values for character variables. Notice that the value to the left of the = sign must also be enclosed in single or double quotes for character variables. If your format value contains any single quotes, enclose it with double quotes.

Once we have defined a set of formats (such as $SEXFMT, $RACE, LIKERT, etc.), we need to associate the formats to the appropriate variables. Just because we have named a format $RACE, for example, does not mean that this format has to be used with the variable we have called RACE. We need another SAS statement that indicates which formats will be used or associated with which variables. Format statements start with the word FORMAT, followed by a single variable or a list of variables, followed by a format name to be used with the preceding variables. The list of variable and format names continues on as many lines as necessary and ends with a semicolon.

SAS software knows the difference between our variable names and our format names since we place a period after each format name in our format statement.

Format statements can be placed within a DATA step or as a statement in a PROC step. If we choose to place our format statement in the DATA step, the formatted values will be associated with the assigned variable(s) for all PROCS that follow. If we place a format statement in a PROC step, the formatted values will be used only for that procedure. In this example, we will place our format statement in the DATA step. Therefore, we will define our formats with PROC FORMAT before we write our DATA step. We will associate the format $SEXFMT with the variable GENDER, $RACE with the variable RACE, and so forth. The variables PRESIDENT, ARMS, and CITIES are all "sharing" the LIKERT format. If you look at the completed questionnaire program shown here, the use of PROC FORMAT and its application in other PROCs should become clear.

```
PROC FORMAT;
   VALUE $SEXFMT '1' = 'Male'
                 '2' = 'Female'
            OTHER    = 'Miscoded';
   VALUE $RACE    '1' = 'White'
                  '2' = 'African Am.'
                  '3' = 'Hispanic'
                  '4' = 'Other';
   VALUE $OSCAR   '1' = 'Single'
                  '2' = 'Married'
                  '3' = 'Widowed'
                  '4' = 'Divorced';
   VALUE $EDUC    '1' = 'High Sch or Less'
                  '2' = 'Two Yr. College'
                  '3' = 'Four Yr. College'
                  '4' = 'Graduate Degree';
   VALUE LIKERT    1 = 'Str Disagree'
                   2 = 'Disagree'
                   3 = 'No opinion'
                   4 = 'Agree'
                   5 = 'Str Agree';
RUN;

DATA QUEST;
   INPUT ID         $ 1-3
         AGE          4-5
         GENDER     $  6
         RACE       $  7
         MARITAL    $  8
         EDUCATION  $  9
         PRESIDENT    10
         ARMS         11
         CITIES       12;
```

[Continued]

```
    LABEL MARITAL    = "Marital Status"
          EDUCATION  = "Education Level"
          PRESIDENT  = "President Doing a Good Job"
          ARMS       = "Arms Budget Increase"
          CITIES     = "Federal Aid to Cities";
    FORMAT GENDER     $SEXFMT.
           RACE       $RACE.
           MARITAL    $OSCAR.
           EDUCATION  $EDUC.
           PRESIDENT ARMS CITIES LIKERT.;
DATALINES;
001091111232
002452222422
003351324442
004271111121
005682132333
006651243425
;
PROC MEANS DATA=QUEST MAXDEC=2 N MEAN STD CLM;
    TITLE "Questionnaire Analysis";
    VAR AGE;
RUN;

PROC FREQ DATA=QUEST;
    TITLE "Frequency Counts for Categorical Variables";
    TABLES GENDER RACE MARITAL EDUCATION
           PRESIDENT ARMS CITIES;
RUN;
```

Selected portions of the output from this program are shown here:

```
Questionnaire Analysis

The MEANS Procedure

                    Analysis Variable : AGE

                                    Lower 95%       Upper 95%
 N        Mean        Std Dev      CL for Mean     CL for Mean

 6        41.50        22.70          17.68           65.32
```

```
Frequency Counts for Categorical Variables

The FREQ Procedure

                                    Cumulative    Cumulative
GENERAL       Frequency     Percent   Frequency     Percent
```

GENDER	Frequency	Percent	Cumulative Frequency	Cumulative Percent
Male	4	66.67	4	66.67
Female	2	33.33	6	100.00

RACE	Frequency	Percent	Cumulative Frequency	Cumulative Percent
White	3	50.00	3	50.00
African Am.	2	33.33	5	83.33
Hispanic	1	16.67	6	100.00

Marital Status

MARITAL	Frequency	Percent	Cumulative Frequency	Cumulative Percent
Single	2	33.33	2	33.33
Married	2	33.33	4	66.67
Widowed	1	16.67	5	83.33
Divorced	1	16.67	6	100.00

Education Level

EDUCATION	Frequency	Percent	Cumulative Frequency	Cumulative Percent
High Sch or Less	2	33.33	2	33.33
Two Yr. College	2	33.33	4	66.67
Four Yr. College	1	16.67	5	83.33
Graduate Degree	1	16.67	6	100.00

President Doing a Good Job

PRESIDENT	Frequency	Percent	Cumulative Frequency	Cumulative Percent
Str Disagree	1	16.67	1	16.67
Disagree	1	16.67	2	33.33
No opinion	1	16.67	3	50.00
Agree	3	50.00	6	100.00

   Notice how much easier it is to read this listing compared to the earlier version that did not contain labels or formats.

E. RECODING DATA

In the previous questionnaire example, we coded the respondents' actual age in years. What if we want to look at the relationship between age and questions 6–8 (opinion questions)? It might be convenient to have a variable that indicated an age group rather than the person's actual age. We will look at two ways of accomplishing this.

Look at the following SAS statements:

```
PROC FORMAT;
    VALUE $SEXFMT '1' = 'Male'
                  '2' = 'Female';
    .
    .
    .
    VALUE AGEFMT 1 = '0-20'
                 2 = '21-40'
                 3 = '41-60'
                 4 = 'Greater than 60';
RUN;

DATA QUEST;
    INPUT ID          $ 1-3
          AGE            4-5
          GENDER    $    6
          RACE      $    7
          MARITAL   $    8
          EDUCATION $    9
          PRESIDENT     10
          ARMS          11
          CITIES        12;
    IF AGE GE 0 AND AGE LE 20 THEN AGEGRP = 1;
    ELSE IF AGE GT 20 AND AGE LE 40 THEN AGEGRP = 2;
    ELSE IF AGE GT 40 AND AGE LE 60 THEN AGEGRP = 3;
    ELSE IF AGE GT 60 THEN AGEGRP= 4 ;
    LABEL MARITAL = 'MARITAL STATUS'
    .
    .
    .
    AGEGRP = 'AGE GROUP';
    FORMAT GENDER     $SEXFMT.
           RACE       $RACE.
           MARITAL    $OSCAR.
           EDUCATION  $EDUC.
           PRESIDENT ARMS CITIES LIKERT.;
DATALINES;
(data lines)
;
PROC FREQ DATA=QUEST;
    TABLES GENDER -- AGEGRP;
RUN;
```

Several new features have been added to the program. The major additions are the IF and ELSE IF statements following the INPUT. The general form of an IF statement is:

```
IF condition THEN statement;
```

If the condition is true, the statement following the word THEN is executed. In these IF and ELSE IF statements, you see the logical operators GE (greater than or equal to), LE (less than or equal to), and LT (less than). A complete list of these comparison operators is shown later. The INPUT statement reads in values from the lines of data, according to the INPUT specifications. Next, the first IF statement is evaluated. If the condition is true, then the variable AGEGRP will be set to 1, and the remaining ELSE statements will be skipped. If the first IF statement is false, the next ELSE statement will be tested, and if it is true, the value of AGEGRP will be set to 2, and so forth. Finally, when the DATALINES statement is reached, an observation is added to the SAS data set. The program returns to the top of the DATA step and this process repeats until all the lines of data have been read. The variables in the data set will include all the variables listed in the INPUT statement as well as the variable AGEGRP. The variable AGEGRP may be used in any PROC just like any of the other variables. Be sure there are no "cracks" in your recoding ranges. That is, make sure you code your IF statements so that there isn't a value of AGE that is not recoded. If that happens, the variable AGEGRP for that person will have a missing value. Notice that the first IF statement is written:

```
IF AGE GE 0 AND AGE LE 20 THEN AGEGRP = 1;
```

and not like this:

```
IF AGE LE 20 THEN AGEGRP = 1;
```

The reason is subtle **but very important**. SAS programs treat missing values as negative infinity for ordering purposes. That is, a missing value will be considered less than any numeric value. Thus, the previous IF statement will be true for missing values as well as the valid ages from 0 to 20. Another way of writing the statement:

```
IF AGE GE 0 AND AGE LE 20 THEN AGEGRP = 1;
```

is:

```
IF AGE LE 20 AND AGE NE . THEN AGEGRP = 1;
```

This last IF statement uses a period to represent a SAS missing numeric value. By the way, while we are on the topic of missing values, SAS missing character values are represented as ' ' (quote, blank, quote—either single or double quotes). In choosing one of the previous IF statements, feel free to choose whichever method makes most sense to you. The other IF statements can also be written using an alternative syntax. For example, you could write the IF statement for AGEGRP = 2 as:

```
ELSE IF 20 LT AGE LE 40 THEN AGEGRP=2;
```

A complete list of the SAS comparison operators is shown here. Either the two-letter abbreviations or the symbols may be used.

Definition	Symbol(s)
Equal to	$=$ or EQ
Greater than	$>$ or GT
Less than	$<$ or LT
Greater than or equal to	$>=$ or GE
Less than or equal to	$<=$ or LE
Not equal to	$^\wedge=$ or NE

(NOTE: ~= is an alternative not equal symbol)

You could have written four IF statements instead of one IF statement followed by three ELSE IF statements. Remember, the effect of the ELSE statements is that when any IF statement is true, all the following ELSE statements are skipped. Thus, the ELSE IF statements have the advantage of reducing computer time (since four IF statements do not have to be tested) and avoiding the following type of problem. Can you see what will happen with the following statements? (Assume X can have values of 1, 2, 3, 4, or 5.)

```
IF X=1 THEN X=5;
IF X=2 THEN X=4;
IF X=4 THEN X=2;
IF X=5 THEN X=1;
```

What happens when X is 1? The first IF statement is true. This causes X to have a value of 5. The next two IF statements are false, but the last IF statement is true. X is back to 1! The ELSE statements, besides reducing computer time, prevent the previous problem. The correct coding is:

```
     IF X=1 THEN X=5;
ELSE IF X=2 THEN X=4;
ELSE IF X=4 THEN X=2;
ELSE IF X=5 THEN X=1;
```

One final note: If all we wanted to do was recode X so that $1 = 5$, $2 = 4$, $3 = 3, 4 = 2$, and $5 = 1$, the statement:

```
X = 6 - X;
```

would be the best way to recode the X values. Check it out.

Notice that we added a line to the LABEL section and to PROC FORMAT to supply a variable label and a format for our new variable.

F.   USING A FORMAT TO RECODE A VARIABLE

There is another way of recoding our AGE variable without creating a new variable. We use a "trick." By defining a format and associating it with a variable, we can have SAS assign subjects to age categories. The formats can be associated with a variable in the DATA step or directly in a PROC. If you include a format statement in the DATA step, the format will remain associated with the variable for all following procedures. If you include a format statement within a procedure, that association remains only for that particular procedure. Also, remember that the actual "internal" value of the original variable has not changed. So, if you place a format statement within a DATA step and associate a format with a variable, all computations regarding that variable still involve the original value. You can still compute means (with PROC MEANS), for example, on AGE even though it has an associated format. It is only in such procedures as PROC FREQ or when used as a CLASS variable that the associated format has an effect. So, to continue with our example, we write:

```
PROC FORMAT;
     VALUE AGROUP LOW-20   = '0-20'
                  21-40    = '21-40'
                  41-60    = '41-60'
                  61-HIGH  = 'Greater than 60';
```

As you can see, instead of single values to the left of the equals sign, we are supplying a range of values. SAS will not let us specify overlapping ranges when defining formats. Thus, LOW-20='Very Young' and 15–30='Young' are not allowed. The special keywords HIGH and LOW are available to indicate the highest and lowest values in the data set (not counting missing values), respectively. Thus, the term LOW-20 refers to all values from the lowest value up to and including 20. Remember, the keyword LOW does not include missing values for numeric variables (it does include missing values for character variables).

One additional keyword, OTHER, can be used to match any data value not included in any of the other format ranges. Thus, you can use the form:

```
VALUE AGROUP LOW-20   = '0-20'
             21-40    = '21-40'
             41-60    = '41-60'
             60-HIGH  = 'Greater than 60'
             .        = 'Did Not Answer'
             OTHER    = 'Out of Range';
```

As you would expect, the `. = 'format label'` allows you to supply a label to numeric missing values.

Once we have defined a format, we can then associate it with a variable, either in the DATA step or in the procedure itself. For example, we can write a TABLES statement on the original variable (such as AGE). By supplying the format information using the format AGROUP for the variable AGE, frequencies will be computed for the formatted values instead of the original AGE values. In this example, we place the format statement in the appropriate PROC rather than in the DATA statement since we want to use the recoded values only for PROC FREQ.

Thus, the SAS statements:

```
PROC FREQ DATA=QUEST;
   TITLE "Using a Format to Group a Numeric Variable";
   TABLES AGE;
   FORMAT AGE AGROUP.;
RUN;
```

produce the following output:

```
Using a Format to Group a Numeric Variable

The FREQ Procedure

                                     Cumulative    Cumulative
        AGE    Frequency    Percent  Frequency      Percent
        ─────────────────────────────────────────────────────
0-20              1         16.67        1          16.67
21-40             2         33.33        3          50.00
41-60             1         16.67        4          66.67
Greater than 60   2         33.33        6         100.00
```

While on the topic of creating formats, we mention a few other ways to present formatted values. You can combine specific codes and code ranges in a single format statement. Suppose we assigned the codes 1 through 3 to the colors 'Red,' 'White,' and 'Blue,' and codes 0, and 4 through 9 to other colors. The codes 0, and 4 through 9 are to be lumped together and called 'Other Colors,' and any other codes are to be labeled as 'Out of Range.' The VALUE statement to create a SAS format called COLOR, which meets these specifications, is shown here:

```
VALUE COLOR 1      = 'Red'
            2      = 'White'
            3      = 'Blue'
            0, 4-9 = 'Other Colors'
            OTHER  = 'Out of Range';
```

A series of values or ranges of values can be separated by commas. In this example, either a 0 or any number from 4 through 9 will be formatted as 'Other Colors.' A good application of this would be to regroup a large number of categories into a smaller number of categories. Suppose we had coded our original questionnaire with five levels of RACE:

```
1 = WHITE   2 = SPANISH AMERICAN   3 = ASIAN   4 = AFRICAN AM.   5 = OTHER
```

Now, for a particular analysis, we want to have only three groups: WHITE, AFRICAN AM., and OTHER. A quick and easy way to do this is to supply a format like the following:

```
VALUE $RACE '1'         = 'WHITE'
            '4'         = 'AFRICAN AM.'
            '2','3','5' = 'OTHER RACES';
```

The advantage of this method of grouping values is that you don't have to create a new data set or make new variables. All that is necessary is to write the PROC FORMAT statements to create a new format and add a format statement to the procedure where you want to use the new grouping.

G.   TWO-WAY FREQUENCY TABLES

Besides computing frequencies on individual variables, we may have occasion to count occurrences of one variable at each level of another variable. An example will make this clear. Suppose we took a poll of presidential preference and also recorded the gender of the respondent. Sample data might look like this:

GENDER	CANDIDATE
M	DEWEY
F	TRUMAN
M	TRUMAN
M	DEWEY
F	TRUMAN
etc.	

We would like to know (1) how many people were for Dewey and how many were for Truman, (2) how many males and females were in the sample, and (3) how many males and females were for Dewey and Truman, respectively.

A previous example of PROC FREQ shows how to perform tasks (1) and (2). For (3) we would like a table that looks like this:

Table of Candidate by Gender

Candidate	Gender		
Frequency	**F**	**M**	**Total**
Dewey	70	40	110
Truman	30	40	70
Total	100	80	180

If this were our table, it would show that females favored Dewey over Truman 70 to 30, while males were split evenly. A SAS program to solve all three tasks is:

```
DATA ELECT;
   INPUT GENDER $ CANDID $ ;
DATALINES;
M DEWEY
F TRUMAN
M TRUMAN
M DEWEY
F TRUMAN
(more data lines)
;
PROC FREQ DATA=ELECT;
   TABLES GENDER CANDID
          CANDID*GENDER;
RUN;
```

Notice that since the variables GENDER and CANDID are coded as character values (alphanumeric), we follow each variable name with a $ in the INPUT statement. Another fact that we have not mentioned thus far is that the VALUES of our character variables also cannot be longer than eight letters in length when using the "list" form of the INPUT statement unless we modify our INPUT statement to indicate this change. So, for the time being, we cannot use this program for the Eisenhower/Stevenson election (without using nicknames), but it's great for Bush/Kerry.

The first two TABLES are one-way frequency tables, the same type we have seen before; the TABLE specification, CANDID*GENDER is a request for a two-way table.

What would a table like the previous one tell us? If it were based on a random sample of voters, we might conclude that gender affected voting patterns. Before we conclude that this is true of the nation as a whole, it would be nice to see how likely it was that these results were simply due to a quirky sample. A statistic called chi-square will do just this.

Consider the table again. There were 180 people in our sample, 110 for Dewey and 70 for Truman; 100 females and 80 males. If there were no gender bias, we would expect the proportion of the population who wanted Dewey (110/180) to be the same for the females and males. Therefore, since there were 100 females, we could expect (110/180) of 100 (approximately 61) females to be for Dewey. Our expectations (in statistics this is called expected values) for all the other cells can be calculated in the same manner. Once we have observed and expected frequencies for each cell (each combination of gender and candidate), the chi-square statistic can be computed. By adding an option for chi-square on our TABLES request, we can have our program compute chi-square and the probability of obtaining a value as large or larger by chance alone. Remembering that statement options follow a slash, the request for chi-square is written as:

```
TABLES CANDID*GENDER / CHISQ;
```

Output from the previous request is shown next:

```
Two-way Table

The FREQ Procedure

Table of Candidate by Gender

Candid      Gender

Frequency|
Percent  |
Row Pct  |
Col Pct  |F        |M        |   Total

Dewey    |      70 |      40 |     110
         |   38.89 |   22.22 |   61.11
         |   63.64 |   36.36 |

Truman   |      30 |      40 |      70
         |   16.67 |   22.22 |   38.89
         |   42.86 |   57.14 |
         |   30.00 |   50.00 |

Total         100        80       180
            55.56     44.44    100.00
```

[Continued]

```
Statistics for Table of Candidate by Gender

Statistic                          DF        Value         Prob

Chi-Square                          1       7.4805       0.0062
Likelihood Ratio Chi-Square         1       7.4930       0.0062
Continuity Adj. Chi-Square          1       6.6626       0.0098
Mantel-Haenszel Chi-Square          1       7.4390       0.0064
Phi Coefficient                             0.2039
Contingency Coefficient                     0.1998
Cramer's V                                  0.2039


        Fisher's Exact Test

Cell (1,1) Frequency (F)              70
Left-sided Pr <= F                0.9981
Right-sided Pr >= F               0.0049


Table Probability (P)             0.0030
Two-sided Pr <= P                 0.0087


Sample Size = 180
```

The key to the table is found in its upper left-hand corner. It tells you what all the numbers in the cells are. By FREQUENCY, we mean the number of subjects in the cell. For example, 70 females favored Dewey for president. The second number in each cell shows the PERCENT of the total population. The third number, labeled ROW PCT, gives the percent of each row. For example, of all the people for Dewey (row 1), 70/110 × 100, or 63.64%, were female. The last number, COL PCT, is the column percent. Of all the females, 70% were for Dewey and 30% were for Truman. In the TABLES request for a two-way cross tabulation, the variable that forms the columns is placed second (e.g., CANDID*GENDER). In our statistical requests, rows come first, then columns.

For our example, chi-square equals 7.48, and the probability of obtaining a chi-square this large or larger by chance alone is .006. Therefore, we can say that, based on our data, there appears to be a gender bias in presidential preference: There is a tendency for females to show greater preference for Dewey than males do or, put another way, for males to prefer Truman.

The number of degrees of freedom (df) in a chi-square statistic is equal to the number of rows minus one multiplied by the number of columns minus one ($(R - 1) \times (C - 1)$). Thus, our 2 × 2 chi-square has 1 df. Whenever a chi-square table has 1 df and the expected value of any cell is less than 5, a "correction for continuity," called Yates' correction, is often applied. SAS software prints out a corrected chi-square value and its associated probability beside the heading Continuity Adj. Chi-Square.

Another alternative, when you have small expected values, is to use Fisher's exact test, which is included in the list of statistics for the table. Remembering that the chi-square test is nondirectional, you will probably want to use the two-tailed Fisher probability. When the degrees of freedom are greater than 1, it is desirable that no more than 20% of the cells have expected values less than 5. The program will print a warning when this condition occurs. This does not mean that you have to throw your data out if you fall into this situation. This is a situation that is complex and about which statisticians don't always agree. Next, we suggest one alternative if your df are greater than 1. If you are in doubt, consult your friendly statistician.

For larger tables (more than four cells), the usual alternative when faced with small expected cell values is to combine, or collapse, cells. If we had four categories of age: 0–20, 21–40, 41–60, and over 60, we might combine 0–20 and 21–40 as one group, and 41–60 and 60+ as another. Another example would be combining categories such as "strongly disagree" and "disagree" on an opinion questionnaire. We can use either method of recoding shown in the previous section to accomplish this.

H.   A SHORTCUT WAY TO REQUEST MULTIPLE TABLES

We can use the questionnaire program at the beginning of this chapter to see another example of a two-way table. Suppose we wanted crosstabulations of AGEGRP against the three variables PRESIDENT, ARMS, and CITIES. We could code:

```
TABLES (PRESIDENT ARMS CITIES)*AGEGRP;
```

This will generate three tables and is a short way of writing:

```
TABLES PRESIDENT*AGEGRP ARMS*AGEGRP CITIES*AGEGRP;
```

We can also have multiple-column variables in a TABLE request. Thus:

```
TABLES (PRESIDENT ARMS) * (AGEGRP GENDER);
```

would produce four tables:

```
PRESIDENT*AGEGRP, PRESIDENT*GENDER, ARMS*AGEGRP, and ARMS*GENDER.
```

When you use this method, be sure to enclose the list of variables within parentheses.

One of the tables generated from this program is shown next:

```
The FREQ Procedure

Table of PRESIDENT by AGEGRP

PRESIDENT(President Doing a Good Job)      AGEGRP(Age Group)
Frequency   |
Percent     |
Row Pct     |
Col Pct     |      1|      2|      3|      4|  Total

Str Disagree|      0|      1|      0|      0|      1
            |   0.00|  16.67|   0.00|   0.00|  16.67
            |   0.00| 100.00|   0.00|   0.00|
            |   0.00|  50.00|   0.00|   0.00|

Disagree    |      1|      0|      0|      0|      1
            |  16.67|   0.00|   0.00|   0.00|  16.67
            | 100.00|   0.00|   0.00|   0.00|
            | 100.00|   0.00|   0.00|   0.00|

No opinion  |      0|      0|      0|      1|      1
            |   0.00|   0.00|   0.00|  16.67|  16.67
            |   0.00|   0.00|   0.00| 100.00|
            |   0.00|   0.00|   0.00|  50.00|

Agree       |      0|      1|      1|      1|      3
            |   0.00|  16.67|  16.67|  16.67|  50.00
            |   0.00|  33.33|  33.33|  33.33|
            |   0.00|  50.00| 100.00|  50.00|

Total             1       2       1       2       6
              16.67   33.33   16.67   33.33  100.00
```

I. COMPUTING CHI-SQUARE FROM FREQUENCY COUNTS

When you already have a contingency table and want to use SAS software to compute a chi-square statistic, there is a WEIGHT statement that makes this task possible. Suppose someone gave you the 2×2 table shown here and wanted to compute chi-square:

		Outcome	
		Alive	**Dead**
Group	Drug	90	10
	Placebo	80	20

We could code this by reading in values for GROUP, OUTCOME, and the number of subjects, COUNT, in the appropriate cell. Thus, we would have:

```
DATA CHISQ;
   INPUT GROUP $ OUTCOME $ COUNT;
DATALINES;
DRUG ALIVE 90
DRUG DEAD 10
PLACEBO ALIVE 80
PLACEBO DEAD 20
;
PROC FREQ DATA=CHISQ;
   TABLES GROUP*OUTCOME / CHISQ;
   WEIGHT COUNT;
RUN;
```

The WEIGHT statement tells the procedure how many subjects there are for each combination of OUTCOME and GROUP. By the way, what would the table look like if we omitted the WEIGHT statement? Remember that without a WEIGHT statement, PROC FREQ counts the number of observations having certain characteristics. Out of the four observations, how many have GROUP equal to DRUG and OUTCOME equal to ALIVE? The answer is one. Likewise for the other three cells. So, without a WEIGHT statement, all four cells would contain a one.

J.   A USEFUL PROGRAM FOR MULTIPLE CHI-SQUARE TABLES

Even though it is a bit early in this book to present more complicated programs, we present a short program that allows you to enter the counts for as many 2 × 2 tables as you wish and compute chi-square statistics for each table. You may copy the following program "as is" and substitute your data lines in place of our sample lines. So, here is the program without any detailed explanations of how it works:

```
***Program to compute Chi-square for any number of 2 x 2 tables
where the data lines consist of the cell frequencies. The order of
the cell counts is upper left, upper right, lower left, and lower
right. To use this program, substitute your cell frequencies for the
sample data lines in this program.;
DATA CHISQ;
   N + 1;
   DO ROW = 1 TO 2;
      DO COL = 1 TO 2;
         INPUT COUNT @;
```

[Continued]

```
        OUTPUT;
      END;
   END;
END;
DATALINES;
3  5  8  6
10 20 30 40
;
PROC FREQ DATA=CHISQ;
   BY N;
   TABLES ROW*COL / CHISQ;
   WEIGHT COUNT;
RUN;
```

K. A USEFUL MACRO FOR COMPUTING CHI-SQUARE FROM FREQUENCY COUNTS

The following is a useful macro for computing chi-square (or other statistics available with PROC FREQ) from frequencies in a 2×2 table. As with the previous program, this is early in the book to be discussing macros, so we will just show the macro and explain how to use it. Since we need to compute chi-square, odds ratios, and relative risks so often from tables containing frequencies, you will find it particularly handy to have this macro in your SAS macro library.

What follows is a listing of the CHISQ macro. You can use it in several ways. First (and best), is to place the macro in your permanent macro library and use the SAS autocall facility to make it available every time you fire up a SAS session. If that is too daunting, you can use the statement:

```
%INCLUDE 'C:\MYMACROS\CHISQ.SAS';
```

where the program CHISQ.SAS is on your hard drive in an appropriate folder. Finally, you can bring the program CHISQ.SAS into the Display Manager and submit it.

```
/**************************************************************
Macro CHISQ
Purpose: To compute chi-square (and any other valid
         PROC FREQ TABLES options) from frequencies in a
         2 x 2 table.
Sample Calling Sequences:
   ***Using the default options;
   %CHISQ(10,20,30,40)

   ***Omitting all options;
   %CHISQ(10,20,30,40,OPTIONS=)

   ***Removing percentages;
   %CHISQ(10,20,30,40,OPTIONS=NOCOL NOROW NOPERCENT)
```

```
***Requesting Odds Ratios and Relative Risk;
%CHISQ(10,20,30,40,OPTIONS=CMH)

***Requesting Chi-square, OR, and RR;
%CHISQ(10,20,30,40,OPTIONS=CHISQ CMH)
************************************************************/
%MACRO CHISQ(A,B,C,D,OPTIONS=CHISQ);
    DATA CHISQ;
        ARRAY CELLS[2,2] _TEMPORARY_ (&A &B &C &D);
        DO ROW = 1 TO 2;
            DO COL = 1 TO 2;
                COUNT = CELLS[ROW,COL];
                OUTPUT;
            END;
        END;
    RUN;
    PROC FREQ DATA=CHISQ;
        TABLES ROW*COL / &OPTIONS;
        WEIGHT COUNT;
    RUN;
%MEND CHISQ;
```

No matter which of these three methods you use, you can then use the macro like this:

```
%CHISQ(10,20,30,40)
```

Notice that you don't need a semicolon following the call to the macro. If you would like any options other than chi-square, ask for them like this:

```
%CHISQ(10,20,30,40,OPTION=CHISQ CMH)
```

You may list any of the valid PROC FREQ TABLES options following the word OPTIONS.

L.   MCNEMAR'S TEST FOR PAIRED DATA

Suppose you want to determine the effect of an anticigarette advertisement on people's attitudes toward smoking. In this hypothetical example, we ask 100 people their attitude toward smoking (either positive or negative). We then show them the anticigarette advertisement and again ask their smoking attitude. This experimental design is called a paired or matched design since the same subjects are responding to a question under two different conditions (before and after an advertisement). Paired designs are also used when a specific person is matched on some criteria, such as age and gender, to another person for purposes of analysis. Suppose you collected the following data in your cigarette study (P = Positive; N = Negative):

Subject	Before	After
001	P	P
002	P	N
003	N	N
.		
.		
.		
100	N	P

The two-way table of BEFORE and AFTER values looks like this:

		After	
		Negative	**Positive**
Before	Negative	32	15
	Positive	30	23

Notice that each cell in this table represents how a person felt before the advertisement compared to after the advertisement. Thus, there were 32 people who were negative before and after the advertisement; 15 who were initially negative but became positive after the advertisement; 30 who were positive before but negative after the advertisement; and 23 who were positive both before and after the advertisement. To analyze a table like this, we do not use the traditional chi-square test, but rather the McNemar test for paired samples. The logic goes like this: We want to only look at the people who changed their minds (called the discordant cells). In this table, there were a total of 45 subjects who changed their minds. Under that null hypothesis that the film had no effect on opinion, you would expect as many Negative-Positives as Positive-Negatives. Therefore, the expected value for the two discordant cells is 45/2 or 22.5. Chi-square can then be computed as $(15 - 22.5)^2/22.5 + (30 - 22.5)^2/22.5 = 5$. We can have SAS compute this for use like this:

```
***Program Name: MCNEMAR.SAS in C:\APPLIED
Purpose: To perform McNemar's Chi-square test for
paired samples;

PROC FORMAT;
   VALUE $OPINION 'P'='Positive'
                  'N'='Negative';
RUN;
DATA MCNEMAR;
   LENGTH BEFORE AFTER $ 1;
   INPUT SUBJECT BEFORE $ AFTER $;
   FORMAT BEFORE AFTER $OPINION.;
DATALINES;
```

```
001 P P
002 P N
003 N N
(more data lines)
100 N P
;
PROC FREQ DATA=MCNEMAR;
   TITLE "McNemar's Test for Paired Samples";
   TABLES BEFORE*AFTER / AGREE;
RUN;
```

Some comments about this program before we study the output. Notice the LENGTH statement immediately before the INPUT statement. By setting the length of the two variables BEFORE and AFTER to 1, we need only a single byte to store the values. Without the LENGTH statement, the default 8 bytes would be used. To make the output more readable, we created a format for the two variables BEFORE and AFTER. Since the title contains a single quote, we are careful to use double quotes to surround the title. Finally, the AGREE option on the TABLES statement produces both McNemar's chi-square as well as a measure of agreement called Kappa. (Coefficient Kappa is often used as a measure of interrater reliability.) Now for the output:

```
McNemar's Test for Paired Samples

The FREQ Procedure

Table of BEFORE by AFTER

BEFORE      AFTER

Frequency|
Percent  |
Row Pct  |
Col Pct  |Negative|Positive|  Total

Negative |    32  |    15  |    47
         |  32.00 |  15.00 |  47.00
         |  68.09 |  31.91 |
         |  51.61 |  39.47 |

Positive |    30  |    23  |    53
         |  30.00 |  23.00 |  53.00
         |  56.60 |  43.40 |
         |  48.39 |  60.53 |

Total        62       38      100
           62.00    38.00   100.00
```

                                                    [Continued]

```
Statistics for Table of BEFORE by AFTER

    McNemar's Test
─────────────────────────────
Statistic (S)      5.0000
DF                      1
Pr > S             0.0253

    Simple Kappa Coefficient
─────────────────────────────
Kappa                   0.1128
ASE                     0.0946
95% Lower Conf Limit   -0.0727
95% Upper Conf Limit    0.2983

Sample Size = 100
```

This output shows us that the McNemar's chi-square statistic is 5.0, with a corresponding p-value of .0253. We conclude that the anticigarette advertisement was effective in changing people's attitudes toward smoking.

If you already had frequency counts for the 2 × 2 table and wanted SAS to compute McNemar's chi-square, you could use the following program:

```
DATA MCNEMAR;
   LENGTH AFTER BEFORE $ 1;
   INPUT AFTER $ BEFORE $ COUNT;
   FORMAT BEFORE AFTER $OPINION.;
DATALINES;
N N 32
N P 30
P N 15
P P 23
;
PROC FREQ DATA=MCNEMAR;
   TITLE "McNemar's Test for Paired Samples";
   TABLES BEFORE*AFTER / AGREE ;
   WEIGHT COUNT;
RUN;
```

M.   COMPUTING THE KAPPA STATISTICS (COEFFICIENT OF AGREEMENT)

Another use of the AGREE option on the TABLES statement is to compute the Kappa statistic. This is a commonly used measure of agreement between two raters. For example, suppose you wanted to determine how accurately a radiologist could determine a malignancy on an X-ray. One way to do this is to have two independent radiologists rate the same series of X-rays and rate each one as positive or negative. As an example, look at the following table, which represents the results of two radiologists rating 83 X-rays:

		Radiologist 2		
	Frequency	No	Yes	Total
	No	25	3	28
Radiologist 1	Yes	5	50	55
	Total	30	53	83

You have to remember that if each radiologist flipped a coin, they would both agree 50% of the time (out of the four outcomes, No/No, Yes/Yes, No/Yes, and Yes/No, each equally likely, there is agreement half the time). The Kappa statistic removes the effect of random agreement and provides a commonly used statistic of agreement. To program SAS to compute Kappa for this table, you would write:

```
DATA X_RAY;
    INPUT RADIOLOGIST_1 $ RADIOLOGIST_2 $ COUNT;
DATALINES;
No No 25
No Yes 3
Yes No 5
Yes Yes 50
;
PROC FREQ DATA=X_RAY;
    TITLE "Computing Coefficient Kappa for Two Observers";
    TABLES RADIOLOGIST_1 * RADIOLOGIST_2 / AGREE;
    WEIGHT COUNT;
RUN;
```

Looking at the following output, we see that Kappa is 0.7881, indicating moderate agreement.

```
Computing Coefficient Kappa for Two Observers

The FREQ Procedure

Table of RADIOLOGIST_1 by RADIOLOGIST_2

RADIOLOGIST_1
          RADIOLOGIST_2

Frequency|
Percent  |
Row Pct  |
Col Pct  |No      |Yes     |  Total

No       |     25 |      3 |     28
         |  30.12 |   3.61 |  33.73
         |  89.29 |  10.71 |
         |  83.33 |   5.66 |

Yes      |      5 |     50 |     55
         |   6.02 |  60.24 |  66.27
         |   9.09 |  90.91 |
         |  16.67 |  94.34 |

Total          30       53        83
            36.14    63.86    100.00

Statistics for Table of RADIOLOGIST_1 by RADIOLOGIST_2

      McNemar's Test
  ─────────────────────────
Statistic (S)      0.5000
DF                      1
Pr > S             0.4795

      Simple Kappa Coefficient
  ─────────────────────────
Kappa                 0.7881
ASE                   0.0710
95% Lower Conf Limit  0.6491
95% Upper Conf Limit  0.9272

Sample Size = 83
```

N. ODDS RATIOS

Suppose we want to determine if people with a rare brain tumor are more likely to have been exposed to benzene than people without a brain tumor. One experimental design used to answer this question is called a case-control design. As the name implies, you first start with cases, people with a disease or condition (in this example, a brain tumor), and find people who are as similar as possible but who do not have brain tumors. Those people are the controls. We provide some data here to demonstrate some features of a case-control study:

		Outcome	
		Case	**Control**
Exposure	Yes	50	20
	No	100	130

Inspection of the table shows a higher percentage of Cases being exposed to benzene than Controls. To test this, we can compute a chi-square statistic. The results of a case-control study are frequently reported by an odds ratio and a 95% confidence interval about the odds ratio. Briefly, the odds of a Case being exposed to benzene is 50/100. The odds of a Control being exposed to benzene is 20/130. Therefore, the odds ratio is

$$\frac{50/100}{20/130} = \frac{.5}{.155} = 3.25.$$

If we run PROC FREQ with the CHISQ and CMH (Cochran-Mantel-Haenszel) options, we obtain a chi-square statistic, the odds ratio, a 95% CI on the odds ratio, and quite a few additional statistics. A program to analyze the previous table and the resulting output is shown next:

```
***Program to compute an Odds Ratio and the 95% CI;
DATA ODDS;
    INPUT OUTCOME $ EXPOSURE $ COUNT;
DATALINES;
CASE 1-YES 50
CASE 2-NO 100
CONTROL 1-YES 20
CONTROL 2-NO 130
;
PROC FREQ DATA=ODDS;
    TITLE "Program to Compute an Odds Ratio";
    TABLES EXPOSURE*OUTCOME / CHISQ CMH;
    WEIGHT COUNT;
RUN;
```

We used an expedient here to ensure that the 'Yes' row came before the 'No' row. We want the 'Yes' group on top so that the odds ratio will be the odds that a case (in the first column) is more (or less) likely to be exposed (i.e., EXPOSURE = 'Yes'). Since SAS, by default, will order the values in a frequency table by the alphabetical order of character variables (or the numerical order of numerical variables), by using the names '1-YES' and '2-NO' for the variable EXPOSURE, we forced the program to place the 'Yes' row on top (since '1' comes before '2' "alphabetically"). Another useful technique is to use formats for the row or column variables, choosing values that result in the desired ordering of rows and columns, and using the ORDER=FORMATTED option with PROC FREQ. This option causes PROC FREQ to use the formatted values rather than the internal raw values when ordering values for tables. The output from this program is as follows:

```
Program to Compute an Odds Ratio

The FREQ Procedure

Table of EXPOSURE by OUTCOME

EXPOSURE        OUTCOME

Frequency|
Percent  |
Row Pct  |
Col Pct  |CASE    |CONTROL |  Total

1-YES    |     50 |     20 |     70
         |  16.67 |   6.67 |  23.33
         |  71.43 |  28.57 |
         |  33.33 |  13.33 |

2-NO     |    100 |    130 |    230
         |  33.33 |  43.33 |  76.67
         |  43.48 |  56.52 |
         |  66.67 |  86.67 |

Total         150      150      300
            50.00    50.00   100.00


Statistics for Table of EXPOSURE by OUTCOME

Statistic                      DF      Value      Prob

Chi-Square                      1    16.7702    <.0001
Likelihood Ratio Chi-Square     1    17.2071    <.0001
Continuity Adj. Chi-Square      1    15.6708    <.0001
Mantel-Haenszel Chi-Square      1    16.7143    <.0001
Phi Coefficient                       0.2364
Contingency Coefficient               0.2301
Cramer's V                            0.2364
```

```
              Fisher's Exact Test
          ─────────────────────────────
Cell (1,1) Frequency (F)          50
Left-sided Pr <= F            1.0000
Right-sided Pr >= F          3.125E-05


Table Probability (P)        2.220E-05
Two-sided Pr <= P            6.249E-05


Sample Size = 300


Program to Compute an Odds Ratio


The FREQ Procedure


Summary Statistics for EXPOSURE by OUTCOME


  Cochran-Mantel-Haenszel Statistics (Based on Table Scores)


Statistic    Alternative Hypothesis    DF    Value     Prob
          ───────────────────────────────────────────────────
    1        Nonzero Correlation        1    16.7143   <.0001
    2        Row Mean Scores Differ     1    16.7143   <.0001
    3        General Association        1    16.7143   <.0001


         Estimates of the Common Relative Risk (Row1/Row2)


Type of Study      Method           Value    95% Confidence Limits
              ──────────────────────────────────────────────────────
Case-Control       Mantel-Haenszel  3.2500    1.8189       5.8070
  (Odds Ratio)     Logit            3.2500    1.8189       5.8070

Cohort             Mantel-Haenszel  1.6429    1.3331       2.0246
  (Col1 Risk)      Logit            1.6429    1.3331       2.0246

Cohort             Mantel-Haenszel  0.5055    0.3432       0.7446
  (Col2 Risk)      Logit            0.5055    0.3432       0.7446


Total Sample Size = 300
```

The chi-square value is 16.770, which is significant at the .001 level. As we calculated earlier, the odds ratio is 3.25. To the right of this value is the 95% confidence interval (1.847 to 5.719). We interpret this to mean that if we took many similar samples from the given population, 95% of the computed confidence intervals would contain the true population odds ratio. More practically, we can say that we are 95% confident that the true population odds ratio is in the interval 1.847 to 5.719 (but don't say it too loudly near a statistician). Also, since the interval does not contain 1, we conclude that the odds ratio of 3.25 is significant at the .05 level. (In this case, we have a chi-square and p-value we can use.) For a case-control study, the relative risks (labeled cohort) are not interpretable and should be ignored. For very low incidence rates, the odds ratio is an acceptable estimate of the relative risk and is discussed next.

O.   RELATIVE RISK

Just as an odds ratio is the appropriate statistic when dealing with case-control studies, relative risk is the appropriate statistic when dealing with cohort studies. As an example, suppose we conducted a prospective cohort study to investigate the effect of cholesterol on heart attacks. A group of patients who have high cholesterol and a group of patients with low cholesterol (the two cohorts) are followed for 5 years. The number of patients developing a heart attack (MI) are observed for both groups. Some fictitious data are presented below:

		Outcome		
		MI	**No MI**	
	High	20	80	100
Cholesterol				
	Low	15	135	150
		35	215	250

We see that out of 100 patients with high cholesterol, 20 had an MI, giving us an incidence rate of 20/100, or .20. For patients with low cholesterol, the incidence rate is 15/150, or .10. The ratio of the incidence rates is called relative risk. In this example it is .20/.10 = 2.00. We can say that the risk of a heart attack for people with high cholesterol is twice that of people with low cholesterol. We can use basically the same program as we used for the previous case-control study to compute the relative risk. Here it is:

```
***Program to compute a Relative Risk and a 95% CI;
DATA RR;
   LENGTH GROUP $ 9;
   INPUT GROUP $ OUTCOME $ COUNT;
DATALINES;
HIGH-CHOL MI 20
HIGH-CHOL NO-MI 80
LOW-CHOL MI 15
LOW-CHOL NO-MI 135
;
PROC FREQ DATA=RR;
   TITLE "Program to Compute a Relative Risk";
   TABLES GROUP*OUTCOME / CMH;
   WEIGHT COUNT;
RUN;
```

Program to Compute a Relative Risk

The FREQ Procedure

Table of GROUP by OUTCOME

GROUP   OUTCOME

Frequency Percent Row Pct Col Pct	MI	NO-MI	Total
HIGH-CHOL	20 8.00 20.00 57.14	80 32.00 80.00 37.21	100 40.00
LOW-CHOL	15 6.00 10.00 42.86	135 54.00 90.00 62.79	150 60.00
Total	35 14.00	215 86.00	250 100.00

Program to Compute a Relative Risk

The FREQ Procedure

Summary Statistics for GROUP by OUTCOME

Cochran-Mantel-Haenszel Statistics (Based on Table Scores)

Statistic	Alternative Hypothesis	DF	Value	Prob
1	Nonzero Correlation	1	4.9635	0.0259
2	Row Mean Scores Differ	1	4.9635	0.0259
3	General Association	1	4.9635	0.0259

Estimates of the Common Relative Risk (Row1/Row2)

Type of Study	Method	Value	95% Confidence Limits	
Case-Control	Mantel-Haenszel	2.2500	1.0905	4.6425
(Odds Ratio)	Logit	2.2500	1.0905	4.6425
Cohort	Mantel-Haenszel	2.0000	1.0761	3.7171
(Col1 Risk)	Logit	2.0000	1.0761	3.7171
Cohort	Mantel-Haenszel	0.8889	0.7950	0.9938
(Col2 Risk)	Logit	0.8889	0.7950	0.9938

Total Sample Size = 250

We did not need to use any labeling tricks to get the rows and columns in the right order since 'HIGH-CHOL' comes before 'LOW-CHOL' and 'MI' comes before 'NO MI' alphabetically. The LENGTH statement assigned a length of 9 for the variable GROUP. Had we not done this, the values of GROUP would have defaulted to 8 characters, and the last letter of the GROUP values would have been chopped off. An alternative approach is to provide an INFORMAT statement or an INFORMAT on the INPUT statement like this:

```
INFORMAT GROUP $ 9.; /* Separate INFORMAT statement */
INPUT GROUP OUTCOME $ COUNT; /* $ after GROUP optional */
```

or

```
INPUT GROUP : $9. OUTCOME $ COUNT; /* Colon modifier */
```

This time we include only the CMH option, so chi-square and associated statistics are not computed. Since we want to see the "risk" of a heart attack (for those with HIGH versus LOW cholesterol), we want to look at the cohort (Col1 Risk) results in the output. The Col1 relative risk is how much more or less likely you are to be in the column 1 category (in this case, MI) if you are in the row 1 group (in this case, high cholesterol). The computed relative risk is 2.00 with the 95% confidence interval of 1.087 to 3.680.

P.   CHI-SQUARE TEST FOR TREND

If the categories in a 2 × N table represent ordinal levels, you may want to compute what is called a chi-square test for trend. That is, are the proportions in each of the N levels increasing or decreasing in a linear fashion? Consider the following table:

| | | Group | | | |
		A	B	C	D
Test Results	Fail	10	15	14	25
	Pass	90	85	86	75
		100	100	100	100

Notice that the proportions of 'Fail' in groups A through D is increasing (except from group B to C). To test if there is a significant linear trend in proportions, we can use the CHISQ option of PROC FREQ and look at the statistic labeled "Mantel-Haenszel chi-square." (A note of caution: Some statisticians believe this test is inappropriate since it tends to result in significant trends even when only relatively minor changes in proportions are present.) A program to enter the table values and compute the chi-square test for trend is:

```
***Chi-square Test for Trend;
DATA TREND;
   INPUT RESULT $ GROUP $ COUNT @@;
DATALINES;
FAIL A 10 FAIL B 15 FAIL C 14 FAIL D 25
PASS A 90 PASS B 85 PASS C 86 PASS D 75
;
PROC FREQ DATA=TREND;
   TITLE "Chi-square Test for Trend";
   TABLES RESULT*GROUP / CHISQ;
   WEIGHT COUNT;
RUN;
```

If you are not familiar with the double "at" sign (@@) notation, see Chapter 12 for more information. Fortunately, the GROUP and RESULT values are already in the proper alphabetical order, so we don't have to resort to any expedients. The output is shown here:

```
Chi-square Test for Trend

The FREQ Procedure

Table of RESULT by GROUP

RESULT     GROUP

Frequency│
Percent  │
Row Pct  │
Col Pct  │ A        │ B        │ C        │ D        │  Total
─────────┼──────────┼──────────┼──────────┼──────────┼
FAIL     │       10 │       15 │       14 │       25 │     64
         │     2.50 │     3.75 │     3.50 │     6.25 │  16.00
         │    15.63 │    23.44 │    21.88 │    39.06 │
         │    10.00 │    15.00 │    14.00 │    25.00 │
─────────┼──────────┼──────────┼──────────┼──────────┼
PASS     │       90 │       85 │       86 │       75 │    336
         │    22.50 │    21.25 │    21.50 │    18.75 │  84.00
         │    26.79 │    25.30 │    25.60 │    22.32 │
         │    90.00 │    85.00 │    86.00 │    75.00 │
─────────┼──────────┼──────────┼──────────┼──────────┼
Total          100        100        100        100      400
             25.00      25.00      25.00      25.00   100.00

                                               [Continued]
```

```
Statistics for Table of RESULT by GROUP

Statistic                        DF      Value      Prob

Chi-Square                        3      9.0774     0.0283
Likelihood Ratio Chi-Square       3      8.7178     0.0333
Mantel-Haenszel Chi-Square        1      7.1844     0.0074
Phi Coefficient                          0.1506
Contingency Coefficient                  0.1490
Cramer's V                               0.1506

Sample Size = 400
```

From this output, you can see that the Mantel-Haenszel (M-H) chi-square test for trend is 7.184 (p = .007). There may be times when your table chi-square is not significant but, since the test for trend is using more information (the order of the columns), it may be significant.

Q. MANTEL-HAENSZEL CHI-SQUARE FOR STRATIFIED TABLES AND META-ANALYSIS

You may have a series of 2×2 tables for each level of another factor. This may be a confounding factor such as age, or you may have a 2×2 table at each site in a multisite study. In any event, one way to analyze multiple 2×2 tables of this sort is to compute a Mantel-Haenszel chi-square. This same technique can also be used to combine results from several studies identified in a literature search on a specific topic. Although the studies may have some minor differences, you may prefer to ignore those differences and combine the results anyway. This technique is sometimes referred to as meta-analysis and is becoming quite popular in medicine, education, and psychology. There are **lots** of cautions concerning this technique, but this is neither the time nor the place to discuss them. The Mantel-Haenszel statistic is also used frequently for item bias research.

It is actually easy to compute a chi-square for stratified tables using SAS software. All that is necessary is to request a three-way table with PROC FREQ and to include the option ALL.

As an example, suppose we have just two 2×2 tables, one for boys and the other for girls. Each table represents the relationship between hours of sleep and the chance of failing a test of physical ability. We want to investigate the risk factor (lack of sleep) on the outcome (failing a test). The hours-of-sleep variable has been dichotomized as 'LOW' (less than 8 hours) and 'HIGH' (more than 8 hours). Here are the two tables:

Test Results, Boys

		Fail	Pass
Sleep	Low	20	100
	High	15	150

Test Results, Girls

		Fail	Pass
Sleep	Low	30	100
	High	25	200

Our SAS program is straightforward:

```
***Program to compute a Mantel-Haenszel Chi-square Test
   for Stratified Tables;
DATA ABILITY;
   INPUT GENDER $ RESULTS $ SLEEP $ COUNT;
DATALINES;
BOYS FAIL 1-LOW 20
BOYS FAIL 2-HIGH 15
BOYS PASS 1-LOW 100
BOYS PASS 2-HIGH 150
GIRLS FAIL 1-LOW 30
GIRLS FAIL 2-HIGH 25
GIRLS PASS 1-LOW 100
GIRLS PASS 2-HIGH 200
;
PROC FREQ DATA=ABILITY;
   TITLE "Mantel-Haenszel Chi-square Test";
   TABLES GENDER*SLEEP*RESULTS/ALL;
   WEIGHT COUNT;
RUN;
```

Since the output from this program is voluminous, we edit it and show you the relevant portions:

```
Mantel-Haenszel Chi-square Test

Statistics for Table 1 of SLEEP by RESULTS
Controlling for GENDER=BOYS

Statistic                     DF      Value      Prob
─────────────────────────────────────────────────────
Chi-Square                     1      3.7013    0.0544

            Estimates of the Relative Risk (Row1/Row2)

Type of Study                 Value      95% Confidence Limits
──────────────────────────────────────────────────────────────
Case-Control (Odds Ratio)    2.0000       0.9777      4.0911
Cohort (Col1 Risk)           1.8333       0.9795      3.4313
Cohort (Col2 Risk)           0.9167       0.8349      1.0064


Table 2 of SLEEP by RESULTS
Controlling for GENDER=GIRLS
```
                                                          [Continued]

```
Statistic                        DF       Value      Prob

Chi-Square                        1      9.0106     0.0027

            Estimates of the Relative Risk (Row1/Row2)

Type of Study                         Value      95% Confidence Limits

Case-Control (Odds Ratio)            2.4000        1.3404      4.2973
Cohort (Col1 Risk)                   2.0769        1.2789      3.3728
Cohort (Col2 Risk)                   0.8654        0.7792      0.9611
```

Summary Statistics for SLEEP by RESULTS
Controlling for GENDER

```
  Cochran-Mantel-Haenszel Statistics (Based on Table Scores)

Statistic   Alternative Hypothesis    DF       Value       Prob

    1        Nonzero Correlation       1      12.4770     0.0004
    2        Row Mean Scores Differ    1      12.4770     0.0004
    3        General Association       1      12.4770     0.0004

         Estimates of the Common Relative Risk (Row1/Row2)

Type of Study    Method               Value      95% Confidence Limits

Case-Control     Mantel-Haenszel     2.2289        1.4185      3.5024
  (Odds Ratio)   Logit               2.2318        1.4205      3.5064

Cohort           Mantel-Haenszel     1.9775        1.3474      2.9021
  (Col1 Risk)    Logit               1.9822        1.3508      2.9087

Cohort           Mantel-Haenszel     0.8891        0.8283      0.9544
  (Col2 Risk)    Logit               0.8936        0.8334      0.9582

         Breslow-Day Test for
Homogeneity of the Odds Ratios

Chi-Square               0.1501
DF                            1
Pr > ChiSq               0.6985

Total Sample Size = 640
```

   As you can see, the results for boys gives us a chi-square of 3.7013 (p = .0544) with a relative risk of 1.8333 (95% CI .9795, 3.4313). (We choose the Col1 Risk value since we want the RR for failing the test, which is the value for column 1.) For girls, the results are a chi-square of 9.0106 (p = .0027) and a relative risk of 2.0769 (95% CI 1.2789, 3.3728).

The results of combining the two tables is found under the heading "SUMMARY STATISTICS." Looking at the Cohort Mantel-Haenszel statistics (col1 risk), we find a p-value of .001 and a relative risk of 1.977 (95% CI 1.355, 2.887). Notice that p-value from the combined genders is lower than that for either the boys or girls alone. The Breslow-Day test for homogeneity of the odds ratio is not significant (p = .698), so we can be comfortable in combining these two tables.

R.   "CHECK ALL THAT APPLY" QUESTIONS

A common problem in questionnaire analysis is "check all that apply" questions. For example, suppose we had a question asking respondents which course or courses they were interested in taking. It might be written:

```
Which course, or courses, would you like to see offered next
semester?

                    (Check ALL that apply)

_____ 1.   Micro-computers        _____ 4.   Job Control Language

_____ 2.   Intro to SAS           _____ 5.   FORTRAN

_____ 3.   Advanced SAS           _____ 6.   PASCAL
```

Insofar as our analysis is concerned, this is not one question with up to six answers, but six yes/no questions. Each course offering is treated as a variable with values of YES or NO (coded as 1 or 0, for example). Our questionnaire would be easier to analyze if it were arranged thus:

```
Please indicate which of the following courses you
would like to see offered next semester:


                           (1) yes      (0) no

a) Micro-computers            ____        ____

b) Intro to SAS               ____        ____

c) Advanced SAS               ____        ____

d) Job Control Language       ____        ____

e) FORTRAN                    ____        ____

f) PASCAL                     ____        ____
```

Our INPUT statement would have six variables (COURSE1–COURSE6 for instance), each with a value of 1 or 0. A format matching 1 = yes and 0 = no would lend readability to the final analysis.

This approach works well when there is a limited number of choices. However, when we are choosing several items from a large number of possible choices, this approach becomes impractical since we need a variable for every possible choice. A common example in the medical field would be the variable "diagnosis" on a patient record.

We might have a list of hundreds or even thousands of diagnosis codes and want to consider a maximum of two or three diagnoses for each patient. Our approach in this case would be to ask for up to three diagnoses per patient, using a diagnosis code from a list of standardized codes. Our form might look like this:

```
Enter up to 3 diagnosis codes for the patient.

Diagnosis 1 _____
Diagnosis 2 _____
Diagnosis 3 _____
```

Our INPUT statement would be straightforward:

```
DATA DIAG1;
    INPUT ID 1-3 … DX1 20-22 DX2 23-25 DX3 26-28;
DATALINES;
(data goes here)
```

Suppose we had the following data:

OBS	ID	DX1	DX2	DX3
1	1	3	4	.
2	2	1	3	7
3	3	5	.	.

Notice that one patient could have a certain diagnosis code as his first diagnosis, whereas another patient might have the same code as his second or third diagnosis. If we want a frequency distribution of diagnosis codes, what can we do? We could try this:

```
PROC FREQ DATA=DIAG1;
    TABLES DX1-DX3;
RUN;
```

Using this method, we would have to add the frequencies from three tables to compute the frequency for each diagnosis code. A better approach would be to create a separate data set that is structured differently. Our goal would be a data set with as many observations per subject as there were nonmissing diagnosis codes, like this:

OBS	ID	DX
1	1	3
2	1	4
3	2	1
4	2	3
5	2	7
6	3	5

The program statements to create the previous data set are:

```
DATA DIAG2;
    SET DIAG1;
    DX = DX1;
    IF DX NE . THEN OUTPUT;
    DX = DX2;
    IF DX NE . THEN OUTPUT;
    DX = DX3;
    IF DX NE . THEN OUTPUT;
    KEEP ID DX;
RUN;
```

Each observation from our original data set (DIAG1) will create up to three observations in our new data set (DIAG2). The SET statement (line 2) reads an observation from our original data set (DIAG1). The first observation in DIAG1 is:

ID	AGE	GENDER	DX1	DX2	DX3
1	23	M	3	4	.

To these values, we add a new variable called DX, and set it equal to DX1. The Program Data Vector now contains:

ID	AGE	GENDER	DX1	DX2	DX3	DX
1	23	M	3	4	.	3

Since DX is not missing, we output (line 4) these values to the first observation in data set DIAG2 (for now, let's ignore the KEEP statement in line 9). Next, DX is set to DX2, and another observation is written to data set DIAG2:

ID	AGE	GENDER	DX1	DX2	DX3	DX
1	23	M	3	4	.	3
1	23	M	3	4	.	4

The values for ID, AGE, GENDER, DX1, DX2, and DX3 are still the same, only DX has changed. Finally, DX is set equal to DX3, but since DX3, and therefore DX, is missing, no observation is written to DIAG2.

We are now at the bottom of the DATA step, and control is returned to the top of the DATA step where the SET statement will read another observation from data set DIAG1. This routine continues until all the observations from DIAG1 have been read. Since we had a KEEP statement in the program, only the variables ID and DX actually exist in data set DIAG2:

OBS	ID	DX
1	1	3
2	1	4
3	2	1
4	2	3
5	2	7
6	3	5

We can now count the frequencies of DX codes using PROC FREQ with DX as the TABLE variable.

This program can be made more compact by using an ARRAY and a DO loop. (See Chapter 15 for a detailed explanation of ARRAYS, and Chapter 16, Section C, for details on restructuring SAS data sets using ARRAYS.) Finally, you can use PROC TRANSPOSE to restructure SAS data sets (see *Longitudinal Data and SAS: A Programmer's Guide* (Cody). Order Number: 58176, ISBN: 1-58025-924).

```
DATA DIAG2;
   SET DIAG1;
   ARRAY D[*] DX1-DX3;
   DO I = 1 TO 3;
      DX = D[I];
      IF D[I] NE . THEN OUTPUT;
   END;
   KEEP ID DX;
RUN;
```

PROBLEMS

Remember, you can download all the data sets and programs for these problems from the web site: www.prenhall.com/cody

3.1 Suppose we have a variable called GROUP that has numeric values of 1, 2, or 3. Group 1 is a control group, group 2 is given aspirin, and group 3 is given ibuprofen. Create a format to be assigned to the GROUP variable.

3.2 You have raw data consisting of ID (character), GENDER (numeric: 1 = Male, 2 = Female), SES (character: L = Low, M = Medium, H = High), DRUG (character: A–H), and AGE (numeric). First create three formats. One for GENDER, one for SES, and one to group ages into three groups: $<= 20$, 21–40, and 41 and older. Write a DATA step to read the following space delimited data and create a data set called QUES2. Format only SES and GENDER here. Also, add labels for SES (Socioeconomic Status), DRUG (Drug Group), and AGE (Age of Subject). Also in the DATA step, create a new variable, COST, with a value of LOW when the value of DRUG is A, C, F, or B and a value of HIGH otherwise. Assume there may be missing values for DRUG. Use PROC PRINT to obtain a listing of the data set. Next, compute frequencies for SES, COST, and AGE, where AGE is formatted into the three age groups mentioned.

Here are some sample data lines to work with:

```
001 1   L B 15
002 2 M     Z   35
003    2 H  F 76
004 1 L C 21
005 2 H . 58
```

3.3 A survey is conducted and data are collected and coded. The data layout is shown here:

Variable	Description	Columns	Coding Values
ID	Subject identifier	1–3	
GENDER		4	M = Male F = Female
PARTY	Political party	5	1 = Republican
			2 = Democrat
			3 = Not registered
VOTE	Did you vote in the last election?	6	0 = No 1 = Yes
FOREIGN	Do you agree with the government's foreign policy?	7	0 = No 1 = Yes
SPEND	Should we increase domestic spending?	8	0 = No 1 = Yes

Collected data are shown next:

```
007M1110
013F2101
137F1001
117 1111
428M3110
017F3101
037M2101
```

(a) Create a SAS data set, complete with labels and formats for this questionnaire.

(b) Generate frequency counts for the variables GENDER, PARTY, VOTE, FOREIGN, and SPEND.

(c) Test if there is a relationship between voting in the last election versus agreement with spending and foreign policy. (Have SAS compute chi-square for these relationships.)

3.4 Run the following program to create a temporary SAS data set called BLOOD. Produce frequencies for WBCs (white blood cell counts) using the following cutoffs: For WBC, Low = 3,000 to 4,000; Medium = 4,001 to 6,000; High = 6,001 to 12,000. Values below 3,000 (but not missing) are to be labeled 'Abnormally Low', and values above 12,000 are to be labeled 'Abnormally High'. Finally, Missing values are to be labeled 'Not Available'. Write the necessary SAS statements to produce frequencies of all of these categories (omit the cumulative frequencies from the tables). Have the frequencies for missing values placed in the main table (use the TABLES option MISSING to do this).

(a) Do the recoding using a DATA step.

(b) Do the recoding using a format.

```
*Program to create data set BLOOD;
DATA BLOOD;
   DO I = 1 TO 500;
      WBC = INT(RANNOR(1368)*2000 + 5000);
      X = RANUNI(0);
      IF X LT .05 THEN WBC = .;
      ELSE IF X LT .1 THEN WBC = WBC - 3000;
      ELSE IF X LT .15 THEN WBC = WBC + 4000;
      OUTPUT;
   END;
   DROP I X;
RUN;
```

3.5 Run the following program to create a SAS data set called DEMOG:

```
DATA DEMOG;
   INPUT WEIGHT HEIGHT GENDER $;
DATALINES;
155 68 M
98 60 F
202 72 M
280 75 M
130 63 F
.   57 F
166 . M
;
```

We want to recode WEIGHT and HEIGHT as follows (assume that WEIGHT and HEIGHT are integer values only):

```
WEIGHT      0-100    = 1
            101-150 = 2
            151-200 = 3
            >200     = 4
HEIGHT      0-70     = 1
            >70      = 2
```

We then want to generate a table of WEIGHT categories (rows) by HEIGHT categories (columns). Recode these variables in two ways: (1) with "IF" statements; (2) with formats. Then write the necessary statements to generate the table.

3.6 A case-control study relating leukemia to exposure to non-ionizing radiation resulted in the following: In the cases (those with leukemia), there were 50 exposed and 500 non-exposed; for the controls, there were 40 exposed and 500 non-exposed. Write the necessary SAS statements to compute the odds ratio for non-ionizing radiation exposure. Make sure you have your first column for cases and the first row for those exposed. What is the 95% CI for the odds ratio and what does this imply?

3.7 A friend gives you some summary data on the relationship between socioeconomic status (SES) and asthma, as follows:

Asthma	Yes	No
Low SES	40	100
High SES	30	130

Create a SAS data set from these data and compute chi-square.

3.8 Using the data from Problem 3.7, what is the relative risk (or risk ratio) for low SES and asthma? What is the 95% CI and what is your interpretation of this. Hint: You can use the PROC FREQ option ORDER=DATA to control the order of the rows and columns. This option uses the data order of the data values to decide the column and row orders.

3.9 A matched-pairs case-control study is conducted. Each case (a person with disease X) is matched to a single control, based on age (plus or minus 2 years) and gender. Each person is then asked if he/she used multivitamins regularly in the past year. The results are:

Case Use of Vitamins	Matched Controls Use of Vitamins	Count
Yes	Yes	100
Yes	No	50
No	Yes	90
No	No	200

Remembering that this is a matched study, compute a McNemar chi-square. Are the cases more or less likely to have used vitamins?

3.10 Two methods of determining carotid stenosis (narrowing of the carotid artery which supplies blood to the brain) are being compared. Method one is an invasive method that is believed to be very accurate. Method two is a noninvasive (MRI) method. One hundred patients were evaluated by these two methods with these results:

Method One	Method Two	Count
Occuled	Occuled	15
Occuled	Non-occuled	8
Non-occuled	Occuled	10
Non-occuled	Non-occuled	67

Compute coefficient Kappa.

3.11 A researcher wants to determine if there is a relationship between use of computer terminals (VDT—video display terminals) and miscarriages. An unpaired, case-control study is conducted. Of the cases (women with miscarriages), there were 30 women who used VDTs and 50 who did not. Among the controls, there were 90 women who used VDTs and 200 who did not. Compute chi-square, the odds ratio, and a 95% confidence interval for the odds ratio.

3.12 You have a SAS data set called CASE_CONTROL in your C:\MYDATA folder (or any other folder of your choice, or even a temporary data set). Variables are: EXPOSURE (character, values of 1-yes, and 2-no), and OUTCOME (values of 1-dead, 2-alive). Write a program to output a 2×2 table along with chi-square and the OR.

```
***Program to create data set CASE_CONTROL;
LIBNAME PERM 'C:\MYDATA';
***Substitute a libname of your choice, or make the data set
   temporary;
      DATA PERM.CASE_CONTROL;
         ARRAY COUNTS[4] _TEMPORARY_ (50 100 150 35);
         DO EXPOSURE = '1-Yes','2-No';
            DO OUTCOME = '1-Dead ','2-Alive';
               I + 1;
               DO N = 1 TO COUNTS[I];
                  SUBJ + 1;
                  OUTPUT;
               END;
            END;
         END;
         DROP I;
      RUN;
```

3.13 A researcher wants to determine if soundproofing a classroom leads to better behavior among the students. In the standard classrooms, there were 30 behavioral problems from a total of 250 children. In the soundproofed classrooms, there were 20 behavioral problems from a total of 300 children. Test if room noise results in an increased number of behavioral problems by computing the relative risk (of noisy classrooms) for producing behavioral problems. Have SAS produce a 95% CI for the relative risk as well.

3.14 Run the following program to generate a SAS data set DOSE, showing the number of patients having significant pain and those not having significant pain for each of three doses of medication. Compute the regular chi-square statistics and a chi-square for a test of trend. Do the proportions of patients having significant pain show a significant downward trend? Notice that the values for PAIN and DOSE are already in the correct order for your table.

```
PROC FORMAT;
   VALUE PAIN 1 = 'YES' 2 = 'NO';
   VALUE DOSE 1='LOW' 2='MEDIUM' 3='HIGH';
RUN;
```

```
DATA DOSE;
    DO DOSE = 1 TO 3;
        DO I = 1 TO 50;
            PAIN = 2 - (RANUNI(135) GT (.6 + .08*DOSE));
            OUTPUT;
        END;
    END;
    FORMAT PAIN PAIN. DOSE DOSE.;
RUN;
```

3.15 A school administrator believes that larger class sizes lead to more discipline problems. Four class sizes (small, medium, large, and gigantic) are tested. The following table summarizes the problems recorded for each of the class sizes. Treating class size as an ordinal variable, test if there is a linear increase in the proportion of class problems. (Be careful to arrange the order of the class size columns so that they range from small to gigantic, or your test for trend will not be valid.)

Class Size	Small	Medium	Large	Gigantic
Problem	3	6	17	80
No Problem	12	22	38	120
Total	15	28	55	200

3.16 A randomized clinical trial compared aspirin to placebo for the prevention of heart attacks (MIs) and strokes. Out of a total of 1,000 subjects on aspirin, there were 80 heart attacks and 65 strokes; out of a total of 2,000 subjects on placebo, there were 240 heart attacks and 165 strokes. Is there a significant benefit for aspirin therapy for heart attacks and strokes? What is the RR for aspirin use for each of these two outcomes? (Careful here, you are given the total number of subjects in each group and the number of complications.)

3.17 The relationship between office temperature and head colds is tested for smokers and nonsmokers. Since smoking is assumed to be a confounding factor in this relationship, use a Mantel-Haenszel chi-square for stratified tables to analyze the two tables here:

		Smokers				**Nonsmokers**	
		Colds	No Colds			Colds	No Colds
Temp Control	Poor	20	100	Temp Control	Poor	30	100
	Good	15	150		Good	25	200

3.18 Three studies were conducted looking at the effects of magnesium sulfate to improve survival of a cardiac arrest. Results from the three studies were as follows:

	Study One		Study Two		Study Three	
	Survived	**Died**	**Survived**	**Died**	**Survived**	**Died**
MgSO$_4$	20	100	25	150	30	200
Placebo	25	155	21	150	28	240

Conduct a meta-analysis to determine the overall effect of magnesium sulfate (MgSO$_4$) on survival. Check the Breslow-Day statistic for the homogeniety of the odds ratio, and determine the overall RR and the 95% CI.

*3.19 A physical exam is given to a group of patients. Each patient is diagnosed as having none, one, two, or three problems from the following code list:

Code	Problem Description
1	Cold
2	Flu
3	Trouble sleeping
4	Chest pain
5	Muscle pain
6	Headaches
7	Overweight
8	High blood pressure
9	Hearing loss

The coding scheme is as follows:

Variable	Description	Column(s)
SUBJ	Subject number	1–2
PROB1	Problem 1	3
PROB2	Problem 2	4
PROB3	Problem 3	5
HR	Heart rate	6–8
SBP	Systolic blood pressure	9–11
DBP	Diastolic blood pressure	12–14

Using the sample data here:

```
Columns 12345678901234
       ----------------------
           11127 78130 80
           1787  82180110
           031   62120 78
           4261  68130 80
           89    58120 76
           9948  82178100
```

(a) Compute the mean HR, SBP, and DBP.

(b) Generate frequency counts for each of the nine medical problems.

*3.20 A survey was conducted where each participant was asked to choose their three favorite classical composers from the following list:

Composer	Number
Bach	1
Handel	2
Scarlatti	3
Hayden	4
Mozart	5
Beethoven	6
Schubert	7
Brahms	8
Schumann	9
Stravinsky	10
Shostakovich	11

The following data was recorded:

Subject	Gender	Choice 1	Choice 2	Choice 3
1	M	6	4	5
2	F	2	4	3
3	M	7	11	9
4	M	5	6	3
5	F	7	8	10
6	F	1	2	4
7	M	6	5	9
8	M	2	9	11
9	F	5	1	2

Run the following program to create the SAS data set CLASSICAL. Using this data set as the starting point, write a SAS program to obtain a single frequency listing showing the composer choices, from the most popular to the least. (Hint: Use the PROC FREQ option ORDER=FREQ to order the list in this way.)

```
***Program to create the data set CLASSICAL;
PROC FORMAT;
   VALUE COMPOSER 1  = 'Bach'
                  2  = 'Handel'
                  3  = 'Scarlatti'
                  4  = 'Hayden'
                  5  = 'Mozart'
```
[Continued]

```
                    6  = 'Beethoven'
                    7  = 'Schubert'
                    8  = 'Brahms'
                    9  = 'Schumann'
                   10  = 'Stravinsky'
                   11  = 'Shostakovich';
RUN;
DATA CLASSICAL;
   INPUT SUBJ $ GENDER : $1. CHOICE1 CHOICE2 CHOICE3;
DATALINES;
1 M 6 4 5
2 F 2 4 3
3 M 7 11 9
4 M 5 6 3
5 F 7 8 10
6 F 1 2 4
7 M 6 5 9
8 M 2 9 11
9 F 5 1 2
;
```

3.21 What's wrong with this program?

```
1     DATA IGOOFED;
2            INPUT
         #1  ID 1-3 GENDER 4 AGE 5-6 RACE 7
             (QUES1-QUES10) (1.)
         #2  @4 (QUES11-QUES25) (1.);
3          FORMAT GENDER SEX. RACE RACE. QUES1-QUES25 YESNO.;
4     DATALINES;
      00112311010011101
      1100111001101011
      002244210111011100
      0111011101111010
      ;
5     PROC FORMAT;
6        VALUE SEX 1='MALE' 2='FEMALE';
7        VALUE RACE 1='WHITE' 2='AFRICAN AM.' 3='HISPANIC';
8        VALUE YESNO 0='NO' 1='YES';
9     RUN;
10    PROC FREQ DATA=IGOOFED;
11       VAR GENDER RACE QUES1-QUES25 / CHISQ;
12    RUN;
13    PROC MEANS MAXDEC=2 N MEAN STD MIN MAX;
14       BY RACE;
15       VAR AGE;
16    RUN;
```

HINTS AND COMMENTS:

The INPUT statement is correct. The pointers (@ signs) and format lists (1.) are explained in Chapter 12. There are four errors.

CHAPTER 4

Working with Date and Longitudinal Data

A. Introduction
B. Processing Date Variables
C. Working with Two-digit Year Values (the Y2K Problem)
D. Longitudinal Data
E. Selecting the First or Last Visit per Patient
F. Computing Differences between Observations in a Longitudinal Data Set
G. Computing the Difference between the First and Last Observation for Each Subject
H. Computing Frequencies on Longitudinal Data Sets
I. Creating Summary Data Sets with PROC MEANS or PROC SUMMARY
J. Outputting Statistics Other Than Means

A.   INTRODUCTION

Working with dates is a task that data analysts frequently face. SAS software contains many powerful resources for working with dates. These include the ability to read dates in almost any form or to compute the number of days, months, or years between any two dates.

Data collected for the same set of subjects at different times are sometimes called longitudinal data. These data require specialized techniques for analysis. Let's begin by seeing how date values are handled with SAS software.

B.   PROCESSING DATE VARIABLES

Suppose you want to read the following information into a SAS data set:

ID	DOB	Admitting Date	Discharge Date	DX	Fee
001	10/21/1946	12/12/2004	12/14/2004	8	8000
002	05/01/1980	07/08/2004	09/05/1990	7	5000

We can arrange this data in columns, thus:

Variable Name	Description	Column(s)
ID	Patient ID	1–3
DOB	Date of birth	4–13
ADMIT	Date of admission	14–23
DISCHRG	Discharge date	24–33
DX	Diagnosis	34
FEE	Hospital fee	35–39

You might be tempted to write an input statement like:

```
INPUT ID      1-3
      DOB     4-13
      ADMIT   14-23
      DISCHRG 24-33
      DX      34
      FEE     35-39;
```

However, you cannot read the 10 date digits and slashes as a number (you would get an error). You could read the dates as character values, but you would not be able to do any calculations on their values (at least directly). So what do we do? SAS software includes extensive provisions for working with date and time variables. The first step in reading date values is to tell the program how the date value is written. Common examples are:

Example	Explanation	SAS INFORMAT
102150	Month – Day – 2-digit Year	MMDDYY6.
10211950	Month – Day – 4-digit Year	MMDDYY8.
10/21/50	Month – Day – 2-digit Year	MMDDYY8.
10/21/1950	Month – Day – 4-digit Year	MMDDYY10.
211050	Day – Month – 2-digit Year	DDMMYY6.
21101950	Day – Month – 4-digit Year	DDMMYY8.
501021	2-digit Year – Month – Day	YYMMDD6.
19501021	4-digit Year – Month – Day	YYMMDD8.
21OCT50	Day, 3-character Month, 2-digit Year	DATE7.
21OCT1950	Day, 3-character Month, 4-digit Year	DATE9.
OCT50	Month and 2-digit Year only	MONYY5.
OCT1950	Month and 4-digit Year only	MONYY7.

SAS programs can read any of these dates, provided we tell it which form of date we are using. Once SAS knows it's reading a date, that date is converted into the number of days from a fixed point in time: January 1, 1960. It doesn't matter if our date comes before or after this date. Dates before January 1, 1960, will be negative numbers. Therefore we can subtract any two dates to find the number of days in-between. We can also convert the SAS internal date value back to any of the allowable SAS date formats for reporting purposes. As we saw in the last chapter, we can associate formats with variables for printing. We can specify how to read values by specifying an INFORMAT to give the program instructions on how to read a data value. The SAS date informat MMDDYY10., for example, is used to read dates in month-day-year order. The 10 in the informat refers to the number of columns occupied by the date. Thus far, we have used space-between-the-numbers and column specifications to instruct SAS how to read our data values.

To read date values, we can use pointers and informats (this is called formatted input). A column pointer (@) first tells the program which column to start reading. We follow this with the variable name and a specification of what type of data we are reading, called an informat. Two very common informats are W.n and $W. W.n is a numeric informat that says to read W columns of data and to place n digits to the right of the decimal point. For example, the informat 6.1 says to read six columns and to place a decimal point before the last digit. If you are not specifying a decimal point, you do not have to specify the value to the right of the period. Thus, an informat of 6. is equivalent to 6.0. The $W. informat is used to read W columns of character data. The informat MMDDYY10. is used for dates in the form 10/21/1950 or 10–21–1950. Using our previous column assignments, a valid INPUT statement would be:

```
INPUT @1    ID              $3.
      @4    DOB       MMDDYY10.
      @14   ADMIT     MMDDYY10.
      @24   DISCHRG MMDDYY10.
      @34   DX              1.
      @35   FEE             5.;
```

The @ signs, referred to as column pointers, tell the program at which column to start reading the next data value. Our three dates are all in the form MMDDYY (month-day-year) and occupy 10 columns, so the MMDDYY10. informat is used. If you had decided not to include the two slashes in the date, the date would occupy eight columns and the MMDDYY8. informat would be used. Remember that all of our formats and informats end with periods (or a period followed by a number) to distinguish them from variable names. All the column pointers in the previous program are not necessary. For example, the program would start reading in column 1 without the @1 pointer. Also, since the ID ends in column 3 and the date of birth starts in column 4, the pointer @4 is redundant. Good programming practice suggests that using a column pointer with every variable is a good idea. (See Chapter 12 for more details on how to use the INPUT statement.)

Let's calculate two new variables from these data. First, we compute our subject's age at the time of admission. Next, we compute the length of stay in the hospital. All we have to do is subtract any two dates to obtain the number of days in-between. Therefore, our completed program looks like:

```
DATA HOSPITAL;
   INPUT @1   ID             $3.
         @4   DOB     MMDDYY10.
         @14  ADMIT   MMDDYY10.
         @24  DISCHRG MMDDYY10.
         @34  DX            1.
         @35  FEE           5.;
   LENGTH_STAY = DISCHRG-ADMIT + 1;
   AGE = ADMIT - DOB;
DATALINES;
00110/21/194612/12/200412/14/20048 8000
00205/01/198007/08/200408/08/2004412000
00301/01/196001/01/200401/04/20043 9000
00406/23/199811/11/200412/25/2004715123
;
```

The calculation for length of stay (LENGTH_STAY) is relatively straightforward. We subtract the admission date from the discharge date and add 1 (we want to count both the admission day and the discharge day in the length-of-stay computation). If we subtract the date of birth (DOB) from the admission date (ADMIT), the result is the age in **days**. Look at a listing of this data set:

```
Listing of Data Set HOSPITAL

                                              LENGTH_
 ID      DOB     ADMIT    DISCHRG   DX    FEE    STAY      AGE

 001    -4820    16417     16419     8   8000     3      21237
 002     7426    16260     16291     4  12000    32       8834
 003        0    16071     16074     3   9000     4      16071
 004    14053    16386     16430     7  15123    45       2333
```

This rather strange listing clearly demonstrates how SAS stores dates. For example, look at the date of birth for subject 003. Notice it is zero. Why? Because this person was born on January 1, 1960, and this is day zero in SAS-land. Well, you wouldn't want to show this listing to your boss or colleagues. How do we make the date values look like the dates we know and love? Just as we used formats in the last chapter to change the way our values printed, we can use formats here to change the appearance of the date values. We don't even have to use PROC FORMAT to create these formats—SAS has already done it for us. Two very popular date formats are MMDDYY10. and DATE9.

To see how this works, let's format the date of birth with the MMDDYY10. format and format the admission and discharge dates using the DATE9. format. We just need to add the single line:

```
FORMAT DOB MMDDYY10. ADMIT DISCHRG DATE9.;
```

to our DATA step. If we do, the listing now looks like this:

```
Listing of Data Set HOSPITAL

                                                    LENGTH_
    ID         DOB       ADMIT     DISCHRG    DX    FEE     STAY    AGE

    001    10/21/1946   12DEC2004  14DEC2004   8    8000      3    21237
    002    05/01/1980   08JUL2004  08AUG2004   4   12000     32     8834
    003    01/01/1960   01JAN2004  04JAN2004   3    9000      4    16071
    004    06/23/1998   11NOV2004  25DEC2004   7   15123     45     2333
```

Notice the result of these two SAS formats. (The DATE9. format used with the admission and discharge dates is particularly useful when you are working with some countries that use the month-day-year format and others that use the day-month-year format.) Next, we need to compute age in years rather than in days. Here is how we do that:

By subtracting the date of birth from the admission date, we have the number of days between these two dates. We could convert this to years by dividing by 365.25 (approximately correct since there is a leap year every 4 years). This would give us:

```
AGE = (ADMIT-DOB) / 365.25;
```

However, a better way to compute the difference in years between two dates is to use a SAS function called YRDIF. You supply the YRDIF function with the first date and the second date, and it computes the number of years between the two dates. For example, the exact age as of the admission date is:

```
AGE = YRDIF(DOB,ADMIT,'ACTUAL');
```

SAS functions (see Chapters 17 and 18) are part of the SAS system and often perform calculations for us. In this example, the three arguments to the YRDIF function are the first date, the second date, and a calculation method. Notice that these arguments (information that is supplied to the function) are separated by commas and are placed in parentheses following the function name.

What if you wanted a person's age as of a particular date, say January 1, 2005? How many days is January 1, 2005, from January 1, 1960? Don't know? We don't either. Let's let SAS compute this for us. You can specify a SAS date by using a date constant, for example, January 1, 2005, is equal to:

```
'01JAN2005'D
```

The form of a SAS date constant is a one- or two-digit day of the month, a three-letter month abbreviation, and a two- or four-digit year (always use a four-digit year) placed in single or double quotes, followed by a lower- or uppercase 'D'. Thus if you wanted to compute a person's age as of January 1, 2005, you would write:

```
AGE = YRDIF(DOB,'01JAN2005'D,'ACTUAL');
```

What if you wanted the age as of the date you ran your SAS program? Well, the TODAY function returns the current date. So, if you wanted a person's age as of the date you ran the program, you would write:

```
AGE = YRDIF(DOB,TODAY(),'ACTUAL');
```

Notice that the TODAY function, even though it does not have any arguments, still has a set of parentheses after it. This is consistent with the rule that **all** SAS functions are followed by a set of parentheses.

We may want to define age so that a person is not considered, say, 18 years old, until his 18th birthday. That is, we want to omit any fractional portion of his/her age in years. A SAS function that does this is the INT (integer) function. We can write:

```
AGE = INT(AGE);
```

to remove the fractional part of the age value. You can "nest" functions (place one inside of another), as shown here, to make this part of your program more compact (and yes, elegant).

```
AGE = INT(YRDIF(DOB,ADMIT,'ACTUAL'));
```

Notice the use of parentheses to keep things straight. If we wanted to round the age to the nearest tenth of a year, we would use the ROUND function. This function has two arguments: the number to be rounded and the roundoff unit. To round to the nearest tenth of a year, we use:

```
AGE = ROUND(YRDIF(DOB,ADMIT,'ACTUAL'),.1);
```

To the nearest year, the function would be:

```
AGE = ROUND(YRDIF(DOB,ADMIT,'ACTUAL'));
```

Note: If you leave off the second argument of the ROUND function, it assumes you want to round to the nearest integer.

It is important to remember that once SAS has converted our dates into the number of days from January 1, 1960, it is stored just like any other numeric value. Therefore, if we print it out (with PROC PRINT, for example), the result will not look like a date. We need to supply SAS with a format to use when printing out dates. This format does not have to be the same one we used to read the date in the first place. Some useful date formats and their results are shown in the following table:

The Date 10/21/1950 Printed with Different Date Formats:

Format	Result
MMDDYY6.	102150
MMDDYY8.	10/21/50
MMDDYY10.	10/21/1950
DATE7.	21OCT50
DATE9.	21OCT1950
WORDDATE.	October 21, 1950

Notice that you can use the informat MMDDYY8. to read dates such as 10/21/50 and 10211950. SAS is smart enough to recognize that an eight-digit date without slashes must be using a four-digit year.

C.   WORKING WITH TWO-DIGIT YEAR VALUES (THE Y2K PROBLEM)

What happens if you have raw data with two-digit years? How does SAS read 10/21/04? As October 21, 1904, or October 21, 2004? SAS has an extremely useful system option called YEARCUTOFF=value. The value you supply is a four-digit year that marks the beginning of a 100-year interval. Any two-digit date will be assumed to fall in that 100-year window. Starting with SAS version 7, the default value for the YEARCUTOFF option is 1920. Thus, any two-digit year would fall between 1920 and 2019. For example, the year 1/1/40 would be read as January 1, 1940; the year 1/1/15 would be read as January 1, 2015. If you want to change the value of the YEARCUTOFF option, use an OPTIONS statement. To set the value back to 1900, you would use the statement:

```
OPTIONS YEARCUTOFF = 1900;
```

Chapter 17, Section D, contains a list of SAS functions that are useful when working with dates. For example, month, day, and year values can be combined to form a SAS date, or you can extract a month or year from a SAS date.

D.   LONGITUDINAL DATA

There is a type of data, often referred to as longitudinal data, that needs special attention. Longitudinal data are collected on a group of subjects over time. CAUTION: This portion of the chapter is difficult and may be "hazardous to your health!"

To examine the special techniques needed to analyze longitudinal data, let's follow a simple example. Suppose we are collecting data on a group of patients. (The same scheme would be applicable to periodic data in business or data in the social sciences with repeated measures.) Each time the patients come in for a visit, we fill out an encounter form. The data items we collect are:

```
PATIENT ID
DATE OF VISIT (Month Day Year)
HEART RATE
SYSTOLIC BLOOD PRESSURE
DIASTOLIC BLOOD PRESSURE
DIAGNOSIS CODE
DOCTOR FEE
LAB FEE
```

Now, suppose each patient's visits are a maximum of four times a year. One way to arrange our SAS data set is as follows (each visit to occupy a separate line):

```
***Note: This is a very poor way to structure your data set. This is
     for instructional purposes only;
DATA HOSP_PATIENTS;
    INPUT #1
        @1   ID         $3.
        @4   DATE1     MMDDYY8.
        @12  HR1          3.
        @15  SBP1         3.
        @18  DBP1         3.
        @21  DX1          3.
        @24  DOCFEE1      4.
        @28  LABFEE1      4.
          #2
        @4   DATE2     MMDDYY8.
        @12  HR2          3.
        @15  SBP2         3.
        @18  DBP2         3.
        @21  DX2          3.
        @24  DOCFEE2      4.
        @28  LABFEE2      4.
          #3
        @4   DATE3     MMDDYY8.
        @12  HR3          3.
        @15  SBP3         3.
        @18  DBP3         3.
        @21  DX3          3.
        @24  DOCFEE3      4.
        @28  LABFEE3      4.
          #4
        @4   DATE4     MMDDYY8.
        @12  HR4          3.
        @15  SBP4         3.
        @18  DBP4         3.
        @21  DX4          3.
        @24  DOCFEE4      4.
        @28  LABFEE4      4.;
    FORMAT DATE1-DATE4 MMDDYY10.;
DATALINES;
007102119830701200800140040015 0
007120119830721300900200050020 0
007
007
009090319830661100701370030000 0
009
009
009
```

```
00507051983074140082013009000000
00501151982080180096014020015000
00506181982070170084014008004000
00507031983064140084014008002000
;
```

The number signs (#) in the INPUT statement signify multiple lines per subject. Since our date is in the month-day-year form, we use the MMDDYY8. informat. We also included an output format for our dates with a FORMAT statement. This FORMAT statement uses the same syntax as the earlier examples in this chapter, where we created our own formats. The output format MMDDYY10. specifies that the date variables be printed in month-day-year form with slahes between the month, day, and year values.

With this method of one line per patient visit, we would need to insert BLANK lines of data for any patient who had less than four visits, in order to fill out four lines per subject. This is not only clumsy but it also occupies a lot of space unnecessarily. If we want to compute within-subject means, we continue (before the DATALINES statement):

```
AVEHR = MEAN(OF HR1-HR4);
AVESBP = MEAN(OF SBP1-SBP4);
AVEDBP = MEAN(OF DBP1-DBP4);
        etc.
```

"MEAN" is one of the SAS built-in functions that will compute the mean of all the variables listed within parentheses. (NOTE: If any of the variables listed as arguments of the MEAN function have missing values, the result will be the mean of the nonmissing values. See Chapter 17, Section B for more details.) Also, since the variable lists in the form VAR1–VAR4 are used, you need to include the word 'OF' before the list. Without the 'OF', SAS would simply subtract the two values. A much better approach is to treat each visit as a separate observation. Our program would then look like:

```
DATA PATIENTS;
   INPUT @1   ID          $3.
         @4   DATE    MMDDYY8.
         @12  HR          3.
         @15  SBP         3.
         @18  DBP         3.
         @21  DX          3.
         @24  DOCFEE      4.
         @28  LABFEE      4.;
   FORMAT DATE MMDDYY10.;
DATALINES;
```

Now we need to include only as many lines of data as there are patient visits; blank lines to fill out four lines per subject are not needed. Our variable names are also simpler since we do not need to keep track of HR1, HR2, etc. A listing of data set PATIENTS is shown here:

```
Listing of Data Set PATIENTS

ID          DATE      HR    SBP    DBP     DX    DOCFEE    LABFEE

007     10/21/1983    70    120    80      14     40        150
007     12/01/1983    72    130    90      20     50        200
009     09/03/1983    66    110    70      137    30        0
005     07/05/1983    74    140    82      13     90        0
005     01/15/1982    80    180    96      14     200       1500
005     06/18/1982    70    170    84      14     80        400
005     07/03/1983    64    140    84      14     80        200
```

How do we analyze this data set?

A simple PROC MEANS on a variable such as HR, SBP, or DOCFEE will not be particularly useful since we are averaging one to four values per patient together. Perhaps the average of DOCFEE would be useful since it represents the average doctor fee per PATIENT VISIT, but statistics for heart rate or blood pressure would be a weighted average, the weight depending on how many visits we had for each patient. How do we compute the average heart rate or blood pressure per patient? The key is to use ID as a CLASS or BY variable.

Here is our program (with sample data):

```
DATA PATIENTS;
    INPUT @1  ID         $3.
          @4  DATE     MMDDYY8.
          @12 HR          3.
          @15 SBP         3.
          @18 DBP         3.
          @21 DX          3.
          @24 DOCFEE      4.
          @28 LABFEE      4.;
    FORMAT DATE MMDDYY10.;
DATALINES;
00710211983070120008001400400150
00712011983072130090020005000200
00909031983066110070137003000000
00507051983074140082013009000000
00501151982080180096014020001500
00506181982070170084014008000400
00507031983064140084014008000200
;
```

```
PROC MEANS DATA=PATIENTS NOPRINT NWAY;
   CLASS ID;
   VAR HR -- DBP DOCFEE LABFEE;
   OUTPUT OUT=STATS MEAN=M_HR M_SBP M_DBP M_DOCFEE M_LABFEE;
RUN;
```

The result is the mean HR, SBP, etc., per patient, which is placed in a new data set called STATS, with variable names M_HR, etc. (See Sections H and I of this chapter for more about using PROC MEANS to create output data sets.) Each variable listed after MEAN = in the OUTPUT statement will be the mean of the variables listed in the VAR statement, in the order they appear. Thus, M_HR in the data set STATS is the mean HR in the data set PATIENTS. In this example, the data set STATS would appear as follows (we can always test this with a PROC PRINT statement):

```
Listing of Data Set STATS

ID    _TYPE_    _FREQ_    M_HR    M_SBP    M_DBP    M_DOCFEE    M_LABFEE
005       1         4       72    157.5    86.5      112.5         525
007       1         2       71    125.0    85.0       45.0         175
009       1         1       66    110.0    70.0       30.0           0
```

This data set contains the mean HR, SBP, etc., per patient. We could analyze this data set with additional SAS procedures to investigate relationships between variables or to compute descriptive statistics where each data value corresponds to a single value (the MEAN) from each patient.

E.   SELECTING THE FIRST OR LAST VISIT PER PATIENT

What if we want to analyze the last (most recent) visit or the first visit for each patient in the previous data set PATIENTS? If we sort the data set in patient-date order, the most recent visit would be the last observation for each patient ID. We can extract these observations with the following SAS program:

```
PROC SORT DATA=PATIENTS;
   BY ID DATE;
RUN;
DATA RECENT;        ①
   SET PATIENTS;    ②
   BY ID;           ③
   IF LAST.ID;      ④
RUN;
```

We first sort the data set by ID and DATE. Even if we believe the observations for each patient are in date order (they aren't in this data set), it is still a good idea to include DATE in the sort. Statement ① creates a new data set called RECENT. As we explained previously, the SET statement ② acts like an INPUT statement except that observations are read one by one from the SAS data set PATIENTS instead of from our original raw data.

We are permitted to use a BY variable ③ following our SET statement, provided our data set has been previously sorted by the same variable (it has). The effect of adding the BY statement is to have SAS create what are called FIRST. and LAST. variables. In this case, since our BY variable is ID, two variables, FIRST.ID and LAST.ID, are automatically created. These variables are available in the DATA step but are not added to the data set (FIRST. and LAST. variables are automatically dropped). The FIRST. and LAST. variables are logical variables; that is, they have values of true (1) or false (0). FIRST.ID will be true (1) whenever we are reading the first observation for a given ID and will be false (0) otherwise; LAST.ID will be true whenever we are reading the last observation for a given ID and will be false (0) otherwise. To clarify this, the following shows our observations and the value of FIRST.ID and LAST.ID. Keep in mind that the two variables FIRST.ID and LAST.ID are **not** in the SAS data set (but they are in the PDV and thus are available to be referenced in the DATA step) and that the data set PATIENTS is now in ID and DATE order.

ID	DATE	HR	SBP	DBP	DX	DOCFEE	LABFEE	FIRST.ID	LAST.ID
5	01/15/82	80	180	96	14	200	1500	1	0
5	06/18/82	70	170	84	14	80	400	0	0
5	07/03/83	64	140	84	14	80	200	0	0
5	07/05/83	74	140	82	13	90	0	0	1
7	10/21/83	70	120	80	14	40	150	1	0
7	12/01/83	72	130	90	20	50	200	0	1
9	09/03/83	66	110	70	137	30	0	1	1

By adding the subsetting IF statement ④, we can select the last visit for each patient (in this case, observations 4, 6, and 7). You can recognize this IF statement as a subsetting IF statement because there is no THEN clause. A subsetting IF statement has the following structure:

```
IF condition;
```

Here is how it works: If the *condition* is true, the program continues to process the statements following the IF statement; if the *condition* is false, the program returns to the top of the DATA step. Specifically in this case, if LAST.ID is true, the program continues and, since this is the bottom of the DATA step, an observation is automatically written out to the data set RECENT. If LAST.ID is not true, the program returns to the top of the DATA step (and an observation is not written to data set RECENT).

Another way to write the subsetting IF statement is:

```
IF LAST.ID = 1;
```

However, since LAST.ID is a logical variable with values of true (1) or false (0), you can omit the equals 1 part.

A listing of data set RECENT (shown here) confirms that we have the last visit for each patient in the PATIENTS data set.

```
Listing of Data Set RECENT

ID          DATE     HR    SBP    DBP    DX    DOCFEE    LABFEE

005      07/05/1983   74    140    82     13      90          0
007      12/01/1983   72    130    90     20      50        200
009      09/03/1983   66    110    70    137      30          0
```

F.   COMPUTING DIFFERENCES BETWEEN OBSERVATIONS IN A LONGITUDINAL DATA SET

Suppose you want to compute the change (difference) between variables, such as heart rate and blood pressure, from visit to visit. With the data set structured with one observation per patient visit, this gets a bit tricky. Two very useful tools for doing computations between observations are the LAG function and retained variables. First let's see how the LAG function works.

You may have come across the term "lagged" values. For example, in a study relating asthma-related doctor visits to ozone levels, you may want to see the relationship between the current day's ozone level and the ozone levels from the day before. This value would be referred to as the ozone level, lagged 24 hours. Now for the definition of the LAG function. The LAG function returns the value of its argument—the last time the LAG function executed. What does this mean? An example will help. Look at the following program:

```
DATA LOOKING_BACK;
    INPUT DAY OZONE;
    OZONE_LAG24 = LAG(OZONE);
    OZONE_LAG48 = LAG2(OZONE);
DATALINES;
1 8
2 10
3 12
3 7
;
PROC PRINT DATA=LOOKING_BACK;
    TITLE "Demonstrating the LAG Function";
RUN;
```

The output is:

```
Demonstrating the LAG Function

                        OZONE_    OZONE_
Obs    DAY    OZONE      LAG24     LAG48

 1      1       8          .         .
 2      2      10          8         .
 3      3      12         10         8
 4      3       7         12        10
```

In this example, the value of OZONE_LAG24 is the value of OZONE from the previous day. It is missing in the first observation since there is no previous day. As you probably figured out by looking at the program and the listing, the LAG2 function returns the value from two days earlier. There is a whole family of LAG functions. Now, you may wonder why the definition seemed so strange. Why didn't we just say that the LAG function returns a value from the previous observation? Well, because it doesn't always. Look carefully at the following program:

```
DATA LAGGARD;
    INPUT X;
    IF X GT 5 THEN LAG_X = LAG(X);
DATALINES;
7
9
1
8
;
PROC PRINT DATA=LAGGARD;
    TITLE "Demonstrating a Feature of the LAG Function";
RUN;
```

Here is a listing of LAGGARD:

```
Demonstrating a Feature of the LAG Function

Obs    X     LAG_X

 1     7       .
 2     9       7
 3     1       .
 4     8       9
```

Let's "play computer" and figure out what is going on. In observation 1, X has a value of 7. The IF condition is true, so the LAG function executes and returns a missing value (since there is no previous X value). In observation 2, X is 9 and the IF condition is true. Therefore, the LAG function returns the value of X the last time the LAG function executed, which was in observation 1, so the value returned is 7. In the third observation, the value of X is 1. The IF condition is not true, so the LAG function does not execute and LAG_X remains a missing value. (Remember that values read from raw data and created in assignment statements in a DATA step are set to missing each time the DATA step iterates.) In observation 4, X is 8 and the IF condition is true. What is the value of X the last time the LAG function executed? Well, it was back in observation 2 and the value of X was 9. That is what the LAG function returns. So, what is the bottom line here? You usually do not want to execute the LAG function conditionally. As long as you execute the LAG (or LAG2 or LAG3, etc.) function for every iteration of the DATA step, you can think of the function as returning a value from the previous observation.

We are now ready to compute differences in heart rates and blood pressures from visit to visit. First the program, then the explanation:

```
*Assume data set PATIENTS is already sorted by ID and DATE;
DATA DIFFERENCE;
   SET PATIENTS;
   BY ID;
   DIFF_HR = HR - LAG(HR);
   DIFF_SBP = SBP - LAG(SBP);
   DIFF_DBP = DBP - LAG(DBP);
   IF NOT FIRST.ID THEN OUTPUT;
RUN;
```

For those readers who are or are becoming "compulsive programmers," you can save a few keystrokes by using the DIF function instead of the LAG function, like this:

```
DATA DIFFERENCE;
   SET PATIENTS;
   BY ID;
   DIFF_HR = DIF(HR);
   DIFF_SBP = DIF(SBP);
   DIFF_DBP = DIF(DBP);
   IF NOT FIRST.ID THEN OUTPUT;
RUN;
```

As you can see, DIF(X) is equivalent to X–LAG(X).

Returning to the original program using the LAG function, we don't want to output an observation for the first visit for each patient since there is no difference to compute. Also, this value would not be correct in any case. Suppose we have just read in values for the first visit for the second patient. The LAG function is actually computing the difference between this patient's heart rate and the last heart rate from the previous patient! Don't panic. This is OK. We are not outputting this observation. Are you tempted to write the assignment statements like this?

```
IF NOT FIRST.PATIENT THEN DIFF_HR = HR - LAG(HR)
```

Well, don't do it! Remember you have to "prime the pump" and execute the LAG function for every observation. As long as we do not output an observation for the first visit, all is well.

Here is the listing of the data set DIFFERENCE:

```
Listing of Data Set DIFFERENCE

ID        DATE HR SBP DBP DX DOCFEE LABFEE DIFF_HR DIFF_SBP DIFF_DBP

005 06/18/1982 70 170  84 14    80     400     -10      -10      -12
005 07/03/1983 64 140  84 14    80     200      -6      -30        0
005 07/05/1983 74 140  82 13    90       0      10        0       -2
007 12/01/1983 72 130  90 20    50     200       2       10       10
```

G.   COMPUTING THE DIFFERENCE BETWEEN THE FIRST AND LAST OBSERVATION FOR EACH SUBJECT

What if you want to see the differences of heart rate and blood pressure from the first visit to the last? You need a way of "remembering" a value from a previous observation. The SAS tool that does this for us is a retained variable. Using a RETAIN statement, we can tell SAS not to set the value in the PDV (program data vector) to missing when the DATA step iterates. So, if you set the value of a retained variable, it stays at that value until you change it. Let's see how we can use this to compute out difference scores. Here is the program:

```
DATA FIRST_LAST;
    SET PATIENTS;
    BY ID;
    ***Data set PATIENTS is sorted by ID and DATE;
    RETAIN FIRST_HR FIRST_SBP FIRST_DBP;   ①
    ***Omit patients with only one visit;
    IF FIRST.ID AND LAST.ID THEN DELETE;   ②
    ***If it is the first visit assign values to the
       retained variables;
```

```
    IF FIRST.ID THEN DO;    ③
       FIRST_HR = HR;
       FIRST_SBP = SBP;
       FIRST_DBP = DBP;
    END;    ④
    IF LAST.ID THEN DO;
       D_HR = HR - FIRST_HR;
       D_SBP = SBP - FIRST_SBP;
       D_DBP = DBP - FIRST_DBP;
       OUTPUT;
    END;
 RUN;
```

We use a RETAIN statement to tell SAS not to set the values of these variables to missing when the DATA step iterates. If a patient has only one visit, we cannot compute a difference between the first and last visit, so we delete that observation (remember, if there is only one visit for a patient, both FIRST.ID and LAST.ID will be equal to one, and the statement ② will be true). When we are reading the first visit for each patient, we set the three retained variables to their respective values. The DO statement ③ can be thought of as an "execute the following statements until you reach the END" statement ④. These values will stay in the PDV since they are not assigned again, and they are not set to missing by SAS. So, when we get to the last visit for each patient, we can subtract the value of our variable from the first visit from the current value. We want to output only one observation per patient with the difference scores, so we include an OUTPUT statement in the DO group. When you include an explicit OUTPUT statement in the DATA step, SAS no longer does its automatic output at the bottom of the DATA step. A listing of data set FIRST_LAST is shown next:

```
Listing of Data Set FIRST_LAST

ID            DATE      HR     SBP     DBP    DX    DOCFEE    LABFEE

005        07/05/1983   74     140     82     13      90        0
007        12/01/1983   72     130     90     20      50       200

             FIRST_    FIRST_
FIRST_HR       SBP       DBP     D_HR   D_SBP   D_DBP

   80         180        96       -6     -40     -14
   70         120        80        2      10      10
```

Compulsive programmers (like one of the authors) are never content to solve a problem just one way. The following program produces an identical result to the previous program. It uses the very unusual trick of executing a LAG function conditionally (remember, we said you almost never want to do this). Please feel free to skip this section if you wish, but it does reinforce exactly how the LAG function works:

```
DATA FIRST_LAST;
   SET PATIENTS;
   BY ID;
   ***Data set PATIENTS is sorted by ID and DATE;
   ***Omit patients with only one visit;
   IF FIRST.ID AND LAST.ID THEN DELETE;
   ***If it is the first or last visit execute the LAG
      function;
   IF FIRST.ID OR LAST.ID THEN DO;
      D_HR = HR - LAG(HR);
      D_SBP = SBP - LAG(SBP);
      D_DBP = DBP - LAG(SBP);
   END;
   IF LAST.ID THEN OUTPUT;
RUN;
```

As you can see, the LAG function only executes when we are reading the first or last visit for each patient. When we read the last visit (LAST.ID is true), the difference is the current value minus the value the last time the LAG function executed—which was the first visit. So, when LAST.ID is true, we output the observation.

H.   COMPUTING FREQUENCIES ON LONGITUDINAL DATA SETS

To compute frequencies for our diagnoses, we use PROC FREQ on our original data set (PATIENTS). We would write:

```
PROC FREQ DATA=PATIENTS ORDER=FREQ;
   TITLE "Diagnoses in Decreasing Frequency Order";
   TABLES DX;
RUN;
```

Notice we use the DATA= option on the PROC FREQ statement to make sure we were counting frequencies from our original data set. The ORDER= option allows us to control the order of the categories in a PROC FREQ output. Normally, the diagnosis categories are listed in sort-sequence order. The ORDER=FREQ option lists the diagnoses in frequency order from the most common diagnosis to the least. While we are on the subject, another useful ORDER= option is ORDER=FORMATTED. This will order the diagnoses in sort sequence by the diagnosis formats (if we had them). Remember that executing a PROC FREQ procedure on our original data set (the one with multiple visits per patient) has the effect of counting the number of times each diagnosis is made. That is, if a patient came in for three visits, each with the same diagnosis, we would add 3 to the frequency count for that diagnosis. If, for some reason, we want to count a diagnosis only once for a given patient, even if this diagnosis is made on subsequent visits, we can sort our data set by ID and DX. We then have a data set such as the following:

ID	DX	FIRST.ID	FIRST.DX
5	13	1	1
5	14	0	1
5	14	0	0
5	14	0	0
7	14	1	1
7	20	0	1
9	137	1	1

If we now use the logical FIRST.DX and FIRST.ID variables, we can accomplish our goal of counting a diagnosis only once for a given patient. The logical variable FIRST.DX will be true each time a new ID-diagnosis combination is encountered. The data set and procedure would look as follows (assume we have previously sorted by ID and DX):

```
DATA DIAG;
   SET PATIENTS;
   BY ID DX;
   IF FIRST.DX;
RUN;
PROC FREQ DATA=DIAG ORDER=FREQ;
   TABLES DX;
RUN;
```

We have accomplished our goal of counting a diagnosis code only once per patient. As you can see, the SAS internal variables FIRST. and LAST. are extremely useful. Think of using them any time you need to do something special for the first or last occurrence of another variable.

I. CREATING SUMMARY DATA SETS WITH PROC MEANS OR PROC SUMMARY

Besides providing a printed output of descriptive statistics, broken down by one or more CLASS (or BY) variables, PROC MEANS or PROC SUMMARY (PROC MEANS with a NOPRINT option and PROC SUMMARY are identical procedures) can produce new SAS data sets containing any of the statistics produced by these procedures. This might be useful, for example, in educational research where the original data were collected on individual students, but you want to use classroom means as the unit of observation (to compare teachers), or it might be useful in business (to compare sales from various quarters when the original data are collected daily). In medicine, you may have clinical data on patient visits, with a different number of visits for each patient, and want to compute patient means for later analysis.

To demonstrate how this is done, suppose we have collected data on several students. We have a student number, gender, the teacher's name, the teacher's age, and two test scores (a pretest and a posttest). We use the following data for our example:

SUBJECT	GENDER	TEACHER	T_AGE	PRETEST	POSTTEST
1	M	Jones	35	67	81
2	F	Jones	35	98	86
3	M	Jones	35	52	92
4	M	Black	42	41	74
5	F	Black	42	46	76
6	M	Smith	68	38	80
7	M	Smith	68	49	71
8	F	Smith	68	38	63
9	M	Hayes	23	71	72
10	F	Hayes	23	46	92
11	M	Hayes	23	70	90
12	F	Wong	47	49	64
13	M	Wong	47	50	63

NOTES: **1.** T_Age is the teacher's age.

2. In a "real" study, we would probably enter the teacher's name and age only once in a separate data set and combine that data set with the student data later on, saving some typing. However, for this example, it is simpler to include the teacher's age for every observation.

As a first step, let's see how we can compute the mean pretest and posttest, and gain scores for each teacher. Look at the following program:

```
DATA SCHOOL;
    LENGTH GENDER $ 1 TEACHER $ 5;  ①
    INPUT SUBJECT
          GENDER   $
          TEACHER  $
          T_AGE
          PRETEST
          POSTTEST;
    GAIN = POSTTEST - PRETEST;
```

```
DATALINES;
1 M JONES 35 67 81
2 F JONES 35 98 86
3 M JONES 35 52 92
4 M BLACK 42 41 74
5 F BLACK 42 46 76
6 M SMITH 68 38 80
7 M SMITH 68 49 71
8 F SMITH 68 38 63
9 M HAYES 23 71 72
10 F HAYES 23 46 92
11 M HAYES 23 70 90
12 F WONG 47 49 64
13 M WONG 47 50 63
;
PROC MEANS DATA=SCHOOL N MEAN STD MAXDEC=2;
   TITLE "Means Scores for Each Teacher";
   CLASS TEACHER;
   VAR PRETEST POSTTEST GAIN;
RUN;
```

This program is straightforward. The DATA step computes a gain score, and PROC MEANS requests statistics for each teacher by including TEACHER as a CLASS variable. The LENGTH statement ① is used to specify how many characters are needed for the character variables GENDER and TEACHER. We include this because, with list input, a default length of eight is used for all character variables. Here is the output:

```
Means Scores for Each Teacher

The MEANS Procedure

                  N
TEACHER          Obs    Variable      N          Mean        Std Dev

BLACK             2     PRETEST       2         43.50           3.54
                       POSTTEST      2         75.00           1.41
                       GAIN          2         31.50           2.12

HAYES             3     PRETEST       3         62.33          14.15
                       POSTTEST      3         84.67          11.02
                       GAIN          3         22.33          22.59

JONES             3     PRETEST       3         72.33          23.46
                       POSTTEST      3         86.33           5.51
                       GAIN          3         14.00          26.00

                                                          [Continued]
```

SMITH	3	PRETEST	3	41.67	6.35
		POSTTEST	3	71.33	8.50
		GAIN	3	29.67	10.79
WONG	2	PRETEST	2	49.50	0.71
		POSTTEST	2	63.50	0.71
		GAIN	2	14.00	1.41

Instead of just printing out the results, we want to create a new data set that has TEACHER as the unit of observation instead of SUBJECT. In our example, we have only five teachers, but we might have 100, and they might be using different teaching methods and be in different schools, etc. To create the new data set, we do the following:

```
PROC MEANS DATA=SCHOOL NOPRINT NWAY;  ①
    CLASS TEACHER;
    VAR PRETEST POSTTEST GAIN;
    OUTPUT OUT=TEACHSUM ②
           MEAN=M_PRE M_POST M_GAIN;
RUN;
*To get a list of what was produced and therefore what
is contained in the data set TEACHSUM, add the following:;
PROC PRINT DATA=TEACHSUM;
    TITLE "Listing of Data Set TEACHSUM";
RUN;
*Hey! This is a good example of why comments
are useful. ;
```

The NOPRINT option on the first line ① tells the program not to print the results of this procedure (since we either already have them from the last run, or the listing would be too large to want to look at). As an alternative, you can use PROC SUMMARY without the NOPRINT option. It is equivalent to PROC MEANS with the NOPRINT option. Take your pick. We want the computed statistics (means in this case) in the new data set. To do this, we include an OUTPUT statement ② in PROC MEANS. The OUTPUT statement creates a new data set. We have to give it a name of our choosing (by saying OUT = TEACHSUM), tell it what statistics to put in it, and what names to give those statistics.

We can output any statistics available with PROC MEANS by using the PROC MEANS options (N, MEAN, STD, etc.) as keywords in the OUTPUT statement. These statistics will be computed for all the variables in the VAR list and will be broken down by the CLASS variable. Since we want only the score means in this new data set, we said, "MEAN = M_PRE M_POST M_GAIN." These new variables represent the means of each of the variables listed in the VAR statement, in the same order the variables are listed. Thus, M_PRE will represent the mean value of PRETEST, M_POST will represent the mean value of POSTTEST, and M_GAIN will represent the mean value of GAIN. You could have named these new variables MANNY, MOE, and JACK. SAS doesn't care what you call them. We used M_PRE, M_POST, and M_GAIN because it helps us remember that they represent Means of PREtest, POSTtest, and GAIN.

Finally, we need to explain the NWAY option in line ①. This tells the procedure to give us only results for each TEACHER (the CLASS variable) and not to include the grand mean in the new data set. Don't forget this. We will later explain what happens if you leave this out. Your new data set (the listing from PROC PRINT) will look like this:

```
Listing of Data Set TEACHSUM

Obs    TEACHER    _TYPE_    _FREQ_     M_PRE     M_POST     M_GAIN

1      BLACK        1          2      43.5000    75.0000    31.5000
2      HAYES        1          3      62.3333    84.6667    22.3333
3      JONES        1          3      72.3333    86.3333    14.0000
4      SMITH        1          3      41.6667    71.3333    29.6667
5      WONG         1          2      49.5000    63.5000    14.0000
```

Let's leave the explanations of the _TYPE_ variable for our next example. The variable _FREQ_ gives us the number of observations (missing or nonmissing) for each value of the CLASS variable. If you go back to the original data values, you will see that teacher BLACK had two students, HAYES had three students, and so forth.

What if you wanted the teacher's age in this new data set (so you could compare age to gain score, for example)? This is easily accomplished by including an ID statement as part of PROC MEANS. So, to include the teacher's age in this data set, you would use the following code:

```
PROC MEANS DATA=SCHOOL NOPRINT NWAY;   ①
    CLASS TEACHER;
    ID T_AGE;
    VAR PRETEST POSTTEST GAIN;
    OUTPUT OUT=TEACHSUM   ②
           MEAN=M_PRE M_POST M_GAIN;
RUN;
```

The resulting data set (TEACHSUM) will now contain the variable T_AGE. As an alternative, you could have included both variables, TEACHER and T_AGE, as CLASS variables with the same result.

We now turn to a more complex example where we create an output data set using PROC MEANS and a CLASS statement with more than one CLASS variable. Yes, folks, hold on to your hats, this gets tricky. First, the raw data:

SUBJ	GENDER	REGION	HEIGHT	WEIGHT
01	M	North	70	200
02	M	North	72	220
03	M	South	68	155
04	M	South	74	210
05	F	North	68	130
06	F	North	63	110
07	F	South	65	140
08	F	South	64	108
09	F	South	.	220
10	F	South	67	130

Next, we create a SAS data set as follows:

```
DATA DEMOG;
    LENGTH GENDER $ 1 REGION $ 5;
    INPUT SUBJ GENDER $ REGION $ HEIGHT WEIGHT;
DATALINES;
01 M North 70 200
02 M North 72 220
03 M South 68 155
04 M South 74 210
05 F North 68 130
06 F North 63 110
07 F South 65 140
08 F South 64 108
09 F South  . 220
10 F South 67 130
;
```

To compute the number of subjects, the mean, and the standard deviation for each combination of GENDER and REGION, include a CLASS statement with PROC MEANS like this:

```
PROC MEANS DATA=DEMOG N MEAN STD MAXDEC=2;
   TITLE "Output from PROC MEANS";
   CLASS GENDER REGION;
   VAR HEIGHT WEIGHT;
RUN;
```

Remember that you do not have to sort your data set when you use a CLASS statement with PROC MEANS. In this example, we have two CLASS variables instead of one. The output from this procedure is shown next:

```
Output from PROC MEANS

The MEANS Procedure

                        N
GENDER   REGION   Obs   Variable   N         Mean      Std Dev

F        North    2     HEIGHT     2        65.50         3.54
                        WEIGHT     2       120.00        14.14

         South    4     HEIGHT     3        65.33         1.53
                        WEIGHT     4       149.50        48.86

M        North    2     HEIGHT     2        71.00         1.41
                        WEIGHT     2       210.00        14.14

         South    2     HEIGHT     2        71.00         4.24
                        WEIGHT     2       182.50        38.89
```

Since we now have two CLASS variables, the requested statistics are computed for each combination of GENDER and REGION.

We first demonstrate what happens when we use PROC MEANS to create an output data set with GENDER and REGION as CLASS variables. Here is the code:

```
PROC MEANS DATA=DEMOG NOPRINT;  ①
   CLASS GENDER REGION;
   VAR HEIGHT WEIGHT;
   OUTPUT OUT=SUMMARY   ②
         MEAN=M_HEIGHT M_WEIGHT;
RUN;
***Add a PROC PRINT to list the observations in SUMMARY;
PROC PRINT DATA=SUMMARY;
   TITLE "Listing of Data Set SUMMARY";
RUN;
```

Don't be confused by the three asterisks in the previous comment. Remember that the first asterisk starts the comment and the semicolon ends it. We added the other two asterisks to make the comment stand out better. (The programming-challenged author thought you might like to know this.)

As before, the NOPRINT option on the first line ① tells the procedure not to print any output. Rather, you want these values in a SAS data set. To do this, you add an OUTPUT statement ② to PROC MEANS. The OUTPUT statement allows you to create a new data set, to select which statistics to place in this data set, and what names to give to each of the requested statistics. The name of the output data set is placed after the OUT= keyword. The request to output means is indicated by the keyword MEAN= ③. The two variable names following the keyword MEAN= are names you choose to represent the mean HEIGHT and WEIGHT, respectively. The order of the names following MEAN= corresponds to the order of the variable names in the VAR statement. In this example, the variable M_HEIGHT will represent the mean height, and the variable M_WEIGHT will represent the mean weight. Other keywords (chosen from the list of statistics available with PROC MEANS found in Chapter 2, Section B) can be used to output statistics such as standard deviation (STD=) or sums (SUM=).

Using a PROC PRINT with DATA=SUMMARY to see the contents of this new data set, we obtain the following listing:

```
Listing of Data Set SUMMARY

Obs     GENDER     REGION     _TYPE_     _FREQ_     M_HEIGHT     M_WEIGHT

 1                                0         10        67.8889      162.300
 2                   North        1          4        68.2500      165.000
 3                   South        1          6        67.6000      160.500
 4         F                      2          6        65.4000      139.667
 5         M                      2          4        71.0000      196.250
 6         F         North        3          2        65.5000      120.000
 7         F         South        3          4        65.3333      149.500
 8         M         North        3          2        71.0000      210.000
 9         M         South        3          2        71.0000      182.500
```

Besides the mean for each combination of GENDER and REGION, we see there are five additional observations and two additional variables, _TYPE_ and _FREQ_. Here's what they're all about. The first observation with a value of 0 for _TYPE_ is the mean of all nonmissing values (9 for HEIGHT and 10 for WEIGHT) and is called the grand mean. The two observations with _TYPE_ equal to 1 are the mean HEIGHT and WEIGHT for each REGION; the next two observations with _TYPE_ equal to 2 are the mean HEIGHT and WEIGHT for each GENDER. Finally, the last four observations with _TYPE_ equal to 3 are the means by GENDER and REGION (sometimes called cell means). This is getting complicated! Relax, there is actually a way to tell which _TYPE_ value corresponds to which breakdown of the data.

Let's look carefully at our CLASS statement. It is written:

```
CLASS GENDER REGION;
```

First, we count in binary (remember, 0, 1, 10, 11, 100, 101, etc.) and place the binary numbers below the CLASS variables like this:

CLASS	GENDER	REGION;		
	Binary		**_TYPE_**	**Interpretation**
	0	0	0	Mean over all GENDERS and REGIONS
	0	1	1	Mean for each value of REGION
	1	0	2	Mean for each value of GENDER
	1	1	3	Mean for each combination of GENDER and REGION (cell means)

Next, we can come up with a simple rule. Whenever the _TYPE_ value, written in binary, gives you a "1" beneath a CLASS variable, the statistics are broken down by that variable. If we look at _TYPE_ = 1, we write that in binary (not too hard) as 01 and realize that the _TYPE_ = 1 statistics represent each REGION and so forth. Still confused? It's OK, this is not easy.

An alternative to interpreting the _TYPE_ variable as a binary number is to use the PROC MEANS (or PROC SUMMARY) option CHARTYPE. When you use this option, the _TYPE_ variable is a **character** variable consisting of a series of 1s and 0s. To see how this works, let's run the previous program with the CHARTYPE option. We have:

```
PROC MEANS DATA=DEMOG NOPRINT CHARTYPE;
    CLASS GENDER REGION;
    VAR HEIGHT WEIGHT;
    OUTPUT OUT=SUMMARY
           MEAN=M_HEIGHT M_WEIGHT;
RUN;
```

The resulting data set SUMMARY now looks like this:

```
Listing of Data Set SUMMARY

GENDER    REGION    _TYPE_    _FREQ_    M_HEIGHT    M_WEIGHT

                     00         10      67.8889     162.300
          North      01          4      68.2500     165.000
                                                 [Continued]
```

	South	01	6	67.6000	160.500
F		10	6	65.4000	139.667
M		10	4	71.0000	196.250
F	North	11	2	65.5000	120.000
F	South	11	4	65.3333	149.500
M	North	11	2	71.0000	210.000
M	South	11	2	71.0000	182.500

Notice that the values of _TYPE_ are now strings of 1s and 0s. You can use this variable to select which means you are interested in. Suppose you wanted a separate data set for each of the _TYPE_ values. You can create several data sets at one time, like this:

```
DATA GRAND REGION GENDER GENDER_REGION;
   SET SUMMARY;
   IF _TYPE_ = '00' THEN OUTPUT GRAND;
   ELSE IF _TYPE_ = '01' THEN OUTPUT REGION;
   ELSE IF _TYPE_ = '10' THEN OUTPUT GENDER;
   ELSE IF _TYPE_ = '11' THEN OUTPUT GENDER_REGION;
RUN;
```

This program demonstrates several things. First, you can create more than one SAS data set in one DATA step. To do this, you list all the data sets you want to create on the DATA statement. Next, you use an OUTPUT statement to force an output at that point in the DATA step. You also need to name the data set you want to output. Otherwise, SAS will output an observation to all the data sets listed in the DATA statement. Finally, you can see how the _TYPE_ variable lets you choose which sets of means you want to output. Using the CHARTYPE option with PROC MEANS really makes the process of choosing the correct value of _TYPE_ much easier. You don't even have to know how to count in binary!

For most applications, you don't even need to look at the _TYPE_ values. Since most applications call for cell means (the values broken down by each of the class variables), you will want the highest value of the _TYPE_ variable. If you include the option NWAY on the PROC MEANS statement, only cell means will be output to the new data set. So, if you only want the mean HEIGHT and WEIGHT for each combination of GENDER and REGION, you would write the PROC MEANS statements like this:

```
PROC MEANS DATA=DEMOG NOPRINT NWAY;
   CLASS GENDER REGION;
   VAR HEIGHT WEIGHT;
   OUTPUT OUT=SUMMARY
          MEAN=M_HEIGHT M_WEIGHT;
RUN;
```

```
PROC PRINT DATA=SUMMARY;
   TITLE "Listing of Data Set SUMMARY with NWAY Option";
RUN;
```

The resulting data set (shown here) contains only the _TYPE_ = 3 values.

```
Listing of Data Set SUMMARY

Obs    GENDER    REGION    _TYPE_    _FREQ_    M_HEIGHT    M_WEIGHT

 1       F       North        3         2      65.5000      120.0
 2       F       South        3         4      65.3333      149.5
 3       M       North        3         2      71.0000      210.0
 4       M       South        3         2      71.0000      182.5
```

The value of the variable _FREQ_ is the number of observations (missing or nonmissing) in each subgroup. For example, there were two females from the North, so _FREQ_ = 2 in observation 1 in the summary data set. If you need to know the number of nonmissing values that were used in computing the requested statistics, include a request for N = in your output data set. Let's demonstrate this with the following code:

```
PROC MEANS DATA=DEMOG NOPRINT NWAY;
   CLASS GENDER REGION;
   VAR HEIGHT WEIGHT;
   OUTPUT OUT  = SUMMARY
          N    = N_HEIGHT N_WEIGHT
          MEAN = M_HEIGHT M_WEIGHT;
RUN;
PROC PRINT DATA=SUMMARY;
   TITLE1 "Listing of Data Set SUMMARY with NWAY Option";
   TITLE2 "with Requests for N= and MEAN=";
RUN;
```

In this program, we have chosen the variable names N_HEIGHT and N_WEIGHT to represent the number of nonmissing observations. The resulting output, shown here, makes the difference between the value of _FREQ_ and the N = variable clear:

```
Listing of Data Set SUMMARY with NWAY Option
with Requests for N= and MEAN=

GENDER  REGION  _TYPE_  _FREQ_  N_HEIGHT  N_WEIGHT  M_HEIGHT  M_WEIGHT

   F    North      3      2        2         2       65.5000    120.0
   F    South      3      4        3         4       65.3333    149.5
   M    North      3      2        2         2       71.0000    210.0
   M    South      3      2        2         2       71.0000    182.5
```

Observe that the value for N_HEIGHT is 3 for females from the South, while the value of _FREQ_ is 4 (there was a missing HEIGHT for a female from the South).

Finally, if you use the NWAY option, there is not much need to keep the _TYPE_ variable in the output data set. You can use a DROP= data set option to omit this variable. The program, modified to do this, is shown here:

```
PROC MEANS DATA=DEMOG NOPRINT NWAY;
   CLASS GENDER REGION;
   VAR HEIGHT WEIGHT;
   OUTPUT OUT  = SUMMARY(DROP=_TYPE_)
          N    = N_HEIGHT N_WEIGHT
          MEAN = M_HEIGHT M_WEIGHT;
RUN;
```

Some lazy programmers sometimes omit the variable list following a request for a statistic when only one statistic is requested. For example, if you only want means for each combination of GENDER and REGION, you would write:

```
PROC MEANS DATA=DEMOG NOPRINT NWAY;
   CLASS GENDER REGION;
   VAR HEIGHT WEIGHT;
   OUTPUT OUT = SUMMARY(DROP=_TYPE_)
          MEAN =;
RUN;
```

Using this method, the variable names in the new summary data set will be the same as those listed on the VAR statement. That is, the variable name representing the mean height will be HEIGHT, and the variable name representing the mean weight will be WEIGHT. This is probably a bad idea since you may get confused and not realize that a variable name represents a summary statistic and not the original value. (Actually, that other author would not even put in (DROP=_TYPE_) since it takes up too much time, and he doesn't mind the extra variable in the printout.)

J.   OUTPUTTING STATISTICS OTHER THAN MEANS

We saw that PROC MEANS can create an output data set using an OUTPUT statement, which contains means for each variable in the VAR list for each level of the variables in the CLASS or BY statement. We used the keyword MEAN= to indicate that we wanted to output means. Any of the options that are available to be used with PROC MEANS (see Chapter 2, Section B) can also be used to create variables in the data set created by PROC MEANS.

Suppose you want the number of nonmissing values, the mean, the median, the minimum, and the maximum value for each combination of GENDER and REGION in the DEMOG data set. The following program would do the trick:

```
PROC MEANS DATA=DEMOG NOPRINT NWAY;
   CLASS GENDER REGION;
   VAR HEIGHT WEIGHT;
   OUTPUT OUT = SUMMARY
          MEAN = MEAN_HEIGHT MEAN_WEIGHT
          N = M_HEIGHT N_WEIGHT
          MEDIAN = MEDIAN_HEIGHT MEDIAN_WEIGHT
          MIN = MIN_HEIGHT MIN_WEIGHT
          MAX = MAX_HEIGHT MAX_WEIGHT;
RUN;
```

Notice that the variable names following the requests for statistics are arbitrary. You can name these variables anything you wish. It makes sense, however, to choose names like MEDIAN_HEIGHT so that it is easy to remember what it represents.

If you would like SAS to name all the summary variables for you, you can use the OUTPUT option AUTONAME (you know, Ford, Chevy, ...). When you use this option, SAS appends an underscore character and the name of the statistic to the variables listed on the VAR statement. To demonstrate this, let's output several statistics from the DEMOG data set and let SAS name them for us. We have:

```
PROC MEANS DATA=DEMOG NOPRINT NWAY;
   CLASS GENDER REGION;
   VAR HEIGHT WEIGHT;
   OUTPUT OUT = SUMMARY(DROP=_TYPE_ RENAME=(_FREQ_ = NUMBER))
          MEAN =
          N =
          MEDIAN =
          MIN =
          MAX = / AUTONAME;
RUN;
```

While we are demonstrating the AUTONAME option, we added the RENAME data set option to show you how to rename the variable _FREQ_ to NUMBER in the SUMMARY DATA SET. Look at the following listing of the data set SUMMARY to see how useful the AUTONAME option can be.

```
Demonstrating the AUTONAME option

                                     HEIGHT_     WEIGHT_
    GENDER     REGION     NUMBER       Mean        Mean     HEIGHT_N     WEIGHT_N

      F        North        2        65.5000      120.0        2            2
      F        South        4        65.3333      149.5        3            4
      M        North        2        71.0000      210.0        2            2
      M        South        2        71.0000      182.5        2            2

    HEIGHT_     WEIGHT_     HEIGHT_     WEIGHT_     HEIGHT_     WEIGHT_
    Median      Median        Min         Min         Max         Max

     65.5        120.0        63          110         68          130
     65.0        135.0        64          108         67          220
     71.0        210.0        70          200         72          220
     71.0        182.5        68          155         74          210
```

PROBLEMS

Remember, you can download all the data sets and programs for these problems from the web site: www.prenhall.com/cody

4.1 We have collected data on a questionnaire as follows:

Variable	Starting Column	Length	Description
ID	1	3	Subject ID
DOB	5	8	Date of birth in MMDDYY format
ST_DATE	13	8	Start date in MMDDYY format
END_DATE	21	8	Ending date in MMDDYY format
SALES	29	5	Total sales

Here is some sample data:

```
         1         2         3
1234567890123456789012345678901234567890  Column Indicators
----------------------------------------------
001 10211946111219801228819887343
002 09131955020219800204419880123
005 06061940031219810312200040000
003 07051944111519801113200009544
```

(a) Write a SAS program to read these data.

(b) Compute age (in years) at time work was started and length of time between ST_DATE and END_DATE (also in years).

(c) Compute the sales per year of work.

(d) Print out a report showing:

ID   DOB   AGE   LENGTH   SALES_YR

where LENGTH is the time at work computed in part (b), and SALES_YR is the sales per year computed in part (c). Use the MMDDYY10. format to print the DOB.

(e) Modify the program to compute AGE as of the last birthday and sales per year rounded to the nearest 10 dollars. Try using the DOLLAR6. format for SALES_YR.

4.2 Run the following DATA step to create a SAS data set called ABC_CORP. Create a new SAS data set called AGES that contains all the variables in ABC_CORP plus three new variables. One is AGE_ACTUAL, which is the exact age from DOB to January 15, 2005. The second is AGE_TODAY, which is the age as of the date the program is run, rounded to the nearest tenth of a year. The third is AGE, with the fractional part dropped, as of the date stored in the variable VISIT_DATE. Print out a listing of this new data set.

```
***Program to create data set ABC_CORP;
DATA ABC_CORP;
   DO SUBJ = 1 TO 10;
      DOB = INT(RANUNI(1234)*15000);
      VISIT_DATE = INT(RANUNI(0)*1000) + '01JAN2000'D;
      OUTPUT;
   END;
   FORMAT DOB VISIT_DATE DATE9.;
RUN;
```

4.3 For each of eight mice, the date of birth, date of disease, and date of death is recorded. In addition, the mice are placed into one of two groups (A or B). Given the following data, compute the time from birth to disease, the time from disease to death, and the age at death. All times can be in days. Compute the mean, standard deviation, and standard error of these three times for each of the two groups. Here are the data:

RAT_NO	DOB	DISEASE	DEATH	GROUP
1	23MAY1990	23JUN1990	28JUN1990	A
2	21MAY1990	27JUN1990	05JUL1990	A
3	23MAY1990	25JUN1990	01JUL1990	A
4	27MAY1990	07JUL1990	15JUL1990	A
5	22MAY1990	29JUN1990	22JUL1990	B
6	26MAY1990	03JUL1990	03AUG1990	B
7	24MAY1990	01JUL1990	29JUL1990	B
8	29MAY1990	15JUL1990	18AUG1990	B

Arrange these data in any columns you wish. HINT: Use the DATE9. informat to read dates in this form.

4.4 Run the following program to create a SAS data set called CLINICAL. Using this data set, compute the difference in weight from one visit to the next.

```
DATA CLINICAL;
    *Use LENGTH statement to control the order of
     variables in the data set;
    LENGTH PATIENT VISIT DATE_VISIT 8;
    RETAIN DATE_VISIT WEIGHT;
    DO PATIENT = 1 TO 25;
        IF RANUNI(135) LT .5 THEN GENDER = 'Female';
        ELSE GENDER = 'Male';
        X = RANUNI(135);
        IF X LT .33 THEN GROUP = 'A';
        ELSE IF X LT .66 THEN GROUP = 'B';
        ELSE GROUP = 'C';
        DO VISIT = 1 TO INT(RANUNI(135)*5);
            IF VISIT = 1 THEN DO;
                DATE_VISIT = INT(RANUNI(135)*100) + 15800;
                WEIGHT = INT(RANNOR(135)*10 + 150);
            END;
            ELSE DO;
                DATE_VISIT = DATE_VISIT + VISIT*(10 + INT(RANUNI(135)*50));
                WEIGHT = WEIGHT + INT(RANNOR(135)*10);
            END;
            OUTPUT;
            IF RANUNI(135) LT .2 THEN LEAVE;
        END;
    END;
    DROP X;
    FORMAT DATE_VISIT DATE9.;
RUN;
```

*4.5 Using the data set (PATIENTS) described in Section D of this chapter (use the program with sample data), write the necessary SAS statements to create a data set (PROB4_5) in which the first visit for each patient is omitted. Then, using that data set, compute the mean HR, SBP, and DBP for each patient. (Patient 9 with only one visit will be eliminated.)

4.6 Using data set CLINICAL from Problem 4.4, create a new SAS data set (CHANGE) with one observation per subject with the difference in WEIGHT between the first and last visit. Include in this data set the number of days between the first and last visit. Do not include any patients who have only one visit.

*4.7 Write a program similar to Problem 4.5, except that we want to include all the data for each patient, excluding any patient who has had only one visit. Instead of having PROC MEANS create printed output, use it to create a SAS data set (PAT_MEAN) containing the mean for each patient (use the AUTONAME output option to name these variables).

4.8 Using data set CLINICAL from Problem 4.4, create a data set with one observation per patient, with the mean, median, minimum and maximum weight.

*4.9 We have a data set called BLOOD that contains from one to five observations per subject. Each observation contains the variables ID, GROUP, TIME, WBC (white blood cells), and RBC (red blood cells). Run the following program to create this data set.

```
***Program to create data set BLOOD;
DATA BLOOD;
   LENGTH GROUP $ 1;
   INPUT ID GROUP $ TIME WBC RBC @@;
DATALINES;
1 A 1 8000 4.5 1 A 2 8200 4.8 1 A 3 8400 5.2
1 A 4 8300 5.3 1 A 5 8400 5.5
2 A 1 7800 4.9 2 A 2 7900 5.0
3 B 1 8200 5.4 3 B 2 8300 5.4 3 B 3 8300 5.2
3 B 4 8200 4.9 3 B 5 8300 5.0
4 B 1 8600 5.5
5 A 1 7900 5.2 5 A 2 8000 5.2 5 A 3 8200 5.4
5 A 4 8400 5.5
;
```

We want to create a data set that contains the mean WBC and RBC for each subject. This new data set should contain the variables ID, GROUP, M_WBC, and M_RBC where M_WBC and M_RBC are the mean values for the subject. Finally, we want to exclude any subjects from this data set who have two or fewer observations in the original data set (assume there are no missing values).

HINT: We will want to use PROC MEANS with a CLASS statement. Since we want both ID and GROUP in the new data set, you can make them both CLASS variables or include an ID statement (ID GROUP;) to cause the variable GROUP to be present in the output data set. Also, remember the _FREQ_ variable that PROC MEANS creates. It will be useful for creating a data set that meets the last condition of excluding subjects with two or fewer observations.

*4.10 Using data set CLINICAL from Problem 4.4, create a summary data set containing the mean, median, and standard deviation broken down by gender and group (use a CLASS statement). Using this summary data set, create four other data sets (in one DATA step) as follows: (1) The grand mean, etc., (2) statistics broken down by gender, (3) statistics broken down by group, and (4) statistics broken down by gender and group. Hint: use the CHARTYPE option to make this problem easier. Use PROC PRINT to list this summary data set.

4.11 Modify the program in Problem 4.9 to include the standard deviation of WBC and RBC for each subject.

4.12 Using the summary data set from Problem 4.10, create four new data sets—GRAND, BY_GENDER, BY_GROUP, and BY_GENDER_GROUP—using the _TYPE_ variable to control which observations go into each of the data sets. Note that you can accomplish this in one DATA step. Use the CHARTYPE option to make this problem easier.

C H A P T E R   5

Correlation and Simple Regression

A. Introduction
B. Correlation
C. Significance of a Correlation Coefficient
D. How to Interpret a Correlation Coefficient
E. Partial Correlations
F. Linear Regression
G. Partitioning the Total Sum of Squares
H. Producing a Scatter Plot and the Regression Line
I. Adding a Quadratic Term to the Regression Equation
J. Transforming Data

A. INTRODUCTION

A common statistic for indicating the strength of a linear relationship existing between two continuous variables is called the Pearson correlation coefficient, or sometimes just correlation coefficient (there are other types of correlation coefficients, but Pearson is the most commonly used). The correlation coefficient is a number ranging from -1 to $+1$. A positive correlation means that as values of one variable increase, values of the other variable also tend to increase. A small or zero correlation coefficient tells us that the two variables are unrelated. Finally, a negative correlation coefficient shows an inverse relationship between the variables: as one goes up, the other goes down.

This chapter also deals with simple regression (one independent variable). Regression bears some similarity to correlation but is useful when you want to predict one variable, given the value of another. Regression also yields a mathematical relationship between the two variables, not just the strength of association. Multiple regression (more than one independent variable) is covered in Chapter 9.

B. CORRELATION

We have recorded the GENDER, HEIGHT, WEIGHT, and AGE of seven people and want to compute the correlations between HEIGHT and WEIGHT, HEIGHT and AGE, and WEIGHT and AGE. The program to accomplish this is as follows:

```
DATA CORR_EG;
    INPUT GENDER $ HEIGHT WEIGHT AGE;
DATALINES;
M 68 155 23
F 61 99 20
F 63 115 21
M 70 205 45
M 69 170 .
F 65 125 30
M 72 220 48
;
PROC CORR DATA=CORR_EG;
    TITLE "Example of a Correlation Matrix";
    VAR HEIGHT WEIGHT AGE;
RUN;
```

The output from this program is:

```
Example of a Correlation Matrix

The CORR Procedure

   3  Variables:    HEIGHT    WEIGHT    AGE

                    Simple Statistics

Variable            N          Mean      Std Dev           Sum

HEIGHT              7      66.85714      3.97612     468.00000
WEIGHT              7     155.57143     45.79613            1089
AGE                 6      31.16667     12.41639     187.00000

            Simple Statistics

Variable        Minimum       Maximum

HEIGHT         61.00000      72.00000
WEIGHT         99.00000     220.00000
AGE            20.00000      48.00000
```

```
            Pearson Correlation Coefficients
               Prob > |r| under H0: Rho=0
                  Number of Observations

                 HEIGHT          WEIGHT            AGE

HEIGHT          1.00000         0.97165         0.86614
                                0.0003          0.0257
                    7               7               6


WEIGHT          0.97165         1.00000         0.92496
                0.0003                          0.0082
                    7               7               6


AGE             0.86614         0.92496         1.00000
                0.0257          0.0082
                    6               6               6
```

PROC CORR gives us some simple descriptive statistics on the variables in the VAR list along with a correlation matrix. If you examine the intersection of any row or column in this matrix, you will find the correlation coefficient (top number), the p-value (the second number), and the number of data pairs used in computing the correlation (third number). If the number of data pairs are the same for every combination of variables, the third number in each group is not printed. Instead, this number is printed at the top of the table.

In this listing, we see that the correlation between HEIGHT and WEIGHT is .97165, and the significance level is .0003; the correlation between HEIGHT and AGE is .86614 (p = .0257); the correlation between WEIGHT and AGE is .92496 (p = .0082). Let's concentrate for a moment on the HEIGHT and WEIGHT correlation. The small p-value computed for this correlation indicates that it is unlikely to have obtained a correlation this large by chance if the sample of seven subjects were taken from a population whose correlation was zero. Remember that this is just an example. Correlations this large are quite rare in social science data.

To generate a correlation matrix (correlations between every pairwise combination of the variables), use the following general syntax:

```
PROC CORR options;
   VAR list-of-variables;
RUN;
```

The term "list-of-variables" should be replaced with a list of variable names, separated by spaces. If no options are selected, Pearson correlation coefficients and simple descriptive statistics are computed. As we discuss later, several nonparametric correlation coefficients are also available. The option SPEARMAN will produce Spearman rank correlations, KENDALL will produce Kendall's tau-b coefficients, and HOEFFDING will produce Hoeffding's D statistic. If you use one of these options, SPEARMAN for example, you must use the PEARSON option as well if you also want Pearson correlations to be computed. If you do not want simple statistics, include the option NOSIMPLE in your list of options.

For example, if you wanted both Pearson and Spearman correlations and did not want simple statistics to be printed, you would submit the following code:

```
PROC CORR DATA=CORR_EG PEARSON SPEARMAN NOSIMPLE;
   TITLE "Example of a Correlation Matrix";
   VAR HEIGHT WEIGHT AGE;
RUN;
```

The CORR procedure will compute correlation coefficients between all pairs of variables in the VAR list. If the list of variables is large, this results in a very large number of coefficients. If you want to see only a limited number of correlation coefficients (the ones with the highest absolute values), include the BEST=number option with PROC CORR. This results in these correlations listed in descending order.

If all you want are the correlations between a subset of variables and another subset of variables in a data set, a WITH statement is available. PROC CORR will then compute a correlation coefficient between every variable in the WITH list against every variable in the VAR list. For example, suppose we had the variables IQ and GPA (grade point average) in a data set called RESULTS. We also recorded a student score on 10 tests (TEST1–TEST10). If we want to see only the correlation between the IQ and GPA versus each of the 10 tests, the syntax is:

```
PROC CORR DATA=RESULTS;
   VAR IQ GPA;
   WITH TEST1-TEST10;
RUN;
```

This notation can save considerable computation time.

C.   SIGNIFICANCE OF A CORRELATION COEFFICIENT

You may ask, "How large a correlation coefficient do I need to show that two variables are correlated?" Each time PROC CORR prints a correlation coefficient, it also prints a probability associated with the coefficient. That number gives the probability of obtaining a sample correlation coefficient as large as or larger than the one obtained by chance alone (i.e., when the variables in question actually have zero correlation).

The significance of a correlation coefficient is a function of the magnitude of the correlation and the sample size. With a large number of data points, even a small correlation coefficient can be significant. For example, with 10 data points, a correlation coefficient of .63 or larger is significant (at the .05 level); with 100 data points, a correlation of .195 would be significant. Note that a negative correlation shows as equally strong a relationship as a positive correlation (although the relationship is inverse). A correlation of −.40 is as strong as one of +.40.

It is important to remember that correlation indicates only the strength of a relationship—it does not imply causality. For example, we would probably find a high positive correlation between the number of hospitals in each of the 50 states versus the number of household pets in each state. Does this mean that pets make people sick and therefore make more hospitals necessary? Doubtful. The most plausible explanation is that both variables (number of pets and number of hospitals) are related to population size.

Being SIGNIFICANT is not the same as being IMPORTANT or STRONG. That is, knowing the significance of the correlation coefficient does not tell us very much. Once we know that our correlation coefficient is significantly different from zero, we need to look further in interpreting the importance of that correlation. Let us digress a moment to explain what we mean by the significance of a correlation coefficient. Suppose we have a population of x and y values in which the correlation is zero. We could imagine a plot of this population as shown in the following graph:

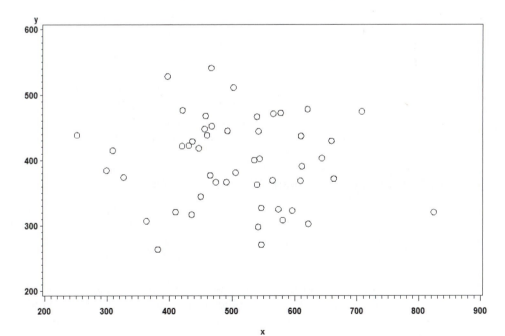

Correlation with Population Correlation = 0

Suppose further that we choose a small sample of 10 points from this population. In the following plot, the black dots represent the x, y pairs we chose "at random" from our population.

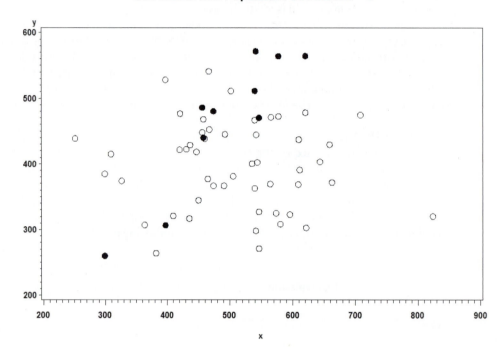

Correlation with Population Correlation = 0

Testing for significance tells us if the correlation found in the sample is large enough to indicate a strong likelihood that there is actually a nonzero correlation in the (larger) population. In this sample, a nonzero correlation would occur, which may or may not be significant and must be tested anew for each correlation.

D.   HOW TO INTERPRET A CORRELATION COEFFICIENT

One of the best ways to interpret a correlation coefficient (r) is to look at the square of the coefficient (r-squared); r-squared can be interpreted as the proportion of variance in one of the variables that can be explained by variation in the other variable. As an example, our height/weight correlation is .97. Thus, r-squared is .94. We can say that 94% of the variation in weights can be explained by variation in height (or vice versa). Also, (1 − .94), or 6%, of the variance of weight, is due to factors other than height variation. With this interpretation in mind, if we examine a correlation coefficient of .4, we see that only .16, or 16%, of the variance of one variable is explained by variation in the other.

Another consideration in the interpretation of correlation coefficients is this: Be sure to look at a scatter plot of the data (using PROC PLOT or PROC GPLOT). It often turns out that one or two extreme data points can cause the correlation coefficient to be much larger than expected. One data-entry error can dramatically alter a correlation coefficient.

An important assumption concerning a correlation coefficient is that each pair of x, y data points is independent of any other pair. That is, each pair of points has to come from a separate subject. Suppose we had 20 subjects, and we measured two variables (say blood pressure and heart rate) for each of the 20 subjects. We could then calculate a correlation coefficient using 20 pairs of data points. Suppose instead that we made 10 measurements of blood pressure and heart rate for each subject. We cannot compute a valid correlation coefficient using all 200 data points (10 points for each of 20 subjects) because the 10 points for each subject are not independent. The best method would be to compute the mean blood pressure and heart rate for each subject and use the mean values to compute the correlation coefficient. Having done so, we'll keep in mind that we correlated mean values when we are interpreting the results.

E. PARTIAL CORRELATIONS

A researcher may wish to determine the strength of the relationship between two variables when the effect of other variables has been removed. One way to accomplish this is by computing a partial correlation. To remove the effect of one or more variables from a correlation, use a PARTIAL statement to list those variables whose effects you want to remove. Using the CORR_EG data set from Section B, we can compute a partial correlation between HEIGHT and WEIGHT with the effect of AGE removed as:

```
PROC CORR DATA=CORR_EG NOSIMPLE;
   TITLE "Example of a Partial Correlation";
   VAR HEIGHT WEIGHT;
   PARTIAL AGE;
RUN;
```

As you can see in the following listing, the partial correlation between HEIGHT and WEIGHT is now lower than before (.91934), although it is still significant (p = .0272).

```
Example of a Partial Correlation

The CORR Procedure

    1 Partial Variables:     AGE
    2        Variables:      HEIGHT    WEIGHT
                                                        [Continued]
```

```
Pearson Partial Correlation Coefficients, N = 6
        Prob > |r| under HO: Partial Rho=0


                    HEIGHT          WEIGHT
HEIGHT             1.00000         0.91934
                                   0.0272


WEIGHT             0.91934         1.00000
                   0.0272
```

F. LINEAR REGRESSION

Given a person's height, what would be his or her predicted weight? How can we best define the relationship between height and weight? By studying the following graph, we see that the relationship is approximately linear. That is, we can imagine drawing a straight line on the graph with most of the data points being only a short distance from the line. The vertical distance from each data point to this line is called a residual.

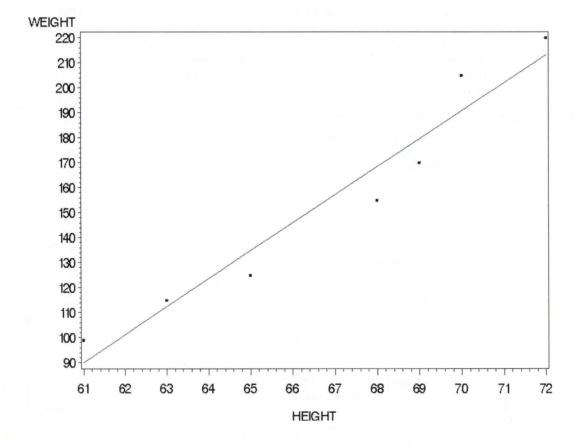

How do we determine the "best" straight line to fit our height/weight data? The method of least squares is commonly used, which finds the line (called the regression line) that minimizes the sum of the squared residuals. A residual is the difference between a subject's predicted score and his or her actual score.

PROC REG (short for regression) has the general form:

```
PROC REG options;
   MODEL dependent variable(s) = independent-variables / options;
RUN;
```

Using our height/weight program, we add the following PROC to give us the equation for the regression line:

```
PROC REG DATA=CORR_EG;
   TITLE "Regression Line for Height-Weight Data";
   MODEL WEIGHT = HEIGHT;
RUN;
```

The output from the previous procedure is as follows:

```
Regression Line for Height-Weight Data


The REG Procedure
Model: MODEL1
Dependent Variable: WEIGHT


Number of Observations Read        7
Number of Observations Used        7


                        Analysis of Variance


                           Sum of         Mean
Source            DF       Squares        Square   F Value  Pr > F


Model              1         11880         11880    84.45   0.0003
Error              5     703.38705     140.67741
Corrected Total    6         12584
```

[Continued]

```
Root MSE                 11.86075    R-Square     0.9441
Dependent Mean          155.57143    Adj R-Sq     0.9329
Coeff Var                 7.62399

                       Parameter Estimates

                     Parameter        Standard
Variable      DF       Estimate          Error     t Value    Pr > |t|
Intercept      1     -592.64458       81.54217       -7.27      0.0008
HEIGHT         1       11.19127        1.21780        9.19      0.0003
```

Look first at the last two lines. There is an estimate for two parameters, INTER-CEPT and HEIGHT. The general equation for a straight line can be written as:

$$y = a + bx$$

where a = intercept, b = slope.

We can write the equation for the "best" straight line defined by our height/weight data as:

$$\text{WEIGHT} = -592.64 + 11.19 \times \text{HEIGHT}$$

Given any height, we can now predict the weight. For example, the predicted weight of a 70-inch-tall person is:

$$\text{WEIGHT} = -592.64 + 11.19 \times 70 = 190.66 \text{ lb.}$$

Under the heading "Parameter Estimates" are columns labeled "Standard Error," "t Value," and "Pr > |t|." The T values and the associated probabilities (Pr > |t|) test the hypothesis that the parameter is actually zero. That is, if the true slope and intercept were zero, what would the probability be of obtaining, by chance alone, a value as large as or larger than the one actually obtained? The standard error can be thought of in much the same way as the standard error of the mean. It reflects the accuracy with which we know the true slope and intercept.

In our case, the slope is 11.19, and the standard error of the slope is 1.22. We can therefore form a 95% confidence interval for the slope by taking two (approximately) standard errors above and below the mean. The 95% confidence interval for our slope is 8.75 to 13.63. Actually, since the number of points in our example is small ($n = 7$), we really should go to a t-table to find the number of standard errors above and below the mean for a 95% confidence interval. (This is true when n is less than 30.) Going to a t-table, we look under degrees of freedom (df) equal to $n - 2$ and level of signifi-cance (two-tail) equal to .05. The value of t for df = 5 and p = .05 is 2.57. Our 95% confidence interval is then 11.19 plus or minus $2.57 \times 1.22 = 3.14$.

Inspecting the Analysis of Variance portion of the output, we see values for Root MSE, R-Square, Dependent Mean, Adj R-Sq, and Coeff Var. Root MSE is the square root of the error variance. That is, it is the standard deviation of the residuals. R-square is the square of the multiple correlation coefficient. Since we have only one independent variable (HEIGHT), r-square is the square of the Pearson correlation coefficient between HEIGHT and WEIGHT. As mentioned in the previous section, the square of the correlation coefficient tells us how much variation in the dependent variable can be accounted for by variation of the independent variable. When there is more than one independent variable, r-square will reflect the variation in the dependent variable accounted for by a linear combination of all the independent variables (see Chapter 9 for a more complete discussion). The Dependent Mean is the mean of the dependent variable: WEIGHT in this case. Coeff Var. is the coefficient of variation—the standard deviation expressed as a percent of the mean. Finally, Adj R-Sq is the squared correlation coefficient corrected for the number of independent variables in the equation. This adjustment has the effect of decreasing the value of r-squared. The difference is typically small but becomes larger and more important when dealing with multiple independent variables (see Chapter 9).

G.   PARTITIONING THE TOTAL SUM OF SQUARES

The top portion of the output from PROC REG presents what is called the analysis-of-variance table for the regression. It takes the variation in the dependent variable and breaks it out into various sources. To understand this table, think about an individual weight. That weight can be thought of as the mean weight for all individuals, plus (or minus) a certain amount, because the individual is taller (or shorter) than average. The regression tells us that taller people are heavier. Finally, there is a portion attributable to the fact that the prediction is less than perfect. So we start at the mean, move up or down due to the regression, and then up or down again due to error (or residual). The analysis-of-variance table breaks these components apart and looks at the contribution of each through the sum of squares.

The total sum of squares is the sum of squared deviations of each person's weight from the grand mean. This total sum of squares (SS) can be divided into two portions: the sum of squares due to regression (or model) and the sum of squares error (sometimes called residual). One portion, called the Sum of Squares (ERROR) in the output, reflects the deviation of each weight from the PREDICTED weight. The other portion reflects deviations between the PREDICTED values and the MEAN. This is called the sum of squares due to the MODEL in the output. The column labeled Mean Square is the sum of squares divided by the degrees of freedom. In our case, there are seven data points. The total degrees of freedom is equal to $n - 1$, or 6. The model here has 1 df. The error degrees of freedom is the total df minus the model df, which gives us 5. We can think of the Mean Square as the respective variance (square of the standard deviation) of each of these two portions of variation. Our intuition tells us that if the deviations about the regression line are small (error mean square) compared to the deviation between the predicted values and the mean (mean square model), then we have a good regression line. To compare these mean squares, the ratio

$$F = \frac{\text{Mean square model}}{\text{Mean square error}}$$

is formed. The larger this ratio, the better the fit (for a given number of data points). The program prints this F value and the probability of obtaining an F this large or larger by chance alone. In the case where there is only one independent variable, the probability of the F statistic is exactly the same as the probability associated with testing for the significance of the correlation coefficient. If this probability is "large" (over .05), then our linear model is not doing a good job of describing the relationship between the variables.

H.   PRODUCING A SCATTER PLOT AND THE REGRESSION LINE

To plot out height/weight data, we can use PROC GPLOT as follows:

```
PROC GPLOT DATA=CORR_EG;
   PLOT WEIGHT*HEIGHT;
RUN;
```

If you want to override the default plotting symbol (a plus sign), you can add a SYMBOL statement before running the PROC GPLOT, like this:

```
SYMBOL VALUE=DOT COLOR=BLACK;
***You can abbreviate VALUE as V= and COLOR as C=;
PROC GPLOT DATA=CORR_EG;
   PLOT WEIGHT*HEIGHT;
RUN;
```

Some other plotting symbols besides DOT (large black dots) that are useful are CIRCLE, SQUARE, TRIANGLE, and PLUS. To see a complete list of plotting symbols, enter SYMBOL in your SAS help window or see the Online Doc™. You can also select VALUE = NONE if you also select an interpolation method (such as JOIN or SMOOTH) and you don't want to see individual data points on the plot.

There are two ways to have SAS show us the data points and the regression line. First, PROC REG allows the use of a PLOT statement. The form is:

```
PLOT y_variable * x_variable / options;
```

Y_variable and *x_variable* are the names of the variables to be plotted on the y- and x-axes, respectively. PROC REG will plot the data points and show the regression line. In addition, there are some special "variable names" that can be used with a PLOT statement. In particular, the names PREDICTED. and RESIDUAL. (the periods are part of these keywords) are used to plot predicted and residual values.

The second way is to use PROC GPLOT with the appropriate SYMBOL statement. Let's first look at the PLOT option of PROC REG. Suppose we want to see a plot of WEIGHT versus HEIGHT and the regression line in the previous PROC REG example. Here are the statements to have PROC REG generate a scatter plot and draw the regression line. While we are at it, we will generate a plot of the residuals versus the x-value:

```
SYMBOL1 VALUE=DOT COLOR=BLACK;
PROC REG DATA=CORR_EG;
   TITLE "Regression and Residual Plots";
   MODEL WEIGHT = HEIGHT;
   PLOT WEIGHT * HEIGHT
        RESIDUAL. * HEIGHT;
RUN;
QUIT;
```

The keyword RESIDUAL. stands for the residual values (the actual y-value minus the predicted value). The SYMBOL statement specifies that we want dots as our plotting symbol and the color to be black. The second plot shows the residuals versus HEIGHT (it uses the symbols defined in SYMBOL1 for this plot). The two plots are shown here:

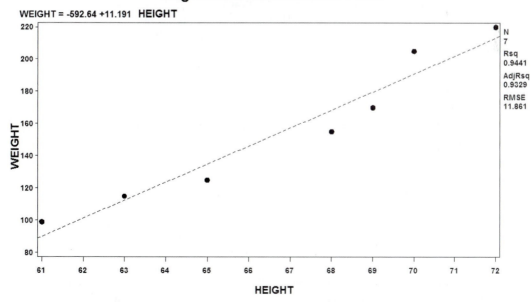

Regression and Residual Plots

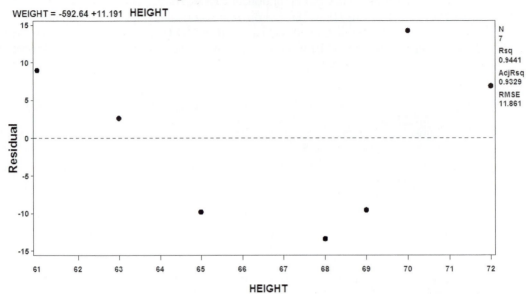

By using the GPLOT procedure, you can show the scatter plot of WEIGHT by HEIGHT, the regression line, and two types of 95% confidence intervals. The most common type of confidence interval about a regression line is an interval that indicates where you believe the mean value of y is most likely to lie. The other, less often used confidence interval, is an interval for individual data points. Limits of this interval determine the likely location of individual y-values. The following PROC GPLOT statements will produce a scatter plot of WEIGHT by HEIGHT and both types of confidence intervals. As you can see, the confidence interval for the mean of y is much narrower than the one for individual data points.

```
GOPTIONS CSYMBOL=BLACK;
SYMBOL1 VALUE=DOT;
SYMBOL2 VALUE=NONE I=RLCLM95;
SYMBOL3 VALUE=NONE I=RLCLI95 LINE=3;
PROC GPLOT DATA=CORR_EG;
   TITLE "Regression Lines and 95% CI's";
   PLOT WEIGHT * HEIGHT = 1
        WEIGHT * HEIGHT = 2
        WEIGHT * HEIGHT = 3 / OVERLAY;
RUN;
QUIT;
```

The first SYMBOL statement requests dots as the plotting symbols. The second and third SYMBOL statements set VALUE equal to 'NONE,' which hides the points. The I = (interpolation) option in SYMBOL2 requests a regression line and the 95% CI about the mean of y (RLCLM95). The SYMBOL3 statement asks for the confidence interval about the individual y-values (CLI95). In addition, this symbol statement uses LINE = 3 to generate a different type of line for this confidence interval. Note the use of the CSYMBOL option on the GOPTIONS statement to set the color to black for all the plotting symbols. The PLOT statement is requesting the overlayed plots, and the = 1, = 2, and = 3 following each of the plot requests indicates which SYMBOL statement to use for each of the plots. In the following plot, notice that the 95% CI about the mean of y is the narrower band (small dashed line), and the 95% CI for individual y-values is represented by the wider larger-dashed lines:

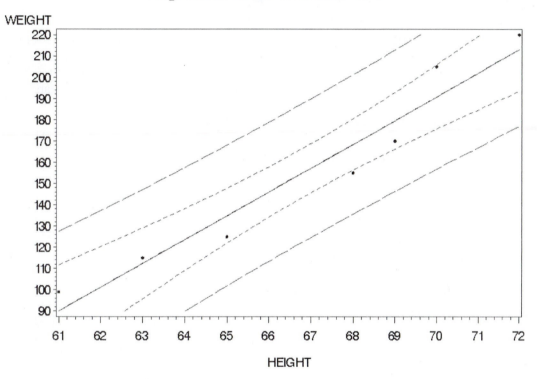

I.   ADDING A QUADRATIC TERM TO THE REGRESSION EQUATION

The plot of residuals, shown in the previous section, suggests that a second-order term (height squared) might improve the model since the points do not seem random but, rather, form a curve that could be fit by a second-order equation. Although this chapter deals mostly with linear regression, let us quickly show you how you might explore this possible quadratic relationship between height and weight. First, we need to add a line in the original DATA step to compute a variable that represents the height squared. After the INPUT statement, include a line such as the following:

```
HEIGHT2 = HEIGHT * HEIGHT;
```

or

```
HEIGHT2 = HEIGHT**2;
```

To see the residuals from this model, you can use the PLOT statement or PROC REG, like this:

```
SYMBOL VALUE=DOT COLOR=BLACK;
PROC REG DATA=CORR_EG;
   MODEL WEIGHT = HEIGHT HEIGHT2;
   PLOT R. * HEIGHT;
   ***Note: R. is short for RESIDUAL.;
RUN;
```

When you run this model, you will get an r-squared of .9743, an improvement over the .9441 obtained with the linear model. The following residual plot shows that the distribution of the residuals is more random than the earlier plot. A caution or two is in order here. First, remember that this is an example with a very small data set and that the original correlation is quite high. Second, although it is possible to also enter cubic terms, etc., one should keep in mind that results need to be interpretable.

Residual Plot with Quadratic Term

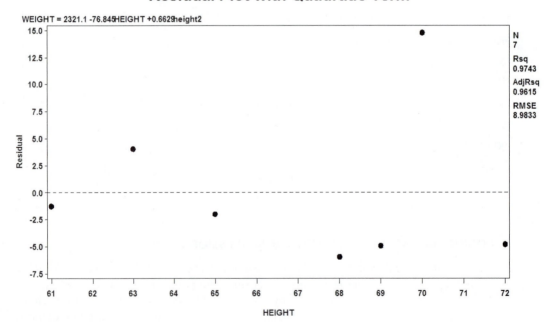

J. TRANSFORMING DATA

Another regression example is provided here to demonstrate some additional steps that may be necessary when doing regression. Shown here are data collected from 10 people:

Subject	Drug Dose	Heart Rate
1	2	60
2	2	58
3	4	63
4	4	62
5	8	67
6	8	65
7	16	70
8	16	70
9	32	74
10	32	73

Let's write a SAS program to define this collection of data and plot drug dose by heart rate.

```
DATA HEART;
    INPUT DOSE HR @@;
    ***The double @ at the end of the INPUT statement allows you to
       read several observations from one data line.  It is an
       instruction not to move the pointer to a new line when you
       reach the bottom of the DATA step (default action);
DATALINES;
2 60 2 58 4 63 4 62 8 67 8 65 16 70 16 70 32 74 32 73
;
SYMBOL VALUE=DOT COLOR=BLACK I=SM;

***I = SM produces a smooth line through the data points. You can
follow SM by a number from 0 to 99 to control how much the line
should try to touch each data point (low values such as SM1 produce
lots of wiggle high values such as SM80, give smoother lines).
Also, you can add an "S" to the end of this option if your x-values
are not sorted (example: I=SM80S);

PROC GPLOT DATA=HEART;
    PLOT HR*DOSE;
RUN;
PROC REG DATA=HEART;
    MODEL HR = DOSE;
RUN;
```

The resulting graph and the PROC REG output are shown here:

```
Investigating the Dose/HR Relationship

The REG Procedure
Model: MODEL1
Dependent Variable: HR
                                                        [Continued]
```

```
Number of Observations Read          10
Number of Observations Used          10

                    Analysis of Variance

                            Sum of       Mean
Source              DF      Squares      Square   F Value   Pr > F

Model                1    233.48441    233.48441    49.01   0.0001
Error                8     38.11559      4.76445
Corrected Total      9    271.60000


Root MSE              2.18276    R-Square    0.8597
Dependent Mean       66.20000    Adj R-Sq    0.8421
Coeff Var             3.29722


                    Parameter Estimates

                   Parameter      Standard
Variable     DF     Estimate         Error    t Value   Pr > |t|

Intercept     1     60.70833       1.04492      58.10    <.0001
DOSE          1      0.44288       0.06326       7.00     0.0001
```

Investigating the Dose/HR Relationship

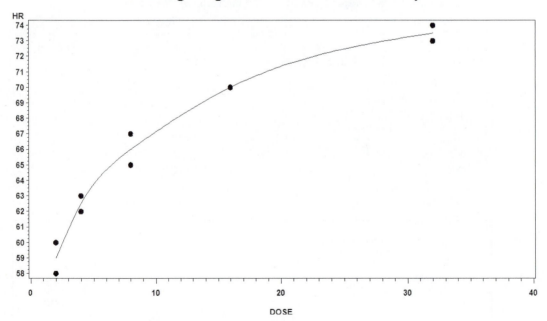

Either by clinical judgment or by careful inspection of the graph, we decide that the relationship is not linear. We see an approximately equal increase in heart rate each time the dose is doubled. Therefore, if we plot log dose against heart rate, we can expect a linear relationship. SAS software has a number of built-in functions such as logarithms and trigonometric functions, described in Chapter 17. We can write mathematical equations to define new variables by placing these statements between the INPUT and DATALINES statements. In SAS programs, we represent addition, subtraction, multiplication, and division by the symbols +, −, *, and /, respectively. Exponentiation is written as **. To create a new variable, which is the log of dose, we write:

```
DATA HEART;
    INPUT DOSE HR @@;
    LDOSE = LOG(DOSE);
    LABEL LDOSE = "Log of Dose";
DATALINES;
2 60 2 58 4 63 4 62 8 67 8 65 16 70 16 70 32 74 32 73
;
```

LOG is a SAS function that yields the natural (base e) logarithm of whatever value is within the parentheses.

We can now plot log dose versus heart rate and compute a new regression line.

```
PROC REG DATA=HEART;
    TITLE "Investigating the Dose/HR Relationship";
    MODEL HR = LDOSE;
RUN;

SYMBOL VALUE=DOT;
PROC GPLOT DATA=HEART;
    PLOT HR*LDOSE;
RUN;
```

Output from the previous statements is shown on the following pages. Approach transforming variables with caution. Keep in mind that when a variable is transformed, one should not refer to the variable as in the untransformed state. That is, don't refer to the "log of dosage" as "dosage." Some variables are frequently transformed: Income, sizes of groups, and magnitudes of earthquakes are usually presented as logs or in some other transformation.

```
Investigating the Dose/HR Relationship

The REG Procedure
Model: MODEL1
Dependent Variable: HR
                                              [Continued]
```

```
Number of Observations Read          10
Number of Observations Used          10

                        Analysis of Variance

                                  Sum of        Mean
Source                   DF       Squares      Square  F Value  Pr > F

Model                     1     266.45000   266.45000   413.90  <.0001
Error                     8       5.15000     0.64375
Corrected Total           9     271.60000


Root MSE              0.80234     R-Square      0.9810
Dependent Mean      66.20000     Adj R-Sq      0.9787
Coeff Var            1.21199

                       Parameter Estimates

                                Parameter    Standard
Variable    Label        DF      Estimate       Error  t Value  Pr > |t|

Intercept   Intercept     1      55.25000     0.59503    92.85   <.0001
LDOSE       Log of Dose   1       5.26584     0.25883    20.34   <.0001
```

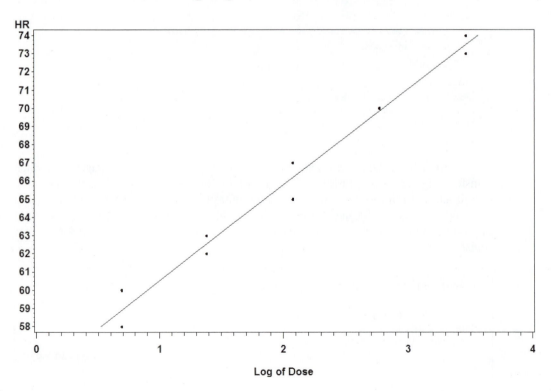

Investigating the Dose/HR Relationship

Notice that the data points are close to the regression line, and the MEAN SQUARE ERROR term is smaller and the r-square is larger, confirming our conclusion that dose versus heart rate fits a logarithmic curve better than a linear one.

PROBLEMS

Remember, you can download all the data sets and programs for these problems from the web site: www.prenhall.com/cody

5.1 Given the following data:

X	Y	Z
1	3	15
7	13	7
8	12	5
3	4	14
4	7	10

(a) Write a SAS program and compute the Pearson correlation coefficient between X and Y; X and Z. What is the significance of each?

(b) Change the correlation request to produce a correlation matrix; that is, the correlation coefficient between each variable versus every other variable.

5.2 Ten students take an eight-item test. The responses to each of the items (scored 0 or 1—incorrect or correct) are stored in variables Q1 to Q8 in the SAS data set EXAM. Run the following program to create this data set.

```
DATA EXAM;
    INPUT (Q1-Q8)(1.);
DATALINES;
10101010
11111111
11110101
01100000
11110001
11111111
11111101
11111101
10110101
00010110
;
```

Starting with this data set, create a new SAS data set with the variables Q1 to Q8, plus the raw score on the test (the sum of Q1 to Q8). Using this data set, determine the correlation of each of the eight questions with the total test score. This number is called the point-biserial correlation coefficient.

5.3 Given the following data:

AGE	SBP
15	116
20	120
25	130
30	132
40	150
50	148

How much of the variance of SBP (systolic blood pressure) can be explained by the fact that there is variability in AGE? (Use SAS to compute the correlation between SBP and AGE.)

5.4 Run the following program to create a SAS data set called SCORES.

```
DATA SCORES;
    DO SUBJECT = 1 TO 100;
        IF RANUNI(1357) LT .5 THEN GROUP = 'A';
        ELSE GROUP = 'B';
        MATH = ROUND(RANNOR(1357)*20 + 550 + 10*(GROUP EQ 'A'));
        SCIENCE = ROUND(RANNOR(1357)*15 + .4*MATH + 300);
        ENGLISH = ROUND(RANNOR(1357)*20 + 500 + .05*SCIENCE +
                    .05*MATH);
        SPELLING = ROUND(RANNOR(1357)*15 + 500 + .1*ENGLISH);
        VOCAB = ROUND(RANNOR(1357)*5 + 400 + .1*SPELLING +
                    .2*ENGLISH);
        PHYSICAL = ROUND(RANNOR(1357)*20 + 550);
        OVERALL = ROUND(MEAN(MATH, SCIENCE, ENGLISH, SPELLING, VOCAB,
                        PHYSICAL));
        OUTPUT;
    END;
RUN;
```

(a) Generate a correlation matrix of all the test scores plus the overall score. Use the option to omit the simple statistics produced by PROC CORR.

(b) PHYSICAL is independent of all the other test scores but is correlated to OVERALL. Why?

(c) Write the statements to correlate each of the scores with OVERALL.

5.5 From the data for X and Y in Problem 5.1:

(a) Compute a regression line (Y on X). Y is the dependent variable, X the independent variable.

(b) What is slope and intercept?

(c) Are they significantly different from zero?

5.6 (a) Using data set SCORES from Problem 5.4, compute a regression equation for predicting the SCIENCE score from the MATH score. Make one plot showing the data points and a regression line and another showing the residuals versus the x-values (use the PLOT statement of PROC REG to do this).

(b) Produce a scatter plot of MATH (x-axis) versus SCIENCE (y-axis). Include the regression line and the 95% CI on the mean of y. Use PROC GPLOT and the appropriate SYMBOL statement to do this.

5.7 Using the data of Problem 5.1, compute three new variables LX, LY, and LZ, which are the natural logs of the original values. Compute a correlation matrix for the three new variables. HINT: The function to compute a natural log is the LOG function (see Chapter 17 for details).

5.8 The following data were collected on DOSE, systolic blood pressure (SBP), and diastolic blood pressure (DBP). Create a SAS data set (DOSE_RESPONSE) from these data and compute two regression equations. One for SBP versus DOSE and one for DBP versus DOSE. Produce one plot of SBP versus DOSE, another of the residuals (SBP − predicted SBP) versus DOSE, one of DBP versus DOSE, and finally, one of residuals for DBP versus DOSE. Use the PLOT statement of PROC REG to do this.

Dose	Systolic Blood Pressure	Diastolic Blood Pressure
4	180	110
4	190	108
4	178	100
8	170	100
8	180	98
8	168	88
16	160	80
16	172	86
16	170	86
32	140	80
32	130	72
32	128	70

5.9 Generate:
(a) A plot of Y versus X (data from Problem 5.1).
(b) A plot of the regression line and the original data on the same set of axes.

5.10 Repeat Problem 5.8 using the natural log of dose (LOG function) instead of DOSE. Compare the fit statistics. Which fit is better for SBP? Which for DBP?

5.11 Given the data set:

COUNTY	POP	HOSPITAL	FIRE_CO	RURAL
1	35	1	2	YES
2	88	5	8	NO
3	5	0	1	YES
4	55	3	3	YES
5	75	4	5	NO
6	125	5	8	NO
7	225	7	9	YES
8	500	10	11	NO

    (a) Write a SAS program to create a SAS data set of the previous data.

    (b) Run PROC UNIVARIATE to check the distributions for the variables POP, HOSPI-TAL, and FIRE_CO.

    (c) Compute a correlation matrix for the variables POP, HOSPITAL, and FIRE_CO. Produce both Pearson and Spearman correlations. Which is more appropriate?

    (d) Recode POP, HOSPITAL, and FIRE_CO so that they each have two levels (use a median cut or a value somewhere near the 50th percentile). Compute crosstabulations between the variable RURAL and the recoded variables.

5.12 Repeat Problem 5.4 (a), except do the analysis separately for GROUP='A' and GROUP='B'. What is the effect on the magnitude of the correlations and the p-values?

5.13 What's wrong with the following program? (NOTE: There may be missing values for X, Y, and Z.)

```
1    DATA MANY-ERR;
2       INPUT X Y Z;
3       IF X LE 0 THEN X=1;
4       IF Y LE 0 THEN Y=1;
5       IF Z LE 0 THEN Z=1;
6       LOGX = LOG(X);
7       LOGY = LOG(Y);
8       LOGZ = LOG(Z);
9    DATALINES;
     1 2 3
     . 7 8
     4 . 10
     7 8 11
     ;
10   PROC CORR DATA=MANY-ERR / PEARSON SPEARMAN;
11      VAR X - LOGZ;
12   RUN;
```

T-tests and Nonparametric Comparisons

A.   INTRODUCTION

Our first topic concerning hypothesis testing is the comparison of two groups. When certain assumptions are met, the popular t-test is used to compare means. When these assumptions are not met, there are several nonparametric methods that can be used. This chapter shows you how to conduct all of these two-group comparisons using SAS. In addition, we show you how to write a simple SAS program to randomly assign subjects to two or more groups.

B.   T-TEST: TESTING DIFFERENCES BETWEEN TWO MEANS

A common experimental design is to assign subjects randomly to a treatment or a control group and then measure one or more variables that would be hypothesized to be affected by the treatment. To determine whether the means of the treatment and control groups are significantly different, we set up what is called a null hypothesis (H_0). It states that the treatment and control groups would have the same mean if we repeated the experiment a large (infinite) number of times and that the differences in any one trial are attributable to the "luck of the draw" in assigning subjects to treatment and control groups. The alternative hypotheses (H_A) to the null hypothesis are that one particular mean will be greater than the other (called a one-tailed test) or that the two means will be different, but the researcher cannot say a priori which will be greater (called a two-tailed test). The researcher can specify either of the two alternative hypotheses.

When the data have been collected, a procedure called the t-test is used to determine the probability that the difference in the means that is observed is due to chance. The lower the likelihood that the difference is due to chance, the greater the likelihood that the difference is due to there being real differences in treatment and control. The following example demonstrates a SAS program that performs a t-test.

Students are randomly assigned to a control or treatment group (where a drug is administered). Their response times to a stimulus is then measured. The times are as follows:

Control	Treatment
(response time in millisec)	
80	100
93	103
83	104
89	99
98	102

Do the treatment scores come from a population whose mean is different from the mean of the population from which the control scores were drawn? A quick calculation shows that the mean of the control group is 88.6, and the mean of the treatment group is 101.6. You may recall the discussion in Chapter 2 about the standard error of the mean. When we use a sample to estimate the population mean, the standard error of the mean reflects how accurately we can estimate the population mean. As you will see when we write a SAS program to analyze this experiment, the standard error of the control mean is 3.26, and the standard error of the treatment mean is .93. Since the means are 11.8 units apart, even if each mean is several standard errors away from its true population mean, they would still be significantly different from each other.

There are some assumptions that should be met before we can apply this test. First, the two groups must be independent. This is ensured by our method of random assignment. Second, the theoretical distribution of sampling means should be normally distributed (this is typically ensured if the sample size is sufficiently large and the distribution of scores is not too "wild"); and third, the variances of the two groups should be approximately equal. This last assumption is checked automatically each time a t-test is computed by SAS. The SAS t-test output contains t-values and probabilities for both the case of equal group variances and unequal group variances. We will see later how to test if the distribution of data values is reasonably close to normal. (Remember that the distribution of data values does not have to be normally distributed—only that the sampling distribution of sample means is normally distributed. If the distribution of data values is reasonable and the sample size is fairly large, the assumption concerning the sampling distribution will be met.) Finally, when the t-test assumptions are not met, there are other procedures that can be used; these are demonstrated later in this chapter.

It's time now to write a program to describe our data and to request a t-test. We know which group each person belongs to and what his or her response time is. The program can thus be written as follows:

```
DATA RESPONSE;
   INPUT GROUP $ TIME;
DATALINES;
C 80
C 93
C 83
C 89
C 98
T 100
T 103
T 104
T 99
T 102
;
PROC TTEST DATA=RESPONSE;
   TITLE "T-test Example";
   CLASS GROUP;
   VAR TIME;
RUN;
```

PROC TTEST uses a CLASS statement to identify the independent variable—the variable that identifies the two groups of subjects. In our case, the variable GROUP has values of C (for the control group) and T (for the treatment group).

The variable or variable list that follows the word VAR identifies the dependent variable(s), in our case, TIME. When more than one dependent variable is listed, a separate t-test is computed for each dependent variable in the list.

Look at the TTEST output here:

```
T-test Example

The TTEST Procedure

                                    Statistics

                             Lower CL           Upper CL    Lower CL
Variable   GROUP        N      Mean      Mean      Mean      Std Dev

TIME       C            5     79.535     88.6     97.665     4.3741
TIME       T            5     99.025    101.6     104.17     1.2424
TIME       Diff (1-2)         -20.83    -13      -5.173      3.6249

                                    Statistics

                                      Upper CL
Variable   GROUP       Std Dev       Std Dev     Std Err   Minimum   Maximum

TIME       C            7.3007       20.979       3.265        80        98
TIME       T            2.0736       5.9587      0.9274        99       104
TIME       Diff (1-2)   5.3666       10.281      3.3941

                                                        [Continued]
```

```
                              T-Tests

    Variable    Method           Variances      DF    t Value    Pr > |t|

    TIME        Pooled           Equal           8     -3.83      0.0050
    TIME        Satterthwaite    Unequal      4.64     -3.83      0.0141

                    Equality of Variances

    Variable    Method        Num DF     Den DF    F Value     Pr > F

    TIME        Folded F        4          4       12.40       0.0318
```

We see the mean value of TIME for the control and treatment groups, along with the standard deviation and standard error. Below this are two sets of t-values, degrees of freedom, and probabilities. One is valid if we have equal variances, the other if we have unequal variances. You will usually find these values very close unless the variances differ widely. The bottom line gives us the probability that the variances are unequal due to chance. Some people use the following rule: If this probability is small (say less than .05), then we are going to reject the hypothesis that the variances are equal. We then use the t-value and probability labeled Unequal. If the PROB>F value is greater than .05, we use the t-value and probability for equal variances.

We point out here that, as with most rules of thumb, there is a weakness in the rule. That is, when the sample sizes are small, the test of equal variances has low power and may fail to reject the null hypothesis of equal variances. This is just the situation where the assumption is most important to performing a proper t-test. Also, when the sample sizes are large (where the assumptions of equal variance are less important to the t-test), the null hypothesis of equal variances is frequently rejected. Therefore, some statisticians simply decide ahead of time if the assumption of equal variance is reasonable or not, and then stay with that decision, regardless of the equality of variance test outcome.

In this example, we look at the end of the t-test output and see that the F ratio (larger variance divided by the smaller variance) is 12.40. The probability of obtaining, by chance alone, a ratio this large or larger is 0.0318. That is, if the two samples came from populations with equal variance, there is a small probability (.0318) of obtaining a ratio of our sample variances of 12.40 or larger by chance. We may therefore decide to use the t-value appropriate for groups with unequal variance.

C.   RANDOM ASSIGNMENT OF SUBJECTS

In our discussion of t-tests, we said that we randomly assigned subjects to either a treatment or control group. This is actually a very important step in our experiment, and we can use SAS to provide us with a method for making the assignments.

We could take our volunteers one by one, flip a coin, and decide to place all the "heads" in our treatment group and all the "tails" in our control group. This is acceptable, but we would prefer a method that ensures an equal number of subjects in each group. One method would be to place all the subjects' names in a hat, mix them up, and pull out half the names for treatment subjects and the others for controls. This is essentially what we will do in our SAS program. The key to the program is a SAS random number function.

The function RANUNI(*seed*) will generate a pseudorandom number in the interval from 0 to 1. The argument of this function, called a seed, can either be a zero or a number of your choice. A zero (or negative) seed specifies that the random number function use the time clock to generate a random seed to initiate the random number sequence. If you use a zero seed, you will obtain a different series of random numbers every time the program is run (unless you run the program at EXACTLY the same time every day). You may want to supply your own seed instead. It can be any number you wish. Each time you run the program with your own seed, you will generate the same series of random numbers, which is sometimes desirable. You will have to decide whether to supply your own seed or use the time clock. It is important to note that the sequence of seed values is determined by the first random number function you use in a DATA step. Regardless of what you use as an argument to any other random function in the same DATA step (RANUNI, RANNOR, etc.), the series of random numbers (time clock or user specified) is determined **only** by the first use of a random function in the DATA step. If you need more control over seed values, take a look at the random number call routines in the SAS Online Doc™ or in *SAS Functions by Example* (Cody, SAS Institute, Cary, NC, 2004).

One way to have SAS perform a random group assignment is to use the RANUNI function to generate a random number between 0 and 1. We can then use this number to decide if a person is in one group or the other. The following short program makes random assignments for 50 subjects:

```
DATA ASSIGN;
    DO SUBJ = 1 TO 50;   ①
        IF RANUNI(123) LE .5 THEN GROUP = 'A';   ②
        ELSE GROUP = 'B';
        OUTPUT;   ③
    END;
RUN;
OPTIONS PS=16 LS=72;
PROC REPORT DATA=ASSIGN PANELS=99 NOWD;
    TITLE "Simple Random Assignment";
    COLUMNS SUBJ GROUP;
    DEFINE SUBJ / WIDTH=4;
    DEFINE GROUP / WIDTH=5;
RUN;
```

In this program, a DO loop ① generates values for SUBJ from 1 to 50. The loop starts with SUBJ equal to 1. A uniform random number from 0 to 1 is then generated ② If this value is less than .5 (which will happen, on average, half the time), GROUP is set to 'A'.

Otherwise, GROUP is set to 'B'. The OUTPUT statement ③ is needed to output an observation every time the DO loop iterates. The END statement at the bottom of the loop causes a return to the top of the DO loop and SUBJ is incremented by 1. This loop continues until the value of SUBJ reaches 51. Since this value is larger than the maximum loop value (50), the loop stops and the program executes the statement following the END statement. But, since this is the end of the DATA step, the program stops.

One other feature of this program is the use of PROC REPORT to produce the listing. You may ignore this and use PROC PRINT instead. The reason we used PROC REPORT here was to demonstrate the really useful PANELS = option. When you set PANELS equal to a large number, PROC REPORT will attempt to fit as many columns across the page as possible. We also set the page size to a small value (16) so that you could see how this feature works. You can find more about PROC REPORT in any one of the many SAS books and manuals or in the Online Doc™.

A listing of data set ASSIGN is show next:

```
Simple Random Assignment

SUBJ  GROUP    SUBJ  GROUP    SUBJ  GROUP    SUBJ  GROUP
   1  B          14  A          27  A          40  B
   2  A          15  B          28  B          41  A
   3  A          16  B          29  B          42  A
   4  B          17  B          30  A          43  B
   5  A          18  B          31  B          44  B
   6  A          19  A          32  A          45  B
   7  B          20  B          33  B          46  B
   8  A          21  B          34  A          47  A
   9  A          22  B          35  B          48  B
  10  A          23  B          36  B          49  B
  11  B          24  A          37  A          50  A
  12  A          25  B          38  A
  13  B          26  B          39  B
```

This program works fine, except that it does not always assign an equal number of subjects to the two groups.

One way to ensure an equal number of subjects in each group is as follows: First, we have SAS generate a random number for each subject in the data set. Then we can assign all those subjects with random numbers below the median to group 'A' and subjects with random numbers above the median to group 'B'. If we want more than two groups, we can divide the random numbers into any number of equal groups that we need (as long as the number of subjects is a multiple of the number of groups). Here is a program to randomly assign 20 subjects equally into two groups:

```
PROC FORMAT;
   VALUE GRPFMT 0='CONTROL' 1='TREATMENT';
RUN;
```

```
DATA RANDOM;
   DO SUBJ = 1 TO 20;
      GROUP=RANUNI(0);
      OUTPUT;
   END;
RUN;
PROC RANK DATA=RANDOM GROUPS=2 OUT=SPLIT;
   VAR GROUP;
RUN;
PROC PRINT DATA=SPLIT NOOBS;
   TITLE "Subject Group Assignments";
   VAR SUBJ GROUP;
   FORMAT GROUP GRPFMT.;
RUN;
```

The key to this program is the RANUNI function, which assigns a random number from 0 to 1 to each subject. The GROUPS = 2 option of PROC RANK divides the subjects into two groups (0 and 1), depending on the value of the random variable GROUP. Values below the median become 0; those at or above the median become 1. The GRPFMT format assigns the labels "CONTROL" and "TREATMENT" using values of 0 and 1, respectively. We can use this program to assign our subjects to any number of groups by changing the "GROUPS=" option of PROC RANK to indicate the desired number of groups. Sample output from this program is shown here:

```
Subject Group Assignments

SUBJ       GROUP

  1        TREATMENT
  2        CONTROL
  3        TREATMENT
  4        TREATMENT
  5        CONTROL
  6        CONTROL
  7        TREATMENT
  8        TREATMENT
  9        CONTROL
 10        CONTROL
 11        TREATMENT
 12        CONTROL
 13        CONTROL
 14        TREATMENT
 15        TREATMENT
 16        CONTROL
 17        CONTROL
 18        TREATMENT
 19        TREATMENT
 20        CONTROL
```

You can verify, using PROC FREQ, that there are equal numbers of Control and Treatment subjects. Although this may seem like a lot of work just to make random assignments of subjects, we recommend that this or an equivalent procedure be used for assigning subjects to groups. Other methods, such as assigning every other person to the treatment group, can result in unsuspected bias.

D.   TWO INDEPENDENT SAMPLES: DISTRIBUTION-FREE TESTS

There are times when the assumptions for using a t-test are not met. One common problem is that the data are not normally distributed, and your sample size is small. For example, suppose we collected the following numbers in a psychology experiment that measured the response to a stimulus:

    0 6 0 5 7 6 9 4 8 0 7 0 5 6 6 0 0

A frequency distribution would look like this:

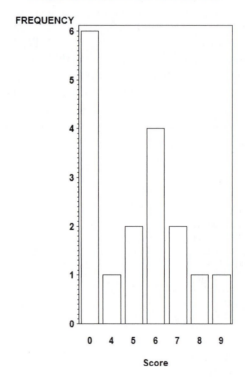

Distribution of Scores

What we are seeing is probably due to a threshold effect. The response is either 0 (the stimulus is not detected), or, once the stimulus is detected, the average response is about 6. Data of this sort would artificially inflate the standard deviation (and thus the standard error) of the sample and make the t-test more conservative. However, we would be safer to use a nonparametric test (a test that does not assume a normal distribution of data).

Another common problem is that the data values may only represent ordered categories. Scales such as 1 = very mild, 2 = mild, 3 = moderate, 4 = strong, 5 = severe reflect the strength of a response, but we cannot say that a score of 4 (strong) is worth twice the score of 2 (mild). Scales like these are referred to as ordinal scales. (Most of the scales we have been using until now have been interval or ratio scales.) We need a nonparametric test to analyze differences in central tendencies for ordinal data. Finally, for very small samples, nonparametric tests are often more appropriate since assumptions concerning distributions are difficult to determine.

SAS provides us with several nonparametric two-sample tests. Among these is the Wilcoxon rank-sum test (equivalent to the Mann-Whitney U-test) for two samples.

Consider the following experiment. We have two groups, A and B. Group B has been treated with a drug to prevent tumor formation. Both groups are exposed to a chemical that encourages tumor growth. The masses (in grams) of tumors in groups A and B are:

```
A:     3.1     2.2     1.7     2.7     2.5

B:     0.0     0.0     1.0     2.3
```

Are there any differences in tumor mass between groups A and B? We will choose a nonparametric test for this experiment because of the absence of a normal distribution and the small sample sizes involved. The Wilcoxon test first puts all the data (groups A and B) in increasing order (with special provisions for ties), retaining the group identity. In our experiment we have:

MASS	0.0	0.0	1.0	1.7	2.2	2.3	2.5	2.7	3.1
GROUP	B	B	B	A	A	B	A	A	A
RANK	1.5	1.5	3	4	5	6	7	8	9

The sums of ranks for the As and Bs are then computed. We have:

$$\text{SUM RANKS A} = 4 + 5 + 7 + 8 + 9 = 33$$
$$\text{SUM RANKS B} = 1.5 + 1.5 + 3 + 6 = 12$$

If there were smaller tumors in group B, we would expect the Bs to be at the lower end of the rank ordering and therefore have a smaller sum of ranks than the As. Is the sum of ranks for group A sufficiently larger than the sum of ranks for group B, so that the probability of the difference occurring by chance alone is small (less than .05)? The Wilcoxon test gives us the probability that the difference in rank sums that we obtained occurred by chance.

For even moderate sample sizes, the Wilcoxon test is almost as powerful as its parametric equivalent, the t-test. Thus, if there is a question concerning distributions or if the data are really ordinal, you should not hesitate to use the Wilcoxon test instead of the t-test.

The program to analyze this experiment using the Wilcoxon test is shown here:

```
DATA TUMOR;
   INPUT GROUP $ MASS @@;
DATALINES;
A 3.1 A 2.2 A 1.7 A 2.7 A 2.5
B 0.0 B 0.0 B 1.0 B 2.3
;
PROC NPAR1WAY DATA=TUMOR WILCOXON;
   TITLE "Nonparametric Test to Compare Tumor Masses";
   CLASS GROUP;
   VAR MASS;
   EXACT WILCOXON;
RUN;
```

As you saw in Chapter 5, the double trailing @ signs allow us to place data for several observations on one line. To review: Normally, each time the DATA step reaches the bottom and loops back to the top to read another line of raw data, an internal pointer moves to the next line (record) in the data file. The two '@' signs at the end of the INPUT statement instruct the program to 'hold the line' and not automatically go to a new line of data for the next observation. This way, we can put data from several observations on one line. By the way, a single @ sign will hold the line for another INPUT statement, but it goes to the next line of data when the DATALINES statement or end of the DATA step is encountered. See Chapter 12 for more details on the use of single and double trailing @ signs.

PROC NPAR1WAY performs the nonparametric tests. The option WILCOXON requests the Wilcoxon rank-sum test (as well as a number of other statistics). The CLASS and VAR statements are identical to the CLASS and VAR statements of the t-test procedure. The EXACT statement causes the program to compute exact p-values (in addition to the asymptotic approximations usually computed) for the tests listed after this statement. In this example, we requested exact p-value computations for the Wilcoxon rank-sum test. We suggest that you include the EXACT statement when you have relatively small sample sizes.

The output from the NPAR1WAY procedure follows:

```
Nonparametric Test to Compare Tumor Masses

The NPAR1WAY Procedure

              Wilcoxon Scores (Rank Sums) for Variable MASS
                    Classified by Variable GROUP

                    Sum of        Expected        Std Dev           Mean
GROUP        N      Scores        Under H0        Under H0          Score
```

GROUP	N	Sum of Scores	Expected Under H0	Std Dev Under H0	Mean Score
A	5	33.0	25.0	4.065437	6.60
B	4	12.0	20.0	4.065437	3.00

```
                  Average scores were used for ties.

          Wilcoxon Two-Sample Test

Statistic (S)                    12.0000

Normal Approximation
Z                                -1.8448
One-Sided Pr <  Z                 0.0325
Two-Sided Pr > |Z|                0.0651

t Approximation
One-Sided Pr <  Z                 0.0511
Two-Sided Pr > |Z|                0.1023

Exact Test
One-Sided Pr <=  S                0.0317
Two-Sided Pr >= |S - Mean|        0.0635

Z includes a continuity correction of 0.5.

       Kruskal-Wallis Test

Chi-Square          3.8723
DF                       1
Pr > Chi-Square     0.0491
```

The sum of ranks for groups A and B are shown, as well as their expected values. The exact two-tailed p-value for this test is .0635, which is quite close to the Normal Approximation value of .0651. This suggests that the mass values for group A are larger than the mass of values for group B, although the p-value is just shy of the magic value of .05.

E.   ONE-TAILED VERSUS TWO-TAILED TESTS

When we conduct an experiment like the tumor example of Section D, we have a choice for stating the alternate hypothesis. In our example, the null hypothesis is that the mass of tumors is the same for groups A and B. The alternative hypothesis is that groups A and B are not the same. We would reject the null hypothesis if A > B or B > A. This type of hypothesis requires a two-tailed test. If we were using a t-test, we would have to consider absolute values of t greater than the critical value. If our alpha level is .05, then we have .025 from each tail of the t-distribution. Most statistical tests are conducted as two-tailed tests. Even though you may hope that the results come out in a particular direction or your preliminary tests indicate a direction, you should still probably conduct a two-tailed test. In medical research, if you already know the direction of the result, it would be unethical to conduct the clinical trial. Clinical trials, as well as hypothesis tests in other fields, are performed when the results have not been proven one way or the other. If in doubt, do a two-tailed test.

If you are conducting a test between, say, two pain relievers (yours and the competition), you may only want to test if one treatment (yours) is better than the competition, so you may choose to conduct a one-tailed test. If you can justify a direction of the alternative hypothesis from substantive issues in the study, you may also choose to conduct a one-tailed test. However, when you choose to conduct a one-tailed test, if the results come out the opposite way you expected, you wind up failing to reject the null hypothesis—in a situation where your result may be even more interesting than if the result went in the direction you expected. With a one-tailed test, the 5% of the curve associated with the .05 alpha level is located in one tail, which increases the power of the study (i.e., makes it more likely of finding a significant difference if, in fact, one exists). If our tumor example had been stated as a one-tailed test, we could have divided the p-value by 2, giving p = .0317 for the Wilcoxon test probability.

The decision to do a one-tailed test should be based on an understanding of the theoretical considerations of the study and not as a method of reducing the p-value below the .05 level.

F.   PAIRED T-TESTS (RELATED SAMPLES)

Our t-test example in Section B had subjects randomly assigned to a control or treatment group. Therefore, the groups could be considered independent.

There are many experimental situations where each subject receives both treatments. For example, each subject in the earlier example could have been measured in the absence of the drug (control value) and after having received the drug (treatment value). The response times for the control and treatment groups would no longer be independent. We would expect a person with a very short response time in the control situation to also have a short response time after taking the drug (compared to the other people who took the drug). We would expect a positive correlation between the control and treatment values.

Our regular t-test cannot be used here since the groups are no longer independent. A variant of the t-test, referred to as a paired t-test, is used instead. The differences between the treatment and control times are computed for each subject. If most of the differences are positive, we suspect that the drug lengthens reaction time. If most are about zero, the drug has no effect. The paired t-test computes a mean and standard error of the differences and determines the probability that the absolute value of the mean difference was greater than zero by chance alone.

Before we program this problem, it should be mentioned that this would be a very poor experiment. What if the response time increased because the subjects became tired? If each subject were measured twice, without being given a drug, would the second value be different because of factors other than the treatment? One way to control for this problem is to take half of the subjects and measure their drug response time first and the control value later, after the drug has worn off. We show a better way to devise experiments to handle this problem in Chapter 8 ("Repeated Measures Designs").

Experiments that measure the same subject under different conditions are sometimes called repeated measures experiments. They do have the problem just stated: one measurement might affect the next. However, if this can be controlled, it is much easier to show treatment effects with smaller samples compared to a regular t-test.

If we used two independent groups of people, we would find that within the control group or treatment group there would be variation in response times because of individual differences. However, if we measure the same subject under two conditions, even if that person has much longer or shorter response times than the other subjects, the difference between the scores should approximate the difference for other subjects. Thus, each subject acts as his or her own control and therefore controls some of the natural variation between subjects.

You can request PROC TTEST to perform a paired t-test by using a PAIRED statement and specifying the two variables holding the paired values. The following is some sample data for which we want to perform a paired t-test:

Subject	Control Value	Treatment Value
1	90	95
2	87	92
3	100	104
4	80	89
5	95	101
6	90	105

The program is written as follows:

```
DATA PAIRED;
   INPUT CTIME TTIME;
DATALINES;
90 95
87 92
100 104
80 89
95 101
90 105
;
PROC TTEST DATA=PAIRED;
   TITLE "Demonstrating a Paired T-test";
   PAIRED CTIME * TTIME;
RUN;
```

The variable names CTIME and TTIME were chosen to represent the response times in the control and treatment conditions, respectively. We place the variable names representing the paired data, with an asterisk between them, on the PAIRED statement. If we have more than one set of paired values, we can enter several pairs on one PAIRED statement. For example, if you have BEFORE1, AFTER1, BEFORE2, AFTER2 and want to compare before to after for each, you would write:

```
PAIRED BEFORE1 * AFTER1   BEFORE2 * AFTER2;
```

Output from the paired program is shown next:

```
Demonstrating a Paired T-test

The TTEST Procedure

                           Statistics

                     Lower CL            Upper CL  Lower CL
Difference       N     Mean    Mean        Mean    Std Dev  Std Dev

CTIME — TTIME    6    −11.67  −7.333     −2.998    2.5787   4.1312

                           Statistics

                 Upper CL
Difference       Std Dev    Std Err    Minimum    Maximum

CTIME — TTIME    10.132     1.6865       −15         −4

                 T-Tests

Difference        DF     t Value    Pr > |t|

CTIME — TTIME      5      −4.35      0.0074
```

In this example, the mean difference (CTIME minus TTIME) is negative (time increased) and equal to −7.333. The probability of the difference occurring by chance is .0074. We can state that response times are longer under the drug treatment compared to the control values. Had the mean difference been positive, we would state that response times were shorter under the drug treatment, because DIFF was computed as treatment time minus control time.

PROBLEMS

Remember, you can download all the data sets and programs for these problems from the web site: www.prenhall.com/cody

6.1 The following table shows the time for subjects to feel relief from headache pain:

Aspirin	Tylenol
(Relief time in minutes)	
40	35
42	37
48	42
35	22
62	38
35	29

Write a SAS program to read these data and perform a t-test. Is either product significantly faster than the other (at the .05 level)?

*6.2 Fourteen subjects were enrolled in the Cody reading program and 14 subjects were enrolled in the Smith reading program. At the conclusion of both programs, reading speed was assessed with the following results (to make this a bit more challenging, write your program so that you have the 14 scores from the Cody program on one line and the 14 scores from the Smith program on another line):

```
Cody Program:   500 450 505 404 555 567 588 577 566 644 511 522 543 578
Smith Program: 355 388 440 600 510 501 502 489 499 489 515 520 520 480
```

Perform a t-test and a Wilcoxon rank-sum test on these data. For the Wilcoxon test, use the EXACT Wilcoxon statement and compare the p-value from the exact solution to the normal approximation. How does this value compare to the t-test results?

6.3 Using the same data as for problem 6.1, perform a Wilcoxon rank-sum test. Include a request for an exact p-value.

6.4 Run the following program to create a data set called QUES6_4, with variables GROUP ('A', 'B', and 'C'), X, and Y.

```
DATA QUES6_4;
    DO GROUP = 'A','B','C';
        DO I = 1 TO 10;
            X = ROUND(RANNOR(135)*10 + 300 +
                    5*(GROUP EQ 'A') - 7*(GROUP EQ 'C'));
            Y = ROUND(RANUNI(135)*100 + X);
            OUTPUT;
        END;
    END;
    DROP I;
RUN;
```

Perform a t-test on X and Y using only groups 'A' and 'C'. Do not use any DATA steps to do this. (HINT: Where did I put that statement?)

6.5 Eight subjects are tested to see which of two medications (A or B) works best for headaches. Each subject tries each of the two drugs (for two different headaches), and the time span to pain relief is measured. Ignoring an order effect, what type of test would you use to test if one drug is faster than the other? The following are some made-up data: Write the SAS statements to run the appropriate analysis.

Subject	Drug A	Drug B
1	20	18
2	40	36
3	30	30
4	45	46
5	19	15
6	27	22
7	32	29
8	26	25

6.6 Each of 12 patients went on the South Beach diet for 5 weeks. Their weights before and after the diet are listed here. Perform a paired t-test comparing before and after values. Analyze these data as if the values were not paired. How do the p-values compare? Note: If you want to make this problem easier, read in the data by subject, with the before and after values. If you would like a challenge, read in all of the before values (perhaps into an array) and then all the after values, and create the data set with each observation having a subject number and the before and after scores.

Subject	1	2	3	4	5	6	7	8	9	10	11	12
Before	300	350	190	400	244	321	330	250	190	160	260	240
After	290	331	200	395	240	300	332	242	185	158	256	220

*6.7 A researcher wants to randomly assign 30 patients to one of three treatment groups. Each subject has a unique subject number (SUBJ). Write a SAS program to assign these subjects to a treatment group and produce a listing of subjects and groups in subject order (ascending).

*6.8 You want to randomly assign 48 subjects into two groups (Placebo and Drug). Write a DATA step to do this in such a way as to ensure there are exactly 24 subjects in each group. Can you think of how to do this so that in each group of eight subjects, there are exactly four Placebo and four Drug assignments? (Hint: Create six blocks of eight subjects each.)

6.9 What's wrong with this program?

```
1   DATA DRUGSTDY;
2       INPUT SUBJ 1-3 DRUG 4 HEARTRATE 5-7 SBP 8-10
3            DBP 11-13;
4   AVEBP=DBP + (SBP-DBP)/3;
5   DATALINES;
    0011064130080
    0021068120076
    0031070156090
    0042080140080
    0052088180092
    0062098178094
    ;
6   PROC NPAR1WAY DATA=DRUGSTDY WILCOXON MEDIAN;
7       TITLE "MY DRUG STUDY";
8       CLASS DRUG;
9       VAR HEARTRATE SBP DBP AVEBP;
10  RUN;
11  PROC T-TEST DATA=DRUGSTDY;
12      CLASS DRUG;
13      VAR HEARTRATE SBP DBP AVEBP;
14  RUN;
```

CHAPTER 7

Analysis of Variance

A.   INTRODUCTION

When you have more than two groups, a t-test (or the nonparametric equivalent) is no longer applicable. Instead, we use a technique called analysis of variance. This chapter covers analysis-of-variance designs with one or more independent variables, as well as more advanced topics such as interpreting significant interactions, unbalanced designs, and analysis of covariance. You will have to wait until the next chapter to see how to analyze repeated measures designs.

B.   ONE-WAY ANALYSIS OF VARIANCE

We have analyzed experiments with two groups using a t-test. Now, what if we have more than two groups? Take the situation where we have three treatment groups: A, B, and C. It was once the practice to use t-tests with this design, comparing A with B, A with C, and B with C. With four groups (ABCD) we would have comparisons AB, AC, AD, BC, BD, and CD. As the number of groups increases, the number of possible comparisons grows rapidly. What is wrong with this procedure of multiple comparisons? Any time we show the means of two groups to be significantly different, we state a probability that the difference occurred by chance alone. Suppose we make 20 multiple comparisons and each comparison is made at the .05 level. Even if there are no real differences between the groups, it is likely that we will find one comparison significant at the .05 level. (The actual probability of at least one significant difference by chance alone is .64.) The more comparisons that are made, the greater the likelihood of finding a pair of means significantly different by chance alone.

The method used today for comparisons of three or more groups is called analysis of variance (ANOVA). This method has the advantage of testing whether there are any differences between the groups with a single probability associated with the test. The hypothesis tested is that all groups have the same mean. Before we present an example, note that there are several assumptions that should be met before an analysis of variance is used.

Essentially, the same assumptions for a t-test need to be met when conducting an ANOVA. That is, we must have independence between groups (unless a repeated measures design is used); the sampling distributions of sample means must be normally distributed; and the groups should come from populations with equal variances (called homogeneity of variance).

The analysis-of-variance technique is said to be "robust," a term used by statisticians that means that the assumptions can be violated somewhat, but the technique can still be used. So, if the distributions are not perfectly normal or if the variances are unequal, we may still use analysis of variance. The judgment of how serious a violation to permit is subjective, and, if in doubt, see your local statistician. (Winer has an excellent discussion of the effect of homogeneity-of-variance violations and the use of analysis of variance.) Balanced designs (those with the same number of subjects under each of the experimental conditions) are preferred to unbalanced designs, especially when the group variances are unequal.

Consider the following experiment:

We randomly assign 15 subjects to the three treatment groups X, Y, and Z (with five subjects per treatment). Each of the three groups has received a different method of speed-reading instruction. A reading test is given, and the number of words per minute is recorded for each subject. The following data are collected:

X	Y	Z
700	480	500
850	460	550
820	500	480
640	570	600
920	580	610

The null hypothesis is that each of the samples come from populations with equal means. In symbols:

$$H_0: \mu_X = \mu_Y = \mu_Z$$

The alternative hypothesis is that the means are not all equal. There is one or more pairs of means (X versus Y, X versus Z, or Y versus Z) that are significantly different. The means of groups X, Y, and Z are 786, 518, and 548, respectively. How do we know if the means obtained are different because of differences in the reading programs or because of random sampling error? By chance, the five subjects we choose for group X might be faster readers than those chosen for groups Y and Z.

In our example, the mean reading speed of all 15 subjects (called the GRAND MEAN) is 617.33. Now, we normally think of a subject's score as whatever it happens to be—580 is 580. But we could also think of 580 as being 37.33 points lower than the grand mean.

We might now ask the question, "What causes scores to vary from the grand mean?" In this example, there are two possible sources of variation, the first source is the training method (X, Y, or Z). If X is a far superior method, then we would expect subjects in X to have higher scores, in general, than subjects in Y or Z. When we say "higher scores in general," we mean something quite specific. We mean that being a member of group X causes one's score to increase by so many points.

The second source of variation is due to the fact that individuals are different. Therefore, within each group there will be variation. We can think of a formula to represent each person's score:

The person's score = The grand mean + An addition or + An addition or
subtraction   subtraction
from the grand   depending on
mean depend-   the individual's
ing on which   variability.
group the
person is in,

Now that we have the ideas down, let's return briefly to the mathematics.

It turns out that the mathematics are simplified if, instead of looking at differences in scores from the grand mean, we look instead at the square of the differences. The sum of all the squared deviations is called the total SUM OF SQUARES or, SS, total.

To be sure this is clear, we calculate the total SS in our example. Subtracting the grand mean (617.33) from each score, squaring the differences (usually called deviations), and adding up all the results, we have:

$$\text{SS total} = (700 - 617.33)^2 + (850 - 617.33)^2 + \cdots + (610 - 617.33)^2$$

As mentioned earlier, we can separate the total variation into two parts: one due to differences in the reading methods (often called SUM OF SQUARES BETWEEN) and the other due to the normal variations between subjects (often called the SUM OF SQUARES ERROR). Note that the word ERROR here is not the same as "mistake." It simply means that there is variation in the scores that we cannot attribute to a specific variable. Some statisticians call this RESIDUAL instead of ERROR.

Intuitively, we know that if there is no difference between the group means in the populations, then, on average, how much the sample group means differ from each other will be related to how much the observations differ from each other. In fact, the logical argument that this is the case is straightforward, but it would take several pages to explain. For our purposes here, take it somewhat on faith that this is so. If we take the "average" sum of squares due to group differences (MEAN SQUARE between) divided by the "average" sum of squares due to subject differences (MEAN SQUARE error), the result is called an F ratio:

$$F = \frac{MS \text{ between}}{MS \text{ error}} = \frac{SS \text{ between}/(k-1)}{SS \text{ error}/(N-k)}, \qquad k = \text{number of groups}$$

If the variation between the groups is large compared to the variation within the groups, this ratio will be larger than 1. If the null hypothesis is true, the expected value for the two mean squares will be equal, and the F statistic will be equal to 1.00. Just how far away from 1.00 is too far away to be attributable to chance is a function of the number of groups and the number of subjects in each group. SAS analysis of variance procedures will give us the F ratio and the probability of obtaining a value of F this large or larger by chance alone, when the null hypothesis is true.

The following box contains a more detailed explanation of how an F ratio is computed. (If you prefer, you may wish to skip the box for now.)

Consider a one-way ANOVA with three groups (A, B, C) and three subjects within each group. The design is as follows:

	A	B	C	
	50	70	20	
	40	80	15	
	60	90	25	
Means	50	80	20	50 Grand Mean

If we want to estimate the within-group variance (also called ERROR variance), we take the deviation of each score from its group mean, square the result, and add up the squared deviations for each of the groups, then divide the result by the degrees of freedom. (The number of degrees of freedom is $N - k$, where N is the total number of subjects and k is the number of groups.) In the previous example, the within-group variance is equal to:

$$[(50 - 50)^2 + (40 - 50)^2 + (60 - 50)^2 + (70 - 80)^2 + \cdots$$
$$+ (25 - 20)^2]/6 = 450/6 = 75.0$$

This within-group variance estimate (75.0) can be compared to the between-group variance, the latter obtained by taking the squared deviations of each group mean from the grand mean, multiplying each deviation by the number of subjects in a group, and dividing by the degrees of freedom $(k - 1)$. In our example, the between-group variance is:

$$[3(50 - 50)^2 + 3(80 - 50)^2 + 3(20 - 50)^2]/2 = 2700$$

If the null hypothesis is true, the between-group variance estimate will be close to the within-group variance and the ratio

$$F = \frac{\text{Between-group Variance}}{\text{Within-group Variance}}$$

will be close to 1. In our example, F = 270/75 = 36.0 with probability of .0005 of obtaining a ratio this large or larger by chance alone.

We can write the following program (using our reading-speed data):

```
DATA READING;
    INPUT GROUP $ WORDS @@;
DATALINES;
X 700    X 850    X 820    X 640    X 920
Y 480    Y 460    Y 500    Y 570    Y 580
Z 500    Z 550    Z 480    Z 600    Z 610
;
PROC ANOVA DATA=READING;
    TITLE "Analysis of Reading Data";
    CLASS GROUP;
    MODEL WORDS = GROUP;
    MEANS GROUP;
RUN;
```

If you are confused by the trailing @ sign, a good explanation is presented in Chapter 12. This style of arranging the data values is not necessary to run PROC ANOVA; we're just saving space. One observation per line will do fine.

Now, look at the section of the program beginning with the statement "PROC ANOVA." The first thing we want to indicate is which variable(s) is/are going to be the independent variable(s). We do this with a "CLASS" statement. We are using the variable named GROUP as our independent variable. It could have been GENDER or TREAT or any other variable in a SAS data set. It is almost always the case that the independent variable(s) in an ANOVA will have a relatively small number of possible values. Next, we want to specify what our MODEL is for the analysis. Basically, we are saying what the dependent and independent variables are for this analysis. Following the word MODEL is our dependent variable or a list of dependent variables, separated by spaces. When we have more than one dependent variable, a separate analysis of variance will be performed on each of them. For the simplest analysis, a one-way ANOVA, following the dependent variable(s) is an equal sign, followed by the independent variable. Be sure to list any of the independent variables in the MODEL statement in the previous CLASS statement. In the next line, MEANS GROUP will give us the mean value of the dependent variable (WORDS) for each level of GROUP. Output from this program is shown here:

```
Analysis of Reading Data

The ANOVA Procedure

   Class Level Information

Class          Levels    Values

GROUP               3    X Y Z
                                              [Continued]
```

```
Number of Observations Read        15
Number of Observations Used        15


Dependent Variable: WORDS

                                   Sum of
Source                     DF      Squares    Mean Square    F Value

Model                       2   215613.3333   107806.6667     16.78

Error                      12    77080.0000     6423.3333

Corrected Total            14   292693.3333

Source                     Pr > F

Model                      0.0003

Error

Corrected Total

R-Square     Coeff Var     Root MSE     WORDS Mean

0.736653     12.98256      80.14570      617.3333

Source                     DF     Anova SS     Mean Square    F Value

GROUP                       2   215613.3333   107806.6667     16.78

Source                     Pr > F

GROUP                      0.0003

Level of          ------------WORDS------------
GROUP         N         Mean            Std Dev

X             5      786.000000       113.929803
Y             5      518.000000        54.037024
Z             5      548.000000        58.051701
```

The output begins by recapitulating the details of the analysis. This is particularly helpful if you are running a number of ANOVAs at the same time. Pay attention to the levels of each CLASS variable to be sure that there are no data errors, which would result in extraneous levels of one or more CLASS variables. Then we are given the number of cases (observations) in the data set and the name of the dependent variable.

Next comes the substantive part of the analysis. There are usually two sections here: an analysis for the MODEL as a whole and a breakdown according to the contribution of each independent variable. Where the ANOVA has only one independent variable (a "one-way ANOVA"), these two sections are quite similar. Let's look at the top section first. We see the terms "Source," "DF," "Sum of Squares," "Mean Square," "F Value," and "Pr > F =." Source tells us what aspect of the analysis we are considering. We have "Model," "Error," and "Corrected Total" as categories here. Model means all of the independent variables and their interactions added together. Error means the residual variation after the Model variation has been removed. In our one-way ANOVA, we have only GROUP to consider. It has two degrees of freedom (df). The sum of squares is 215613.33. The mean square (SS/df) is 107806.67. The next row contains the same information for the error term. In the third line we see the df and SUM OF SQUARES for the CORRECTED TOTAL (which just means the total sum of squares about the grand mean). To the right, we find the F statistic and the probability of it having occurred by chance. Below this is the R-SQUARE for the Model, the coefficient of variation (CV), and the mean and standard deviation for the dependent variable.

The next section uses the same terms as previously described, only now each independent variable or interaction is listed separately. Since we have only one independent variable, the results look identical to the ones previous. In this example, we would therefore reject the null hypothesis, since our F (with 2 and 12 degrees of freedom) is 16.78 and the p-value is .0003, and conclude that the reading-instruction methods were not all equivalent.

Now that we know the reading methods are different, we want to know what the differences are. Is X better than Y or Z? Are the means of groups Y and Z so close that we cannot consider them different? In general, methods used to find group differences after the null hypothesis has been rejected are called post hoc, or multiple-comparison tests. SAS provides us with a variety of these tests to investigate differences between levels of our independent variable. These include Duncan's multiple-range test, the Student-Newman-Keuls' multiple-range test, least-significant-difference test, Tukey's studentized range test, Scheffe's multiple-comparison procedure, and others. To request a post hoc test, place the SAS option name for the test you want, following a slash (/) on the MEANS statement. The SAS names for the post hoc tests previously listed are DUNCAN, SNK, LSD, TUKEY, and SCHEFFE, respectively. In practice, it is easier to include the request for a multiple-comparison test at the same time we request the analysis of variance. If the analysis of variance is not significant, WE SHOULD NOT LOOK FURTHER AT THE POST HOC TEST RESULTS. (This is our advice. Some statisticians may not agree, especially when certain post hoc tests are used.) Our examples will use the Student-Newman-Keuls (SNK) test for post hoc comparisons. You may use any of the available methods in the same manner. Winer (see Chapter 1) is an excellent reference for analysis of variance and experimental design. A discussion of most of these post hoc tests can be found there.

For our example we have:

```
MEANS GROUP / SNK;
```

Unless we specify otherwise, the differences between groups are evaluated at the .05 level. Alpha levels of .1 or .01 may be specified by following the post hoc option name with ALPHA = .1 or ALPHA = .01. For example, to specify an alpha level of .1 for a Scheffe test, we would have:

```
MEANS GROUP / SCHEFFE ALPHA=.1;
```

Here is the output from the SNK procedure in our example:

```
Analysis of Reading Data

The ANOVA Procedure

Student-Newman-Keuls Test for WORDS

NOTE: This test controls the Type I experimentwise error rate under the
complete null hypothesis but not under partial null hypotheses.

Alpha                          0.05
Error Degrees of Freedom         12
Error Mean Square        6423.333

Number of Means                   2                 3
Critical Range            110.44134         135.22484

Means with the same letter are not significantly different.

S
N
K

G
r
o
u
p
i
n
g          Mean        N     GROUP

A         786.00       5     X

B         548.00       5     Z
B
B         518.00       5     Y
```

This listing uses the following method to show group differences:

On the right are the group identifications. The order is determined by the group means, from highest to lowest. At the far left is a column labeled "SNK Grouping." Any groups that are not significantly different from one another will have the same letter in the Grouping column. In our example, the Y and Z groups both have the letter 'B' in the GROUPING column and are therefore not significantly different. The letter 'B' between GROUP Z and GROUP Y is there for visual effect. It helps us realize that groups Y and Z are not significantly different (at the .05 level). Group X has an A in the grouping column and is therefore significantly different ($p < .05$) from the Y and Z groups.

From this SNK multiple-comparison test, we conclude that

1. Method X is superior to both methods Y and Z.
2. Methods Y and Z are not significantly different.

How would we describe the statistics used and the results of this experiment in a journal article? Although there is no "standard" format, we suggest one approach here. The key is clarity. Here is our suggestion:

Method. We compared three reading methods: (1) Smith's Speed-Reading Course, (2) Standard Method, and (3) Evelyn Tree's Institute. Fifteen subjects were randomly assigned to one of the three methods. At the conclusion of training, a standard reading test (Cody Count the Words Test version 2.1) was administered.

Results. The mean reading speed for the three methods is:

Method	Reading Speed (words per minute)
1. Smith's	786
2. Standard	518
3. Tree's	548

A one-way analysis of variance was performed. The F-value was 16.78 (df = 2, 12, p = .0003). A Student-Newman-Keuls test (p = .05) shows that Smith's method is significantly superior to either Tree's or the Standard method. Tree's and the Standard method are not significantly different from each other at the .05 level.

The results of the SNK multiple-comparison test can readily be described in words when there are only three groups. With four or more groups, especially if the results are complicated, we can use another method. Consider the following results of an SNK test on an experiment with four treatment groups:

Grouping	Mean	N	Group
A	80	10	1
A			
A B	75	10	3
B			
B	72	10	2
C	60	10	4

To begin, notice that the groups are now ordered from the highest mean to the lowest mean, 1, 3, 2, and 4. We see that groups 1 and 3 are not significantly different. (They both have the letter "A" in the Grouping column). Neither are groups 2 and 3. Remember that "not significantly different" does not mean "equal." But 1 and 2 are significantly different!

Finally, group 4 is significantly different from all the other groups ($p < .05$). We can describe these results in a journal article as previously represented, or like this:

	Group			
	1	3	2	4
Mean	<u>80</u>	<u>75</u>	<u>72</u>	<u>60</u>

SNK multiple-comparison example

Any two groups with a common underscore are not significantly different ($p < .05$).

A final possibility is to simply put the findings within the text of the article (e.g., "group 4 is different from 1, 2, and 3; group 1 is different from group 2").

C. COMPUTING CONTRASTS

Now back to our reading example. Suppose you decide, before you perform this experiment, that you want to make some specific comparisons. For example, if method X is a new method and methods Y and Z are more traditional methods, you may decide to compare method X to the mean of method Y and method Z to see if there is a difference between the new and traditional methods. You may also want to compare method Y to method Z to see if there is a difference between the two traditional methods. These comparisons are called contrasts, planned comparisons, or a priori comparisons. Note that the use of CONTRASTS does not give you any protection against a type I error, as do the various multiple-comparison methods described earlier; however, they allow you to make comparisons among groups of treatments.

To specify comparisons using SAS, you need to use PROC GLM (general linear model) instead of PROC ANOVA. PROC GLM is similar to PROC ANOVA and uses many of the same options and statements. However, PROC GLM is a more generalized program and can be used to compute contrasts or to analyze unbalanced designs (see Section G).

Here are the CONTRAST statements for making the planned comparisons previously described:

```
PROC GLM DATA=READING;
   TITLE "Analysis of Reading Data - Planned Comparions";
   CLASS GROUP;
   MODEL WORDS = GROUP;
   CONTRAST 'X VS. Y AND Z' GROUP    -2 1 1;
   CONTRAST 'METHOD Y VS Z' GROUP     0 1 -1;
RUN;
```

For this one-way design, the syntax of a CONTRAST statement is the word CONTRAST, followed by a label for this contrast (placed in quotes), the independent (CLASS) variable name, and a set of k coefficients (where k is the number of levels of the class variable).

The rules are simple: (1) The sum of the coefficients must add to zero. (2) The order of the coefficients must match the alphanumeric order of the levels of the CLASS variable if it is not formatted. If you associated a format with the CLASS variable, the order is determined by the formatted values (you can override this by specifying the PROC GLM option, ORDER=DATA, which will order the levels of the CLASS variable by the data values, not the formatted values). (3) A zero coefficient means that you do not want to include the corresponding level in the comparison. (4) Levels with negative coefficients are compared to levels with positive coefficients.

The first CONTRAST statement in the previous program gives you a comparison of method X against the mean of methods Y and Z. The second CONTRAST statement will only perform a comparison between methods Y and Z.

Here is a portion of the output showing the results of the comparisons requested:

Contrast	DF	Contrast SS	Mean Square	F Value
X VS. Y AND Z	1	213363.3333	213363.3333	33.22

Contrast	Pr > F			
X VS. Y AND Z	<.0001			

Contrast	DF	Contrast SS	Mean Square	F Value
METHOD Y VS Z	1	2250.0000	2250.0000	0.35

Contrast	Pr > F			
METHOD Y VS Z	0.5649			

Notice that method X is shown to be significantly different from methods Y and Z combined, and there is no difference between methods Y and Z at the .05 level.

D.   ANALYSIS OF VARIANCE: TWO INDEPENDENT VARIABLES

Suppose we ran the same experiment for comparing reading methods, but using 15 male and 15 female subjects. In addition to comparing reading-instruction methods, we could compare male versus female reading speeds. Finally, we might want to see if the effects of the reading methods are the same for males and females.

This experimental design is called a two-way analysis of variance. The "two" refers to the fact that we have two independent variables: GROUP and GENDER. We can picture this experiment as follows:

GROUP

	X	Y	Z
Male	700	480	500
	850	460	550
	820	500	480
	640	570	600
	920	580	610
Female	900	590	610
	880	540	660
	899	560	525
	780	570	610
	899	555	645

GENDER

This design shows that we have each of the three reading-instruction methods (specified by the variable GROUP) for each level of GENDER (male/female). Designs of this sort are called factorial designs. The combination of GROUP and GENDER is called a cell. For this example, males in group X constitute a cell. In general, the number of cells in a factorial design would be the number of levels of each of the independent variables multiplied together. In this case, three levels of GROUP times two levels of GENDER equals six cells.

The total sum of squares is now divided or partitioned into four components. We have the sum of squares due to GROUP differences and the sum of squares due to GENDER differences. The combination of GROUP and GENDER provides us with another source of variation (called an interaction), and finally, the remaining sum of squares is attributed to error. We discuss the interaction term later in this chapter.

Since this design has the same number of subjects in each cell, the design is said to be "balanced" (some statisticians call this "orthogonal"). When we have more than one independent variable in our model, we cannot use PROC ANOVA if our design is unbalanced. For unbalanced designs, PROC GLM (general linear model) is used instead. The programming of our balanced design experiment is similar to the one-way analysis of variance. Here is the program:

```
DATA TWOWAY;
   INPUT GROUP $ GENDER $ WORDS @@;
DATALINES;
X M 700   X M 850   X M 820   X M 640   X M 920
Y M 480   Y M 460   Y M 500   Y M 570   Y M 580
Z M 920   Z M 550   Z M 480   Z M 600   Z M 610
X F 900   X F 880   X F 899   X F 780   X F 899
Y F 590   Y F 540   Y F 560   Y F 570   Y F 555
Z F 520   Z F 660   Z F 525   Z F 610   Z F 645
;
```

```
PROC ANOVA DATA=TWOWAY;
   TITLE "Analysis of Reading Data";
   CLASS GROUP GENDER;
   MODEL WORDS = GROUP | GENDER;
   MEANS GROUP | GENDER / SNK;
RUN;
```

As before, following the word CLASS is a list of independent variables. The vertical line between GROUP and GENDER in the MODEL and MEANS statements indicates that we have a factorial design (also called a crossed design). If we don't include the vertical line, none of the interaction terms will be estimated. Some computer terminals may not have the "|" symbol on the keyboard. In this case, the term GROUP | GENDER can be written as:

```
GROUP GENDER GROUP*GENDER
```

The "|" symbol is especially useful when we have higher-order factorial designs such as GROUP | GENDER | DOSE. Written the long way, this would be

```
GROUP GENDER DOSE GROUP*GENDER GROUP*DOSE
GENDER*DOSE GROUP*GENDER*DOSE
```

That is, each variable, and every two- and three-way interaction term has to be specified.

Let's study the output of the previous example carefully to see what conclusions we can draw about our experiment. The first portion of the output is shown here:

```
Analysis of Reading Data

The ANOVA Procedure

   Class Level Information

Class          Levels    Values

GROUP               3    X Y Z

GENDER              2    F M

Number of Observations Read        30
Number of Observations Used        30

Dependent Variable: WORDS

                                 Sum of
Source                   DF       Squares    Mean Square  F Value
Model                     5   531436.1667   106287.2333    23.92
Error                    24   106659.2000     4444.1333
Corrected Total          29   638095.3667

                                                  [Continued]
```

```
Source                    Pr > F

Model                     <.0001

Error

Corrected Total

R-Square      Coeff Var        Root MSE     WORDS Mean

0.832848      10.31264         66.66433      646.4333


Source                    DF       Anova SS     Mean Square   F Value

GROUP                      2     503215.2667    251607.6333    56.62
GENDER                     1      25404.3000     25404.3000     5.72
GROUP*GENDER               2       2816.6000      1408.3000     0.32

Source                    Pr > F

GROUP                     <.0001
GENDER                    0.0250
GROUP*GENDER              0.7314
```

The top portion labeled "Class Level Information" indicates our two independent variables and the levels of each. The analysis-of-variance table shows us the sum of squares and mean square for the entire model and the error. This overall F value (23.92) and the probability $p = .0001$ shows us how well the model (as a whole) explains the variation about the grand mean. This could be very important in certain types of studies where we want to create a general, predictive model. In this case, we are more interested in the detailed sources of variation (GROUP, GENDER, and GROUP*GENDER).

Each source of variation in the table has an F value and has the probability of obtaining by chance a value of F this large or larger. In our example, the GROUP variable is significant at .0001 and GENDER at .0250. Since there are only two levels of GENDER, we do not need the SNK test to claim that males and females are significantly different with respect to reading speed ($p = .025$).

In a two-way analysis of variance, when we look at GROUP effects, we are comparing GROUP levels without regard to GENDER. That is, when the groups are compared, we combine the data from both genders. Conversely, when we compare males to females, we combine data from the three treatment groups.

The term GROUP*GENDER is called an interaction term. If group differences were not the same for males and females, we would have a significant interaction. For example, if males did better with method A compared to method B, while females did better with B compared to A, we would expect a significant interaction. In our example, the interaction between GROUP and GENDER was not significant ($p = .73$). (Our next example shows a case where there is a significant interaction.)

The portion of the output resulting from the "MEANS GROUP | GENDER / SNK" request is shown next:

```
Student-Newman-Keuls Test for WORDS


NOTE: This test controls the Type I experimentwise error rate under
the complete null hypothesis but not under partial null hypotheses.


Alpha                          0.05
Error Degrees of Freedom         24
Error Mean Square        4444.133


Number of Means                  2                  3
Critical Range           61.53152       74.452114


Means with the same letter are not significantly different.


S
N
K

G
r
o
u
p
i
n
g           Mean        N     GROUP


A          828.80      10     X


B          570.00      10     Z
B
B          540.50      10     Y


Student-Newman-Keuls Test for WORDS


NOTE: This test controls the Type I experimentwise error rate under
the complete null hypothesis but not under partial null hypotheses.


Alpha                          0.05
Error Degrees of Freedom         24
Error Mean Square        4444.133


Number of Means                  2
Critical Range           50.240276
```

[Continued]

```
Means with the same letter are not significantly different.

S
N
K

G
r
o
u
p
i
n
g         Mean      N    GENDER

A        675.53     15    F


B        617.33     15    M
```

Level of GROUP	Level of GENDER	N	-----------WORDS----------- Mean	Std Dev
X	F	5	871.600000	51.887378
X	M	5	786.000000	113.929803
Y	F	5	563.000000	18.574176
Y	M	5	518.000000	54.037024
Z	F	5	592.000000	66.011363
Z	M	5	548.000000	58.051701

The first comparison shows group X as being significantly different ($p < .05$) from Y and Z. The second table shows that females have significantly higher reading speeds than males. We already know this because GENDER is a significant main effect ($p = .025$), and there are only two levels of GENDER. Following the two SNK tests are the mean reading speeds (and standard deviations) for each combination of GROUP and GENDER. These values are the means of the six cells in our experimental design.

E. INTERPRETING SIGNIFICANT INTERACTIONS

Now, consider an example that has a significant interaction term. We have two groups of children. One group is considered normal; the other, hyperactive. (Hyperactivity is often referred to as attention-deficit hyperactivity disorder, or ADHD; we simply use the term hyperactive.) Each group of children is randomly divided, with one-half receiving a placebo and the other a drug called Ritalin. A measure of activity is determined for each of the children, and the following data are collected:

	PLACEBO	RITALIN
	50	67
NORMAL	45	60
	55	58
	52	65
	70	51
HYPERACTIVE	72	57
	68	48
	75	55

We name the variables in this study GROUP (NORMAL or HYPER), DRUG (PLACEBO or RITALIN), and ACTIVITY (activity score). Since the design is balanced (same number of subjects per cell), we can use PROC ANOVA. The DATA step and PROC statements are written like this:

```
DATA RITALIN;
   DO GROUP = 'NORMAL','HYPER ';
      DO DRUG = 'PLACEBO','RITALIN';
         DO SUBJ = 1 TO 4;
            INPUT ACTIVITY @;
            OUTPUT;
         END;
      END;
   END;
DATALINES;
50 45 55 52 67 60 58 65 70 72 68 75 51 57 48 55
;
PROC ANOVA DATA=RITALIN;
   TITLE "Activity Study";
   CLASS GROUP DRUG;
   MODEL ACTIVITY=GROUP | DRUG;
   MEANS GROUP | DRUG;
RUN;
```

Before discussing the ANOVA results, let us precede that with a few words about the DATA step. One way to read the data values is a straightforward INPUT DRUG $ GROUP $ ACTIVITY; statement. Instead we use nested DO loops, more as a demonstration of SAS programming rather than to shorten the program. Notice two things: One, you can write DO loops in a SAS DATA step with character values for the loop variable. This is a very useful and powerful feature of the language. Second, when we nest DO loops (place one inside the other), the rule is to complete all the iterations of an inner loop before dropping back to an outer loop. So, in this example, we set the value of GROUP to 'NORMAL', the value of DRUG to 'PLACEBO', and then read in four values of ACTIVITY. If you have trouble understanding this DATA step, you can always resort to the simpler INPUT statement without loops.

This ANOVA design is another example of a two-way analysis-of-variance factorial design. The vertical bar between GROUP and DRUG in the MODEL and MEANS statements indicates that we have a factorial design and GROUP and DRUG are crossed. Notice that we do not need to request a multiple-comparison test since there are only two levels of each independent variable.

A portion of the output is shown here:

```
Source              DF        Anova SS      Mean Square    F Value

GROUP                1     121.0000000     121.0000000       8.00
DRUG                 1      42.2500000      42.2500000       2.79
GROUP*DRUG           1     930.2500000     930.2500000      61.50

Source              Pr > F

GROUP               0.0152
DRUG                0.1205
GROUP*DRUG          <.0001
```

The first thing to notice is that there is a strong GROUP*DRUG interaction term ($p < .0001$). When this occurs, we must be careful about interpreting any of the main effects (GROUP and DRUG in our example). That is, we must first understand the nature of the interactions before we examine main effects.

By looking more closely at the interaction between GROUP and DRUG, we see why the main effects shown in the analysis-of-variance table can be misleading. The best way of explaining a two-way interaction is to take the cell means and plot them. These means can be found in the portion of the output from the MEANS request. The portion of the output containing the cell means is shown here:

```
Level of      Level of              ----------ACTIVITY----------
GROUP         DRUG         N           Mean           Std Dev

HYPER         PLACEBO      4        71.2500000        2.98607881
HYPER         RITALIN      4        52.7500000        4.03112887
NORMAL        PLACEBO      4        50.5000000        4.20317340
NORMAL        RITALIN      4        62.5000000        4.20317340
```

We can use this set of means to plot an interaction graph. We choose one of the independent variables (we choose DRUG) to go on the x-axis and then plot means of the dependent variable at each level of the other independent variable (GROUP). We can either do this by hand or have SAS plot it for us. To have SAS plot the interaction graph, we first have to use PROC MEANS to create a data set containing the cell means. The SAS statements to create a data set of cell means is shown next:

```
PROC MEANS DATA=RITALIN NWAY NOPRINT;
   CLASS GROUP DRUG;
   VAR ACTIVITY;
   OUTPUT OUT=MEANS MEAN=M_HR;
RUN;
```

Notice that we use GROUP and DRUG as CLASS variables and the NWAY option of PROC MEANS since this will restrict the output data set to the highest order interaction (the cell means). Next, we use PROC GPLOT (or PROC PLOT if you don't have SAS Graph™ installed) to plot the interaction graph. We can choose to place either of the independent variables on the x-axis and plot a separate graph for each level of the other independent variable. We choose DRUG to be the x-axis variable and plot a separate graph for each level of GROUP. We will use values of GROUP as the plotting symbols. Here is the program:

```
SYMBOL1 V=SQUARE COLOR=BLACK I=JOIN;
SYMBOL2 V=CIRCLE COLOR=BLACK I=JOIN;
PROC GPLOT DATA=MEANS;
    TITLE "Interaction Plot";
    PLOT M_HR * DRUG = GROUP;
RUN;
```

If you choose to use PROC PLOT instead of PROC GPLOT, you will omit the SYMBOL statements, and the resulting graph will use the first character of GENDER as the plotting symbols. Using PROC GPLOT, you supply two SYMBOL statements to specify which plotting symbols to use for each value of GENDER and an additional instruction to connect the points with straight lines. In these SYMBOL statements, we used the abbreviations for VALUE (V) and INTERPOLATE (I).

The resulting graph is shown here:

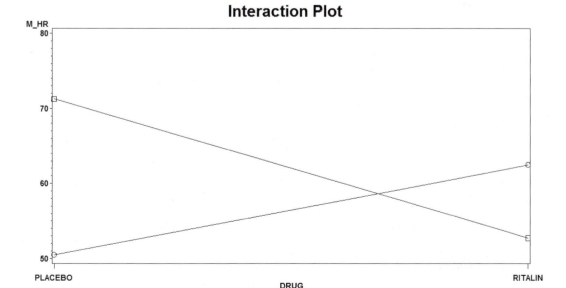

The graph shows that normal children increase their activity when given Ritalin, while hyperactive children are calmed by Ritalin. In the analysis of variance, the comparison of placebo to Ritalin values is done by combining the data from normal and hyperactive children. Since these values tend to cancel each other, the average activity with placebo and Ritalin is about the same. What we have found here is that it is not possible to understand the activity level of children just by knowing whether they had Ritalin. One must also know whether they are hyperactive. This is why it is critical to understand the interaction before looking at main effects. If we really want to study the effect of Ritalin, we should look separately at normal and hyperactive children. For each of these groups, we have two levels of the DRUG. We can therefore do a t-test between placebo and Ritalin within the normal and hyperactive groups. As we know from Chapter 6, this is accomplished by first sorting the data set by GROUP and then including a BY variable in the t-test request. We have:

```
PROC SORT DATA=RITALIN;
   BY GROUP;
RUN;
PROC TTEST DATA=RITALIN;
   TITLE "Drug Comparisons for Each Group Separately";
   BY GROUP;
   CLASS DRUG;
   VAR ACTIVITY;
RUN;
```

Portions of the output from these statements is shown here:

```
Drug Comparisons for Each Group Separately

GROUP=HYPER

The TTEST Procedure

                            Statistics

                            Lower CL           Upper CL  Lower CL
Variable  DRUG         N       Mean    Mean       Mean   Std Dev

ACTIVITY  PLACEBO      4     66.498   71.25     76.002    1.6916
ACTIVITY  RITALIN      4     46.336   52.75     59.164    2.2836
ACTIVITY  Diff (1-2)         12.362   18.5      24.638    2.2859

                            T-Tests

Variable  Method        Variances     DF   t Value   Pr > |t|

ACTIVITY  Pooled        Equal          6      7.38     0.0003
ACTIVITY  Satterthwaite Unequal      5.53     7.38     0.0005
```

```
GROUP=NORMAL

The TTEST Procedure

                          Statistics

                          Lower CL           Upper CL  Lower CL
  Variable  DRUG          N    Mean   Mean       Mean   Std Dev

  ACTIVITY  PLACEBO       4  43.812   50.5     57.188    2.3811
  ACTIVITY  RITALIN       4  55.812   62.5     69.188    2.3811
  ACTIVITY  Diff (1-2)       -19.27    -12     -4.728    2.7085

                          T-Tests

  Variable  Method        Variances     DF   t Value   Pr > |t|

  ACTIVITY  Pooled        Equal          6     -4.04     0.0068
  ACTIVITY  Satterthwaite Unequal        6     -4.04     0.0068
```

Notice that in both groups the two drug means are significantly different ($p < .05$). However, in the normal group, the Ritalin mean is higher than the placebo mean, while in the hyperactive group, the reverse is true. So, watch out for those interactions!

An alternative to the previous t-tests is to break down the two-way ANOVA into a one-way ANOVA by creating an independent (CLASS) variable that has a level for each combination of the original independent variables. In our case, we create a variable (let's call it CONDITION) that has a level for each combination of DRUG and GROUP. Thus, we have NORMAL-PLACEBO, NORMAL-RITALIN, HYPER-PLACEBO, and HYPER-RITALIN as levels of our CONDITION variable. A quick and easy way to create this variable is to concatenate (join) the two original independent variables. In the SAS system, concatenation is accomplished with the concatenation operator '||' (two exclamation marks '!!' are equivalent). To create the CONDITION variable, we add a single line to our data step like this:

```
CONDITION = GROUP || DRUG;
```

This line creates a new variable (CONDITION), which has four values (HYPER PLACEBO, HYPER RITALIN, NORMAL PLACEBO, and NORMAL RITALIN). We can remove the extra spaces between the words if we wish, using the TRIM function described in Chapter 18 or even place a dash between the GROUP and DRUG values like this:

```
CONDITION = TRIM(GROUP) || '-' || DRUG;
```

Or, if you are using SAS version 9.1 or higher, you can use the CATX function, which concatenates two or more strings, trims each of the strings, and allows you to select a separator (the first argument of the function) to be placed between each of the strings to be joined. It would look like this:

```
CONDITION = CATX('-',GROUP,DRUG); /* Version 9 and higher */
```

We now have a one-way design with the single factor (CONDITION) having four levels. The SAS statements to produce the one-way ANOVA are:

```
PROC ANOVA DATA=RITALIN;
    TITLE "One-way ANOVA Ritalin Study";
    CLASS CONDITION;
    MODEL ACTIVITY = CONDITION;
    MEANS CONDITION / SNK;
RUN;
```

Portions of the results of running this procedure are:

```
One-way ANOVA Ritalin Study

The ANOVA Procedure

                    Class Level Information

Class        Levels  Values

CONDITION         4  HYPER-PLACEBO HYPER-RITALIN NORMAL-PLACEBO
                     NORMAL-RITALIN

Dependent Variable: ACTIVITY

Source               DF      Anova SS   Mean Square  F Value

CONDITION             3  1093.500000    364.500000    24.10

Source              Pr > F

CONDITION           <.0001

Student-Newman-Keuls Test for ACTIVITY

NOTE: This test controls the Type I experimentwise error rate under
the complete null hypothesis but not under partial null hypotheses.

Means with the same letter are not significantly different.

S
N
K

G
r
o
u
p
i
n
g         Mean     N    CONDITION

A       71.250     4    HYPER-PLACEBO

B       62.500     4    NORMAL-RITALIN

C       52.750     4    HYPER-RITALIN
C
C       50.500     4    NORMAL-PLACEBO
```

Notice that this analysis tells us more than the two t-tests. Besides verifying that PLACEBO is different from RITALIN within each GROUP (NORMAL and HYPER), we can also see the other pairwise comparisons. Finally, this analysis uses all the data to estimate the error term.

There is another way of comparing the two drugs within each level of GROUP without using either of the two methods just described. If we run PROC GLM instead of PROC ANOVA, we can issue two CONTRAST statements that will make the two within-group comparisons for us. This method is considered more correct statistically than the two t-tests by some statisticians since it uses all the data to estimate the error variance. It is, however, equivalent to the previous one-way analysis. We present it here without too much explanation. It is difficult, and you will need to seek help beyond this book. First, the program:

```
PROC GLM DATA=RITALIN;
   TITLE "Demonstrating the CONTRAST Statement of GLM";
   CLASS GROUP DRUG;
   MODEL ACTIVITY = GROUP | DRUG / SS3;
   CONTRAST 'Hyperactive only' DRUG 1 -1
                           GROUP*DRUG 1 -1 0 0;
   CONTRAST 'Normals only'    DRUG 1 -1
                           GROUP*DRUG 0 0 1 -1;
RUN;
```

The first contrast compares PLACEBO to RITALIN only for the hyperactive children. The second contrast only does this for the normal children. The "real" methodology for writing these CONTRAST statements is quite complicated. We present a simple algorithm just for this simple two-way design. The order of the variables in the CLASS statement is very important. You must also recognize that the order of the levels of each CLASS variable will be determined by the formatted values (if a format is provided) or by the actual raw data values (if a format is not supplied). In this example, we have:

GROUP (HYPER - NORMAL) and DRUG (PLACEBO - RITALIN)

Note that HYPER is listed before NORMAL because, alphabetically, 'H' comes before 'N'. For each level of GROUP, we list all the levels of DRUG like this:

HYPER-PLACEBO HYPER-RITALIN NORMAL-PLACEBO NORMAL-RITALIN

We pick the first level of GROUP and then list all the levels of DRUG before going to the next value of GROUP. Now, since we want to compare drugs, we first list DRUG in the CONTRAST statement as 1 −1. Next, for the interaction term GROUP*DRUG, we place our 1 and −1 in the locations that we want to compare. Thus, to compare HYPER-PLACEBO to HYPER-RITALIN, we code 1 −1 0 0. To compare NORMAL-PLACEBO to NORMAL-RITALIN, we code 0 0 1 −1. If this doesn't make any sense, use one of the two methods we presented earlier or consult your friendly statistician. Here is the result of running these two contrasts:

```
The GLM Procedure

Dependent Variable: ACTIVITY

Contrast                 DF     Contrast SS    Mean Square    F Value

Hyperactive only          1     684.5000000    684.5000000      45.26
Normals only              1     288.0000000    288.0000000      19.04

Contrast                 Pr > F

Hyperactive only         <.0001
Normals only             0.0009
```

We could obtain the identical results if we used a CONTRAST statement in the one-way model just described, where we created a variable (CONDITION) that had four levels representing all combinations of GROUP and DRUG. We would use PROC GLM instead of PROC ANOVA and add two CONTRAST statements following the MODEL statement, thus:

```
PROC GLM DATA=RITALIN;
    TITLE "One-way ANOVA Ritalin Study";
    CLASS CONDITION;
    MODEL ACTIVITY = CONDITION;
    CONTRAST 'Hyperactive only'  CONDITION   1  -1   0   0;
    CONTRAST 'Normals only'      CONDITION   0   0   1  -1;
RUN;
```

This may be simpler and easier to understand than the CONTRAST statements for two-way designs. Take your choice.

F.   N-WAY FACTORIAL DESIGNS

The method we used to perform a two-way analysis of variance can be extended to cover any number of independent variables. An example with three independent variables (GROUP GENDER DOSE) is shown here (we didn't actually create a data set called THREEWAY):

```
PROC ANOVA DATA=THREEWAY;
   TITLE "Three-way Analysis of Variance";
   CLASS GROUP GENDER DOSE;
   MODEL ACTIVITY = GROUP | GENDER | DOSE;
   MEANS GROUP | GENDER | DOSE;
RUN;
```

With three independent variables, we have three main effects (GROUP GENDER DOSE), three two-way interactions (GROUP*GENDERGROUP*DOSE GENDER *DOSE), and one three-way interaction (GROUP*GENDER*DOSE). One usually hopes that the higher-order interactions are not significant since they complicate the interpretation of the main effects and the lower-order interactions. (See Winer for a more complete discussion of this topic.)

It clearly becomes difficult to perform factorial design experiments with a large number of independent variables without expert assistance. The number of subjects in the experiment also must be large so that there are a reasonable number of subjects per cell.

G.   UNBALANCED DESIGNS: PROC GLM

As we mentioned before, designs with an unequal number of subjects per cell are called unbalanced designs. (There is no particular name given to experiments with equal cell sizes conducted by unbalanced researchers, however.) For all designs that are unbalanced (except for one-way designs), we should not use PROC ANOVA; PROC GLM is used instead. CLASS, MEANS, and MODEL statements for PROC GLM are identical to those used with PROC ANOVA. The only difference between the procedures is the mathematical methods used for each and some additional information that is computed when PROC GLM is used. The real differences come in interpreting results.

Here is an example of a two-way analysis of variance that is unbalanced:

A pudding company wants to test-market a new product. Three levels of sweetness and two flavors are produced. Each subject is given a pudding to taste and is asked to rate the taste on a scale from 1 to 10. The following data are collected:

	Sweetness Level		
	1	**2**	**3**
Vanilla	9	8	6
	7	7	5
	8	8	7
	7		
Chocolate	9	8	4
	9	7	5
	7	6	6
	7	8	4
	8		4

The SAS DATA step to create this data set is:

```
DATA PUDDING;
    LENGTH FLAVOR $ 9;
    INPUT FLAVOR $ SWEET RATING @ @;
DATALINES;
VANILLA 1 9   VANILLA 2 8   VANILLA 3 6
VANILLA 1 7   VANILLA 2 7   VANILLA 3 5
VANILLA 1 8   VANILLA 2 8   VANILLA 3 7
VANILLA 1 7
CHOCOLATE 1 9   CHOCOLATE 2 8   CHOCOLATE 3 4
CHOCOLATE 1 9   CHOCOLATE 2 7   CHOCOLATE 3 5
CHOCOLATE 1 7   CHOCOLATE 2 6   CHOCOLATE 3 6
CHOCOLATE 1 7   CHOCOLATE 2 8   CHOCOLATE 3 4
CHOCOLATE 1 8                   CHOCOLATE 3 4
;
```

Since the number of subjects in each cell is unequal, we use PROC GLM.

```
PROC GLM DATA=PUDDING;
    TITLE   "Pudding Taste Evaluation";
    TITLE3 "Two-way ANOVA - Unbalanced Design";
    TITLE4 "--------------------------------";
    CLASS SWEET FLAVOR;
    MODEL RATING = SWEET | FLAVOR / SS3;
    LSMEANS SWEET | FLAVOR / PDIFF ADJUST=TUKEY;
RUN;
```

Notice a new statement, LSMEANS. With unbalanced designs, the MEANS statement will give us unadjusted means; LSMEANS will produce least-square, adjusted means for main effects. We added the PDIFF option, which computes probabilities for all pairwise differences. These pairwise comparisons are essentially a series of t-tests and have all the problems of running multiple t-tests unless we add the ADJUST=TUKEY option, which provides an adjustment for multiple comparisons.

Portions of the output are shown here:

```
Pudding Taste Evaluation

Two-way ANOVA - Unbalanced Design
---------------------------------

The GLM Procedure

Dependent Variable: RATING
```

Source	DF	Sum of Squares	Mean Square	F Value
Model	5	39.96666667	7.99333333	9.36
Error	18	15.36666667	0.85370370	
Corrected Total	23	55.33333333		

Source	Pr > F
Model	0.0002

Source	DF	Type III SS	Mean Square	F Value
SWEET	2	29.77706840	14.88853420	17.44
FLAVOR	1	1.56666667	1.56666667	1.84
SWEET*FLAVOR	2	2.77809264	1.38904632	1.63

Source	Pr > F
SWEET	<.0001
FLAVOR	0.1923
SWEET*FLAVOR	0.2241

The GLM Procedure
Least Squares Means
Adjustment for Multiple Comparisons: Tukey-Kramer

SWEET	RATING LSMEAN	LSMEAN Number
1	7.87500000	1
2	7.45833333	2
3	5.30000000	3

Least Squares Means for effect SWEET
Pr > |t| for H0: LSMean(i)=LSMean(j)

Dependent Variable: RATING

i/j	1	2	3
1		0.6550	<.0001
2	0.6550		0.0009
3	<.0001	0.0009	

The GLM Procedure
Least Squares Means
Adjustment for Multiple Comparisons: Tukey-Kramer

FLAVOR	RATING LSMEAN	H0:LSMean1= LSMean2 Pr > |t|
CHOCOLATE	6.61666667	0.1923
VANILLA	7.13888889	

[Continued]

225

```
The GLM Procedure
Least Squares Means
Adjustment for Multiple Comparisons: Tukey-Kramer


                           RATING         LSMEAN
    SWEET    FLAVOR         LSMEAN         Number

    1        CHOCOLATE     8.00000000         1
    1        VANILLA       7.75000000         2
    2        CHOCOLATE     7.25000000         3
    2        VANILLA       7.66666667         4
    3        CHOCOLATE     4.60000000         5
    3        VANILLA       6.00000000         6


           Least Squares Means for effect SWEET*FLAVOR
               Pr > |t| for H0: LSMean(i)=LSMean(j)


                   Dependent Variable: RATING

    i/j        1         2         3         4         5         6
      1               0.9984    0.8265    0.9958    0.0002    0.0758
      2     0.9984              0.9700    1.0000    0.0009    0.1819
      3     0.8265    0.9700              0.9904    0.0052    0.5065
      4     0.9958    1.0000    0.9904              0.0029    0.2810
      5     0.0002    0.0009    0.0052    0.0029              0.3424
      6     0.0758    0.1819    0.5065    0.2810    0.3424
```

Notice that we have chosen the TYPE III sum of squares (the SS3 option on the MODEL statement). Without this option, GLM would produce both TYPE I and TYPE III sums of squares. When designs do not have equal cell sizes, the TYPE I and TYPE III sums of squares may not be equal for all variables. The difference between TYPE I and TYPE III sum of squares is that TYPE I lists the sums of squares for each variable as if it were entered one at a time into the model, in the order they are specified in the MODEL statement. Hence they can be thought of as incremental sums of squares. If there is any variance that is common to two or more variables, the variance will be attributed to the variable that is entered first. This may or may not be desirable. The TYPE III sum of squares gives the sum of squares that would be obtained for each variable if it were entered last into the model. That is, the effect of each variable is evaluated after all other factors have been accounted for. In any given situation, whether you want to look at TYPE I or TYPE III, sum of squares will vary; however, for most analysis-of-variance applications, you will want to use TYPE III sum of squares.

We see that only sweetness was significant in this model (p < .0001) and the sweetness/flavor interaction was not significant. As you will see in the comparison of cell means a bit farther on in the listing, the sweetness difference was predominately in the dislike for sweet chocolate. The probabilities for FLAVOR and the interaction between FLAVOR and SWEETNESS were .1923 and .2241, respectively.

Just to keep you on your toes, we have added to the program a new form of the TITLE statement. As you probably can guess, TITLE3 provides a third title line; TITLE4, a fourth. Since TITLE2 is missing, line 2 will be blank. In general, TITLEn will be the nth title line on the SAS output, where n is an integer. Note that TITLE is equivalent to TITLE1.

The LSMEANS with the PDIFF option shows us the adjusted means (AD-JUST=TUKEY) for SWEET as well as a 3-by-3 matrix showing us all the pairwise multiple comparisons. You can see here that sweetness levels 1 versus 3 and 2 versus 3 are both significant. Looking farther down on the listing, you can see that the two flavors were not significantly different. Notice that the p-value here (.1923) is the same p-value for the main effect of FLAVOR in the ANOVA table near the top of the listing. Finally, LSMEANS provides the adjusted cell means (every combination of SWEET and FLAVOR) and another matrix comparing each of the cells in a pairwise fashion. You have to use the column labeled LSMEAN Number (1 to 6) to know which two cells are being compared in the 6-by-6 matrix just below these cell means. Here we see that 1 versus 5 (SWEET = 1, FLAVOR = CHOCOLATE versus SWEET = 3, FLAVOR = CHOCOLATE), 2 versus 5 (SWEET = 1, FLAVOR = VANILLA versus SWEET = 3, FLAVOR = CHOCOLATE), 3 versus 5 (SWEET = 2, FLAVOR = CHOCOLATE versus SWEET = 3, FLAVOR = CHOCOLATE), and 4 versus 5 (SWEET = 2, FLAVOR = VANILLA versus SWEET = 3, FLAVOR = CHOCOLATE) are all significant at $p < .05$ (they really didn't like sweet chocolate).

H.   ANALYSIS OF COVARIANCE

If you have a variable, IQ for example, that may be correlated with your dependent measure, you may want to adjust for possible differences due to the confounding variable before analyzing your dependent variable. Consider the following data set:

Group			
A		**B**	
Math Score	**IQ**	**Math Score**	**IQ**
260	105	325	126
325	115	440	135
300	122	425	142
400	125	500	140
390	138	600	160

We want to compare groups A and B on math scores. However, we notice that there seems to be a relationship between math scores and IQ and that group B seems to have higher IQ scores than group A. We can test the relationship between math score and IQ by computing a correlation coefficient, and we can see if there is a significant difference in IQ scores between groups A and B with a t-test. Here is a program that does just that:

```
DATA COVAR;
    LENGTH GROUP $ 1;
    INPUT GROUP MATH IQ @@;
DATALINES;
A  260  105   A  325  115   A  300  122   A  400  125   A  390  138
B  325  126   B  440  135   B  425  142   B  500  140   B  600  160
;
PROC CORR DATA=COVAR NOSIMPLE;
    TITLE "Covariate Example";
    VAR MATH IQ;
RUN;
PROC TTEST DATA=COVAR;
    CLASS GROUP;
    VAR IQ MATH;
RUN;
```

Notice that we requested a t-test for math scores as well (while we were at it). Here are the results (edited):

```
Covariate Example

The CORR Procedure

  2  Variables:    MATH     IQ

Pearson Correlation Coefficients, N = 10
        Prob > |r| under H0: Rho=0

                MATH              IQ

MATH        1.00000          0.92456
                             0.0001

IQ          0.92456          1.00000
            0.0001
```

Covariate Example

The TTEST Procedure

Statistics

Variable	GROUP	N	Lower CL Mean	Mean	Upper CL Mean	Lower CL Std Dev
IQ	A	5	105.82	121	136.18	7.3256
IQ	B	5	125.1	140.6	156.1	7.4784
IQ	Diff (1-2)		−37.62	−19.6	−1.581	8.3454
MATH	A	5	261.02	335	408.98	35.697
MATH	B	5	332.25	458	583.75	60.68
MATH	Diff (1-2)		−244.2	−123	−1.82	56.123

Variable	Method	Variances	DF	t Value	Pr > |t|
IQ	Pooled	Equal	8	−2.51	0.0365
IQ	Satterthwaite	Unequal	8	−2.51	0.0365
MATH	Pooled	Equal	8	−2.34	0.0474
MATH	Satterthwaite	Unequal	6.47	−2.34	0.0547

We see that IQ and math scores are highly correlated (r = .92456, p = .0001) and that there is a significant difference in IQ (p = .0365) and math scores (p = .0474) between groups. We want to correct for the IQ mismatch by running an analysis of covariance.

The first step is to test an important assumption that must be verified before running an analysis of covariance—homogeneity of slope for the covariate versus the dependent variable. For this example, the relationship between the covariate (IQ) and the dependent variable (MATH) must be the same in GROUP A and GROUP B. This can be checked with simple regression, modeling MATH as a function of IQ at the two different levels of GROUP and then testing the interaction of IQ and GROUP. The modeling and comparison of regression coefficients can all be done using PROC GLM and the following MODEL statement:

```
PROC GLM DATA=COVAR;
   CLASS GROUP;
   MODEL MATH = IQ GROUP IQ*GROUP / SS3;
RUN;
```

The term IQ*GROUP will test if there are different regression coefficients for the two groups. Running this, we get (partial listing):

```
Covariate Example

Source                    DF      Type III SS      Mean Square    F Value

IQ                         1      41278.21495      41278.21495      25.81
GROUP                      1       3634.41141       3634.41141       2.27
IQ*GROUP                   1       3816.96372       3816.96372       2.39

Source                  Pr > F

IQ                      0.0023
GROUP                   0.1824
IQ*GROUP                0.1734
```

We see from this that there is no significant difference in the MATH/IQ relationship as a function of GROUP (from the IQ*GROUP interaction term: $F = 2.39$, $p = .1734$). We can go ahead and run the analysis of covariance as follows:

```
PROC GLM DATA=COVAR;
    CLASS GROUP;
    MODEL MATH = IQ GROUP / SS3;
    LSMEANS GROUP;
RUN;
```

The results (following) show that when we adjust for IQ, there are no longer any differences between the groups on math scores. Notice the LSMEANS output shows the math scores for the two groups adjusted for IQ:

```
Covariate Example

The GLM Procedure

Dependent Variable: MATH

Source                    DF      Type III SS      Mean Square    F Value

IQ                         1      41815.68621      41815.68621      21.82
GROUP                      1         96.59793         96.59793       0.05
```

```
Source                  Pr > F

IQ                      0.0023
GROUP                   0.8288

Covariate Example

The GLM Procedure
Least Squares Means

GROUP       MATH LSMEAN

A           392.345889
B           400.654111
```

PROBLEMS

Remember, you can download all the data sets and programs for these problems from the web site: www.prenhall.com/cody

7.1 Problems 7.1 and 7.3 were inspired by one of the authors (Cody) watching the French Open Tennis tournament while working on problem sets. (McEnroe versus Lendl (1984). Lendl won in five sets.)

Three brands of tennis shoes are tested to see how many months of playing would wear out the soles. Eight pairs of brands A, N, and T are randomly assigned to a group of 24 volunteers. The following table shows the results of the study:

	Brand A	Brand N	Brand T
	8	4	12
	10	7	8
Wear time,	9	5	10
in months	11	5	10
	10	6	11
	10	7	9
	8	6	9
	12	4	12

Are the brands equal in wear quality? Write a SAS program to solve this problem, using ANOVA. Include the statements to perform an SNK multiple comparison.

7.2 Two cholesterol-lowering medications (statins) and a placebo were given to each of 10 volunteers with total cholesterol readings of 240 or higher. After 6 weeks, the following total cholesterol values were recorded:

Statin A: 220 190 180 185 210 170 178 200 177 189
Statin B: 160 168 178 200 172 155 159 167 185 199
Placebo: 240 220 246 244 198 238 277 255 190 188

Create a SAS data set by reading these data. As an interesting exercise, try reading three lines like this:

220 190 180 185 210 170 178 200 177 189
160 168 178 200 172 155 159 167 185 199
240 220 246 244 198 238 277 255 190 188

That is, have the program create the treatment variable (call it TREAT) with values of 'A', 'B', and 'Placebo'. Next run a oneway ANOVA followed by a Student-Newman-Keuls multiple-comparison test.

7.3 Tennis balls are tested in a machine to see how many bounces they can withstand before they fail to bounce 30% of their dropping height. Two brands of balls (W and P) are compared. In addition, the effect of shelf life on these brands is tested. Half of the balls of each brand are 6 months old, the other half, fresh. Using a two-way analysis of variance, what conclusions can you reach? The data are shown here:

		Brand W	Brand P
	New	67	75
		72	76
		74	80
		82	72
		81	73
Age			
	Old	46	63
		44	62
		45	66
		51	62
		43	60

7.4 Four different methods of preparing for a college entrance exam were compared. They are labeled 'A', 'B', 'C', and 'D'. The following exam scores were obtained for each of the four programs:

Program

A	560	520	530	525	575	527	580	620
B	565	522	520	530	510	522	600	590
C	512	518	555	502	510	520	516	
D	505	508	512	520	543	523	517	

Compare the four preparation methods. Use a multiple-comparison method of your choice to make pairwise comparisons. Create two contrasts: One to compare methods A and B to C and D, the other to compare method D to the other three.

7.5 A taste test is conducted to rate consumer preference between brands C and P of a popular soft drink. In addition, the age of the respondent is recorded (1 = less than 20, 2 = 20 or more). Preference data (on a scale of 1–10) is as follows:

		Brand C	Brand P
	<20	7	9
		6	8
		6	9
		5	9
		6	8
Age			
	>=20	9	6
		8	7
		8	6
		9	6
		7	5
		8	
		8	

(a) Write a SAS program to analyze these data with a two-way analysis of variance. (Careful: Is the design balanced?) Note: Go ahead and treat these data as interval data even though some analysts would prefer that you use a nonparametric analysis.

(b) Use SAS to plot an interaction graph.

(c) Follow up with a t-test comparing brand C to brand P for each age group separately.

7.6 Two groups of patients, those with a genetic deficiency and the other a "normal" group, are given either a drug to cure depression or a placebo. A standardized depression survey is given to all the subjects with the following results:

	Anti-Depression Drug	Placebo
Genetic Deficiency	9	9
	11	6
	10	6
	10	7
Normal	5	12
	4	11
	7	10
	7	11

Perform a two-way analysis of variance to test for drug and genetic differences.

7.7 A manufacturer wants to reanalyze the data of problem 7.1, omitting all data for brand N. Run the appropriate analysis.

7.8 Run the following program to create the data set CO_VARY. Compute the correlation between IQ and SCORE and conduct a t-test comparing the two groups on IQ and SCORE. Next, test if the slope of IQ versus SCORE is homogeneous across groups. Finally, conduct an analysis of covariance, testing if the mean SCORE is different by GROUP after you have adjusted for IQ.

```
DATA CO_VARY;
   DO I = 1 TO 20;
      DO GROUP = 'A','B';
         SUBJ + 1;
         IQ = INT(RANNOR(124)*10 + 120 + 15*(GROUP EQ 'A'));
         SCORE = INT(.7*IQ + RANNOR(0)*10 + 100 + 10*(GROUP EQ 'B'));
         OUTPUT;
      END;
   END;
   DROP I;
RUN;
PROC PRINT DATA=CO_VARY NOOBS;
   TITLE "Listing of CO_VARY";
RUN;
```

7.9 What's wrong with this program?

```
1    DATA TREE;
2       INPUT TYPE $ LOCATION $ HEIGHT;
3    DATALINES;
     PINE NORTH 35
     PINE NORTH 37
     PINE NORTH 41
     PINE NORTH 41
     MAPLE NORTH 44
     MAPLE NORTH 41
     PINE SOUTH 53
     PINE SOUTH 55
     MAPLE SOUTH 28
     MAPLE SOUTH 33
     MAPLE SOUTH 32
     MAPLE SOUTH 22
     ;
4    PROC ANOVA DATA=TREE;
5       CLASS TYPE LOCATION;
6       MODEL HEIGHT = TYPE | LOCATION;
7       MEANS TYPE LOCATION TYPE*LOCATION;
8    RUN;
```

7.10 You are given data on the speed with which rats can negotiate a maze. The rats are grouped into three age groups and two genetic strains. Conduct a two-way ANOVA with AGE and STRAIN as the independent variables and SPEED as the dependent variable. The collected data are shown here:

```
Rat Maze Data

  Age    Strain  Speed      Age    Strain  Speed      Age    Strain Speed
  ─────────────────────      ─────────────────────      ─────────────────────
  3 Mo.    A      12         3 Mo.    B      18         9 Mo.    A      14
  3 Mo.    A      14         6 Mo.    A      22         9 Mo.    A      14
  3 Mo.    A       9         6 Mo.    A      20         9 Mo.    A      10
  3 Mo.    A      17         6 Mo.    A      12         9 Mo.    A      15
  3 Mo.    A      10         6 Mo.    A      12         9 Mo.    A      17
  3 Mo.    A      11         6 Mo.    A      17         9 Mo.    A      12
  3 Mo.    A       9         6 Mo.    A      14         9 Mo.    A      19
  3 Mo.    A      10         6 Mo.    A      17         9 Mo.    B      27
  3 Mo.    B      24         6 Mo.    B      23         9 Mo.    B      29
  3 Mo.    B      17         6 Mo.    B      26         9 Mo.    B      27
  3 Mo.    B      22         6 Mo.    B      34         9 Mo.    B      23
  3 Mo.    B      16         6 Mo.    B      20
```

*7.11 You want to determine if the mean score on a standardized math test is different among three groups of schoolchildren ranging in age from 12 to 18. Although the test covers only simple math topics easily understood by a 12-year-old, you want to perform the analysis with and without an adjustment based on age.

(a) Using the following sample data, first perform a one-way ANOVA comparing the math scores and ages among the three groups.

(b) Test if the relationship between age and math score is homogeneous among the three groups.

(c) If the test in part B permits, perform an analysis of covariance, comparing the math scores among the three groups, adjusted for age.

Math Scores and Ages for Groups A, B, and C

Group A		Group B		Group C	
Math Score	**Age**	**Math Score**	**Age**	**Math Score**	**Age**
90	16	92	18	97	18
88	15	88	13	92	17
72	12	76	12	88	16
82	14	78	14	92	17
65	12	90	17	99	17
74	13	68	12	82	14

Repeated Measures Designs

A. Introduction
B. One-factor Experiments
C. Using the REPEATED Statement of PROC ANOVA
D. Using PROC MIXED to Compute a Mixed (Random Effects) Model
E. Two-factor Experiments with a Repeated Measure on One Factor
F. Two-factor Experiments with Repeated Measures on Both Factors
G. Three-factor Experiments with a Repeated Measure on the Last Factor
H. Three-factor Experiments with Repeated Measures on Two Factors

A.   INTRODUCTION

This chapter covers the analysis of repeated measures designs. First, a few words about terminology before we begin this topic. Our use of the term "repeated" is based on a common use in the medical field and described in the text *Statistical Principles in Experimental Design*, 2d edition, by B. J. Winer (1991). We use the term "repeated" in this chapter to mean any factor where each subject is measured at every level for that factor. For example, if a runner is timed running on two different types of track surfaces, we are considering "surface" as a repeated measure factor. Other authors use the term "repeated" to refer only to factors that cannot be assigned in random order, such as time. When treatments are randomized, the interpretation of a significant effect can be attributed to treatments and not to the order of presentation. This is often referred to as a split-plot or within-subjects design. If you use the latter meaning of "repeated," feel free to substitute your design terminology in the examples in this chapter. These designs often fall into the category that statisticians call mixed designs, or designs with within-subjects factors.

Designs in this chapter involve a repeated measurement on the unit of analysis (usually subjects) in one or more of the independent variables. For example, an experiment where each subject receives each of four drugs or an experiment where each subject is measured each hour for 5 hours would need a repeated measures design.

With the introduction of version 6 of SAS software, a REPEATED statement was added to the analysis-of-variance procedures (ANOVA and GLM), which greatly simplified the coding of repeated measures designs. As you will see, there are times when you will want to analyze your data using the REPEATED statement, and there will be times when you will choose not to. For each of the repeated measures designs in this chapter, we demonstrate both methods of analysis.

Factors such as TIME or SUBJECT are usually considered as random factors (as compared to fixed factors such as treatment or gender). Although the programs in this chapter will produce the correct F values (and the associated p-values), you should consider using PROC MIXED for all these models to provide better estimates of the standard errors of the estimates. We feel that the discussion of mixed models goes well beyond the scope of this book, and we refer you to several excellent books on mixed models:

Littell, Ramon C., Milliken, George A., Stroup, Walter, W., and Wolfinger, Russell D. 1996. *SAS® System for Mixed Models*. Cary, NC: SAS Institute Inc.

SAS/STAT 9.1 User's Guide. Cary, NC: SAS Institute. 2004.

B. ONE-FACTOR EXPERIMENTS

Consider the following experiment. We have four drugs (1, 2, 3, and 4) that relieve pain. Each subject is given each of the four drugs. The subject's pain tolerance is then measured. Enough time is allowed to pass between successive drug administrations so that we can be sure there is no residual effect from the previous drug. In clinical terms, this is called a "wash-out" period.

The null hypothesis is

$$\text{MEAN(1)} = \text{MEAN(2)} = \text{MEAN(3)} = \text{MEAN(4)}$$

If the analysis of variance is significant at $p < .05$, we want to look at pairwise comparisons of the drugs using a multiple-range test or other post hoc procedure.

Notice how this experiment differs from a one-way analysis of variance without a repeated measure. With the designs we have discussed thus far, we would have each subject receive only one of the four drugs. In this design, each subject is measured under each of the drug conditions. This has several important advantages.

First, each subject acts as his own control. That is, drug effects are calculated by recording deviations between each drug score and the average drug score for each subject. The normal subject-to-subject variation can thus be removed from the error sum of squares. Let's look at a table of data from the pain experiment:

Subject	Drug 1	Drug 2	Drug 3	Drug 4
1	5	9	6	11
2	7	12	8	9
3	11	12	10	14
4	3	8	5	8

To analyze this experiment, we consider the subject to be an independent variable. We therefore have SUBJECT and DRUG as independent variables.

One way of arranging our data and writing our INPUT statement would be like this:

```
DATA PAIN;
    INPUT SUBJ DRUG PAIN;
DATALINES;
1 1 5
1 2 9
1 3 6
1 4 11
2 1 7
(more data lines)
```

It is usually more convenient to arrange all the data for each subject on one line:

SUBJ	DRUG 1	DRUG 2	DRUG 3	DRUG 4
1	5	9	6	11
2	7	12	8	9
3	11	12	10	14
4	3	8	5	8

We can read the data as previously arranged but restructure it to look as if we had read it with the first program as follows:

```
DATA PAIN;
    INPUT SUBJ @;          ①
    DO DRUG = 1 to 4;      ②
        INPUT PAIN @;      ③
        OUTPUT;            ④
    END;     ⑤
DATALINES;
1    5    9    6    11
2    7   12    8     9
3   11   12   10    14
4    3    8    5     8
;
```

The first INPUT statement ① reads the subject number. The "@" sign following SUBJ is an instruction to keep reading from the same line of data. (See Chapter 12

for a more detailed discussion of the trailing @ sign.) Statement ② starts an iterative loop. The value of DRUG is first set to 1. Next, the input statement ③ is executed. Again, if the "@" were omitted, the program would go to the next data line to read a value (which we don't want). The OUTPUT statement ④ causes an observation to be written to the SAS data set. Look at our first line of data. We would have as the first observation in the SAS data set SUBJ = 1, DRUG = 1, and PAIN = 5. When the END statement ⑤ is reached, the program flow returns to the DO statement ② where DRUG is set to 2. A new PAIN value is then read from the data (still from the first line because of the trailing @), and a new observation is added to the SAS data set. This continues until the value of DRUG = 4. Normally, when a DATA step reaches the end, signalled by a DATALINES or RUN statement, an observation is automatically written out to the SAS data set being created. However, since we included an OUTPUT statement in the DATA step to write out our observations, the automatic writing out of an observation at the end of the DATA step does not occur. (If it did, we would get a duplicate of the observation where DRUG = 4.) Since the DO loop has completed, the program control returns to the top of the DATA step to line ① The program will read the subject number from the next line of data.

The general form of a DO statement is

```
DO variable = start TO end BY increment;
   (SAS Statements)
END;
```

Where "*start*" is the initial value for "variable," "*end*" is the ending value and "*increment*" is the increment. If "*increment*" is omitted, it defaults to 1.

The first few observations in the SAS data set created from this program are listed here:

OBS	SUBJ	DRUG	PAIN
1	1	1	5
2	1	2	9
3	1	3	6
4	1	4	11
5	2	1	7
6	2	2	12
7	2	3	8
8	2	4	9
9	3	1	11

etc.

We can make one small modification to the program and, by so doing, avoid having to enter the subject numbers on each line of data. The new program looks as follows:

```
DATA PAIN;
    SUBJ + 1;     ①
    DO DRUG = 1 TO 4;
        INPUT PAIN @;
        OUTPUT;
    END;
DATALINES;
    5    9    6   11
    7   12    8    9
   11   12   10   14
    3    8    5    8
;
```

Statement ① creates a variable called SUBJ, which starts at 1 and is incremented by 1 each time the statement is executed. Here's why: A SAS statement in the form *variable + increment*; is called a SUM statement. Notice that this is not an assignment statement (there is no equal sign). SAS processes SUM statements as follows: The *variable* is retained and initialized at 0. (If the variable was not retained, it would be set back to a missing value for each iteration of the DATA step.) SUM statements also ignore missing values of the *increment*.

We are ready to write our PROC statements to analyze the data. With this design, there are several ways to write the MODEL statement. One way is like this:

```
PROC ANOVA DATA=PAIN;
    TITLE "One-way Repeated Measures ANOVA";
    CLASS SUBJ DRUG;
    MODEL PAIN = SUBJ DRUG;
    MEANS DRUG / SNK;
RUN;
```

Notice that we are not writing SUBJ | DRUG. We are indicating that SUBJ and DRUG are each main effects and that there is no interaction term between them. Once we have accounted for variations from the grand mean due to subjects and drugs, the remaining deviations will be our source of error.

The following is a portion of the output from the one-way repeated measures experiment:

```
One-way Repeated Measures ANOVA

The ANOVA Procedure

    Class Level Information

Class      Levels    Values

SUBJ          4     1 2 3 4

DRUG          4     1 2 3 4

Number of Observations Read        16
Number of Observations Used        16

Dependent Variable: PAIN

                          Sum of
Source            DF      Squares    Mean Square   F Value   Pr > F

Model             6    120.5000000   20.0833333     13.64    0.0005

Error             9     13.2500000    1.4722222

Corrected Total  15    133.7500000

R-Square       Coeff Var        Root MSE        PAIN Mean

0.900935       14.06785         1.213352        8.625000

Source            DF     Anova SS    Mean Square   F Value   Pr > F

SUBJ              3    70.25000000   23.41666667    15.91    0.0006
DRUG              3    50.25000000   16.75000000    11.38    0.0020

Student-Newman-Keuls Test for PAIN

NOTE: This test controls the Type I experimentwise error rate under the
      complete null hypothesis but not under partial null hypotheses.

Alpha                      0.05
Error Degrees of Freedom      9
Error Mean Square          1.472222
                                                      [Continued]
```

```
Number of Means              2              3              4
Critical Range        1.9407923      2.3954582      2.6784122


Means with the same letter are not significantly different.


S
N
K

G
r
o
u
p
i
n
g          Mean        N     DRUG

A       10.5000       4      4
A
A       10.2500       4      2


B        7.2500       4      3
B
B        6.5000       4      1
```

What conclusions can we draw from these results? Looking at the very bottom of the analysis-of-variance table, we find an F value of 11.38 with an associated probability of .0020. We can therefore reject the null hypothesis that the means are equal. Another way of saying this is that the four drugs are not equally effective for reducing pain. Notice that the SUBJ term in the analysis-of-variance table also has an F value and a probability associated with it. This merely tells us how much variability there was from subject to subject. It is not really interpretable in the same fashion as the drug factor. We include it as part of the model because we don't want the variability associated with it to go into the ERROR sum of squares. Most researchers would treat SUBJ as a random effect rather than a fixed effect. That is, we are not actually interested in these particular four subjects. Rather, they are really a random sample of an infinite number of possible subjects. We will show you later how to treat this design as a mixed model using PROC MIXED (although without any detailed explanation—see the last paragraph of the Introduction).

Now that we know that the drugs are not equally effective, we can look at the results of the Student-Newman-Keuls (SNK) test. This shows two drug groupings. Assuming that a higher mean indicates greater pain, we can say that drugs 1 and 3 were more effective in reducing pain than were drugs 2 and 4. We cannot, at the .05 level, claim any differences between drugs 1 and 3 or between drugs 2 and 4.

Looking at the error SS and the SS due to subjects, we see that SUBJECT SS (70.25) is large compared to the ERROR SS (13.25). Had this same set of data been the result of assigning 16 subjects to the four different drugs (instead of repeated measures), the error SS would have been 83.5 (13.25 + 70.25). The resulting F and p-values for the DRUG effect would have been 2.41 and .118, respectively. (Note that the degrees of freedom for the error term would be 12 instead of 9.)

We see, therefore, that controlling for between-subject variability can greatly reduce the error term in our analysis of variance and can allow us to identify small treatment differences with relatively few subjects.

C.   USING THE REPEATED STATEMENT OF PROC ANOVA

This same design can be analyzed using the REPEATED option, first introduced with version 6 of SAS software. When the REPEATED statement is used, we need our data set in the form:

SUBJ PAIN1 PAIN2 PAIN3 PAIN4

where PAIN1–PAIN4 are the pain levels at each drug treatment. Notice that the data set does not have a DRUG variable. The ANOVA statements to analyze this experiment are:

```
DATA REPEAT1;
    INPUT SUBJ PAIN1-PAIN4;
DATALINES;
1    5    9    6   11
2    7   12    8    9
3   11   12   10   14
4    3    8    5    8
;
PROC ANOVA DATA=REPEAT1;
    TITLE "One-way ANOVA Using the REPEATED Statement";
    MODEL PAIN1-PAIN4 = / NOUNI;
    REPEATED DRUG 4 (1 2 3 4);
RUN;
```

Notice several details. First, there is no CLASS statement; our data set does not have an independent variable. We specify the four dependent variables to the left of the equal sign in the MODEL statement. Since there is no CLASS variable, we have nothing to place to the right of the equals sign. The option NOUNI (no univariate) is a request not to conduct a separate analysis for each of the four PAIN variables. Later, when we have both repeated and nonrepeated factors in the design, this option will be especially important. The repeated statement indicates that we want to call the repeated factor DRUG. The "4" following the variable name indicates that DRUG has four levels. This is optional. Had it been omitted, the program would have assumed as many levels as there were dependent variables in the MODEL statement. The number of levels needs to be specified only when we have more than one repeated measure factor. Finally, "(1 2 3 4)" indicates the labels we want printed for each level of DRUG. The labels also act as spacings when polynomial contrasts are requested. (See the SAS/STAT manual for more details on this topic.) We omit the complete output from running this procedure but will show you some excerpts and leave the details of the output to the next model.

```
One-way ANOVA Using the REPEATED Statement

The ANOVA Procedure

Number of Observations Read          4
Number of Observations Used          4

Repeated Measures Analysis of Variance

          Repeated Measures Level Information

Dependent Variable      PAIN1     PAIN2     PAIN3     PAIN4

    Level of DRUG           1         2         3         4

              MANOVA Test Criteria and Exact F Statistics
                 for the Hypothesis of no DRUG Effect
                    H = Anova SSCP Matrix for DRUG
                      E = Error SSCP Matrix

                    S=1      M=0.5     N=-0.5

Statistic                       Value  F Value  Num DF  Den DF  Pr > F

Wilks' Lambda               0.00909295    36.33       3       1  0.1212
Pillai's Trace              0.99090705    36.33       3       1  0.1212
Hotelling-Lawley Trace    108.97530864    36.33       3       1  0.1212
Roy's Greatest Root       108.97530864    36.33       3       1  0.1212

Univariate Tests of Hypotheses for Within Subject Effects

Source              DF       Anova SS    Mean Square    F Value

DRUG                 3    50.25000000    16.75000000      11.38
Error(DRUG)          9    13.25000000     1.47222222

                                    Adj Pr > F
Source            Pr > F      G - G      H - F

DRUG              0.0020     0.0123     0.0020
Error(DRUG)

Greenhouse-Geisser Epsilon      0.5998
Huynh-Feldt Epsilon             1.4433
```

The F value and probabilities (F = 11.38, p = .002) are identical to those in the previous output. Notice that two additional p-values are included in this output. The adjusted p-values (labeled 'Adj. Pr > F') shown to the right are more conservative, the G-G representing the Greenhouse-Geisser correction, and the H-F representing the Huynh-Feldt value. (See Kirk for an explanation; reference in Chapter 1.) Here is a brief explanation: There are some assumptions in repeated measures designs that are rather complicated. You may see these mentioned as symmetry tests or as sphericity tests. Somewhat simplified, what is being tested is the assumption that the variances and correlations are the same among the various dependent variables. It is more or less an extension of the assumption of equal variances in t-tests or ANOVA or equal slopes in ANCOVA. The previous SAS output shows the unadjusted p-values along with the G-G and H-F adjusted values, the more conservative adjustment being the Greenhouse-Geisser.

When we use the REPEATED statement, we cannot use a MEANS statement with the repeated factor name. The only way to compute pairwise comparisons in this case is to use the keyword CONTRAST(n) with the REPEATED statement. The form is:

```
REPEATED factor_name CONTRAST(n);
```

where *factor-name* is a valid SAS name (not a variable in your data set) and *n* is a number from 1 to k, with k being the number of levels of the repeated factor. CONTRAST(1) compares the first level of the factor with each of the other levels. If the first level were a control value, for example, the CONTRAST(1) statement would compare the control to each of the other drugs. If we want all of the pairwise contrasts, we need to write k − 1 repeated statements. In our DRUG example, where there are four levels of DRUG, we write:

```
PROC ANOVA DATA=REPEAT1;
   TITLE "One-way ANOVA Using the Repeated Statement";
   MODEL PAIN1-PAIN4 = / NOUNI;
   REPEATED DRUG 4 CONTRAST(1)/ NOM SUMMARY;
   REPEATED DRUG 4 CONTRAST(2)/ NOM SUMMARY;
   REPEATED DRUG 4 CONTRAST(3)/ NOM SUMMARY;
RUN;
```

These three CONTRAST statements produce all the two-way comparisons. (CONTRAST(1) gives us 1 versus 2, 3, 4; CONTRAST(2) gives us 2 versus 1, 3, 4; and CONTRAST(3) gives us 3 versus 1, 2, 4.) The contrasts are equivalent to multiple t-tests between the levels, and you may want to protect yourself against a type I error with a Bonferroni correction or some other method. The option NOM asks that no multivariate statistics be printed; the option SUMMARY requests analysis-of-variance tables for each contrast defined by the repeated factor.

D.   USING PROC MIXED TO COMPUTE A MIXED (RANDOM EFFECTS) MODEL

Looks like we are getting a bit carried away with this "simple" design. The last approach we will take is to use PROC MIXED so that we can treat the subject variable as a random effect (thought by most to be the proper thing to do). The syntax of PROC MIXED for this model is:

```
PROC MIXED DATA=PAIN;
    TITLE "One Factor Experiment - Mixed Model";
    CLASS SUBJ DRUG;
    MODEL PAIN = DRUG;
    RANDOM SUBJ;
RUN;
QUIT;
```

It looks a lot like the ANOVA model except that we list SUBJ as a random effect (as opposed to a repeated effect). Since the use and abuse of mixed models goes far beyond the scope of this book, we will show you the output and will refer you to one of the references listed in the Introduction of this chapter. Here is the listing (edited):

```
One Factor Experiment - Mixed Model

The Mixed Procedure

                    Model Information

Data Set                    WORK.PAIN
Dependent Variable          PAIN
Covariance Structure        Variance Components
Estimation Method           REML
Residual Variance Method    Profile
Fixed Effects SE Method     Model-Based
Degrees of Freedom Method   Containment

              Class Level Information

Class     Levels     Values

SUBJ        4        1 2 3 4
DRUG        4        1 2 3 4

                  Dimensions

Covariance Parameters          2
Columns in X                   5
Columns in Z                   4
Subjects                       1
Max Obs Per Subject           16

              Number of Observations

Number of Observations Read          16
Number of Observations Used          16
Number of Observations Not Used       0
```

```
                        Iteration History

Iteration     Evaluations     -2 Res Log Like      Criterion

     0              1            62.87898203
     1              1            52.54100308      0.00000000
```

Convergence criteria met.

Covariance Parameter
    Estimates

Cov Parm   Estimate

SUBJ   5.4861
Residual   1.4722

        Fit Statistics

-2 Res Log Likelihood   52.5
AIC (smaller is better)   56.5
AICC (smaller is better)   57.9
BIC (smaller is better)   55.3

        Type 3 Tests of Fixed Effects

             Num   Den
Effect   DF   DF   F Value   Pr > F

DRUG   3   9   11.38   0.0020

Notice that the p-value for the fixed effect (DRUG) is the same as that obtained by using the fixed effects models.

E. TWO-FACTOR EXPERIMENTS WITH A REPEATED MEASURE ON ONE FACTOR

One very popular form of a repeated measures design is the following:

		PRE	POST
	SUBJ		
Control	1		
	2		
	3		
Treatment	4		
	5		
	6		

Subjects are randomly assigned to a control or treatment group. Then, each subject is measured before and after treatment. Obviously, in this case, the TIME factor (PRE and POST) cannot be randomized. The "treatment" for the control group can either be a placebo or no treatment at all. The goal of an experiment of this sort is to compare the pre/post changes of the control group to the pre/post changes of the treatment group. This design has a definite advantage over a simple pre/post design where one group of subjects is measured before and after a treatment (such as having only a treatment group in our design). Simple pre/post designs suffer from the problem that we cannot be sure if it is our treatment that causes a change (e.g., TIME may have an effect). By adding a pre/post control group, we can compare the pre/post control scores to the pre/post treatment scores and thereby control any built-in, systematic, pre/post changes.

A simple way to analyze our design is to compute a difference score (post minus pre) for each subject. We then have two groups of subjects with one score each (the difference score). Then we use a t-test to look for significant differences between the difference scores of the control and treatment groups. With more than two levels of time, however, we will need to use analysis of variance.

Here are some sample data and a SAS program that calculates difference scores and computes a t-test:

	SUBJ	PRE	POST
Control	1	80	83
	2	85	86
	3	83	88
Treatment	4	82	94
	5	87	93
	6	84	98

```
DATA PREPOST;
   INPUT SUBJ GROUP $ PRETEST POSTEST;
   DIFF = POSTEST-PRETEST;
DATALINES;
1   C   80   83
2   C   85   86
3   C   83   88
4   T   82   94
5   T   87   93
6   T   84   98
;
PROC TTEST DATA=PREPOST;
   TITLE "T-test on Difference Scores";
   CLASS GROUP;
   VAR DIFF;
RUN;
```

Results of this analysis show the treatment mean difference to be significantly different from the control mean difference (p = .045). See following:

```
T-test on Difference Scores

The TTEST Procedure

                                Statistics

                            Lower CL            Upper CL   Lower CL
 Variable   GROUP         N      Mean    Mean       Mean    Std Dev

 DIFF       C            3     -1.968       3     7.9683     1.0413
 DIFF       T            3    0.3244  10.667     21.009     2.1677
 DIFF       Diff (1-2)        -15.07  -7.667     -0.263     1.9568

                                Statistics

                            Upper CL
 Variable   GROUP     Std Dev  Std Dev  Std Err   Minimum   Maximum

 DIFF       C             2    12.569   1.1547         1         5
 DIFF       T        4.1633   26.165   2.4037         6        14
 DIFF       Diff (1-2)  3.266    9.385   2.6667

                                T-Tests

 Variable   Method          Variances    DF   t Value    Pr > |t|

 DIFF       Pooled          Equal         4     -2.88      0.0452
 DIFF       Satterthwaite   Unequal    2.88     -2.88      0.0671

                       Equality of Variances

 Variable   Method     Num DF   Den DF   F Value   Pr > F

 DIFF       Folded F        2        2      4.33   0.3750
```

If we choose to assume equal variances in the two groups, the group differences are significantly different at p = .0452. We can alternatively treat this design as a two-way analysis of variance (GROUP × TIME) with TIME as a repeated measure. This method has the advantage of analyzing designs with more than two levels on one or both factors.

We first write a program using the REPEATED statement of ANOVA. No changes in the data set are necessary. The ANOVA statements are:

```
PROC ANOVA DATA=PREPOST;
    TITLE1 "Two-way ANOVA with a Repeated Measure on One Factor";
    CLASS GROUP;
    MODEL PRETEST POSTEST = GROUP / NOUNI;
    REPEATED TIME 2 (0 1);
    MEANS GROUP;
RUN;
```

The REPEATED statement indicates that we want to call the repeated factor TIME, that it has two levels, and that we want to label the levels 0 and 1.

Output (edited) from these procedure statements are shown here:

```
Two-way ANOVA with a Repeated Measure on One Factor

                MANOVA Test Criteria and Exact F Statistics
                   for the Hypothesis of no TIME Effect
                   H = Anova SSCP Matrix for TIME
                      E = Error SSCP Matrix


                    S=1      M=-0.5     N=1


Statistic                      Value   F Value  Num DF  Den DF  Pr > F

Wilks' Lambda               0.13216314    26.27      1       4  0.0069
Pillai's Trace              0.86783686    26.27      1       4  0.0069
Hotelling-Lawley Trace      6.56640625    26.27      1       4  0.0069
Roy's Greatest Root         6.56640625    26.27      1       4  0.0069

                MANOVA Test Criteria and Exact F Statistics
                for the Hypothesis of no TIME*GROUP Effect
                   H = Anova SSCP Matrix for TIME*GROUP
                      E = Error SSCP Matrix


                    S=1      M=-0.5     N=1


Statistic                      Value   F Value  Num DF  Den DF  Pr > F

Wilks' Lambda               0.32611465     8.27      1       4  0.0452
Pillai's Trace              0.67388535     8.27      1       4  0.0452
Hotelling-Lawley Trace      2.06640625     8.27      1       4  0.0452
Roy's Greatest Root         2.06640625     8.27      1       4  0.0452
```

```
Tests of Hypotheses for Between Subjects Effects

Source                   DF      Anova SS    Mean Square   F Value   Pr > F

GROUP                     1    90.75000000   90.75000000    11.84    0.0263
Error                     4    30.66666667    7.66666667
```

```
Univariate Tests of Hypotheses for Within Subject Effects

Source                   DF      Anova SS    Mean Square   F Value   Pr > F

TIME                      1    140.0833333   140.0833333    26.27    0.0069

TIME*GROUP                1     44.0833333    44.0833333     8.27    0.0452
Error(TIME)               4     21.3333333     5.3333333
```

Level of		----------PRETEST----------		----------POSTEST----------	
GROUP	N	Mean	Std Dev	Mean	Std Dev
C	3	82.6666667	2.51661148	85.6666667	2.51661148
T	3	84.3333333	2.51661148	95.0000000	2.6457513

Finally, it should be noted that these data can also be analyzed through an analysis of covariance using the pretest as a covariate. The choice of analysis depends in part on the precise nature of the question being asked. Bock (1975) discusses this issue.

We will discuss the output from PROC ANOVA after we show an alternative method of analyzing this experiment. However, a portion of the previous output is unique and is discussed here. Notice the rows labeled Wilks' Lambda, Pillai's Trace, Hotelling-Lawley Trace, and Roy's Greatest Root. These are multivariate statistics that are of special interest when more than one dependent variable is indicated. Unlike univariate statistics, when you use multivariate procedures, there is no single test analogous to the F-test. Instead, there are about half a dozen. A question arises as to which one to use. Multivariate statisticians spend many pleasant hours investigating this question. The answer to the question is unambiguous: It depends. Bock (1975; see Chapter 1 for references) presents a nice discussion of the options. Our advice is: When there are only very small differences among the p-values, it doesn't really matter which one you use. If in doubt, we suggest two alternatives: (1) Use Wilks' Lamda. It is a likelihood ratio test that is often appropriate, or (2) when there are differences among the p-values, find a consultant.

We now analyze the same experiment as a two-factor analysis of variance without using the REPEATED statement of PROC ANOVA. We may want to do this so that we can use the "built-in" multiple-comparison tests. (You also may not want to do this if you feel you need the "protection" of the more conservative F values computed in the multivariate model.) To do this, we must first create a new variable—say TIME—that will have two possible values: PRE or POST. Each subject will then have two observations, one with TIME = PRE and one with TIME = POST.

As with our one-way, repeated measures design, the method of creating several observations from one is with the OUTPUT statement.

We can add the following SAS statements to the end of the previous program:

```
DATA TWOWAY;
    SET PREPOST;     ①
    LENGTH TIME $ 4;
    TIME = 'PRE';    ②
    SCORE = PRETEST;    ③
    OUTPUT;     ④
    TIME = 'POST';    ⑤
    SCORE = POSTEST;    ⑥
    OUTPUT;    ⑦
    KEEP SUBJ GROUP TIME SCORE;    ⑧
RUN;
```

This section of the program creates a SAS data set called TWOWAY, which has variables SUBJ, GROUP, TIME, and SCORE. The first few observations in this data set are:

SUBJ	GROUP	TIME	SCORE
1	C	PRE	80
1	C	POST	83
2	C	PRE	85
2	C	POST	86

Let's follow this portion of the SAS program step by step to see exactly how the new data set is created.

The SET statement ① causes observations to be read from the original data set, PREPOST. The first observation is

```
SUBJ=1 GROUP=C PRETEST=80 POSTEST=83 DIFF=3.
```

Statement ② creates a new variable called TIME and sets the value of TIME to 'PRE'. Note that a LENGTH statement was added so that the length of TIME would be 4. Without this statement, SAS would set a length of 3 for TIME since SAS determines the length of character variables by the manner in which this variable is first processed in a DATA step. Thus, if you did not have a LENGTH statement and SAS processed line ②, it would see that there were three characters between the single quotes and would set a length of 3 for the variable TIME. A somewhat lazy way to write this program is to place a space after the 'E' in PRE, causing SAS to set a length of 4 for TIME. One further note on this, since it's really important: If you placed the LENGTH statement after the assignment of TIME equal to 'PRE,' it would have no effect. Statement ③ creates a new variable, SCORE, which is equal to the PRETEST value. When the OUTPUT statement ④ is executed, the first observation of the data set called TWOWAY becomes:

```
SUBJ=1 GROUP=C PRETEST=80 POSTEST=83 DIFF=3 TIME=PRE SCORE=80.
```

However, since we included a KEEP statement ⑥, the first observation in data set TWOWAY only contains the variables SUBJ, GROUP, TIME, and SCORE.

Next, ⑤ sets TIME='POST', and ⑥ sets the variable SCORE to the POSTEST value. A new observation is added to the data set TWOWAY with the second OUTPUT statement ⑦. The second observation has SUBJ=1 GROUP=C TIME=POST SCORE=83.

The RUN statement ends the DATA step and control returns to the top of the DATA step where a new observation is read from data set PREPOST.

We are now ready to write our ANOVA statements. Unlike any of our previous examples, we have to specify all the terms, including the sources of error, in the MODEL statement. This is necessary because our main effects and interaction terms are not tested by the same error term. Therefore, we need to specify each of these terms in the MODEL statement so they can be used later in tests of our hypotheses. In this design, we have one group of subjects assigned to a control group and another group assigned to a treatment group. Within each group, each subject is measured at TIME = PRE and TIME = POST. In this design, the subjects are said to be nested within the GROUP. In SAS programs, the term subjects nested within group is written:

```
SUBJ(GROUP)
```

Since the model statement defines ALL sources of variation about the grand mean, the ERROR SUM OF SQUARES printed in the ANOVA table will be zero. To specify which error term to be used to test each hypothesis in our design, we use TEST statements following the MODEL specification. A TEST statement consists of a hypothesis to be tested (H =) and the appropriate error term (E =). The entire ANOVA procedure looks as follows:

```
PROC ANOVA DATA=TWOWAY;
    TITLE "Two-way ANOVA with TIME as a Repeated Measure";
    CLASS SUBJ GROUP TIME;
    MODEL SCORE = GROUP SUBJ(GROUP) TIME
                  GROUP*TIME TIME*SUBJ(GROUP);
    MEANS GROUP|TIME;
    TEST H=GROUP              E=SUBJ(GROUP);
    TEST H=TIME GROUP*TIME    E=TIME*SUBJ(GROUP);
RUN;
```

Notice that the error term for GROUP is SUBJ(GROUP) (subject nested within group), and the error term for TIME and the GROUP*TIME interaction is TIME*SUBJ(GROUP).

The following are portions of the PROC ANOVA output:

```
Two-way ANOVA with TIME as a Repeated Measure

The ANOVA Procedure

        Class Level Information

Class        Levels   Values

SUBJ            6      1 2 3 4 5 6

GROUP           2      C T

TIME            2      AFTER BEFORE

Number of Observations Read        12
Number of Observations Used        12

Dependent Variable: SCORE

                               Sum of
Source                 DF      Squares    Mean Square  F Value  Pr > F

Model                  11    326.9166667   29.7196970     .        .

Error                   0      0.0000000      .

Corrected Total        11    326.9166667

R-Square    Coeff Var      Root MSE     SCORE Mean

1.000000        .              .         86.91667

Source                 DF    Anova SS    Mean Square  F Value  Pr > F

GROUP                   1     90.7500000   90.7500000    .        .
SUBJ(GROUP)             4     30.6666667    7.6666667    .        .
TIME                    1    140.0833333  140.0833333    .        .
GROUP*TIME              1     44.0833333   44.0833333    .        .
SUBJ*TIME(GROUP)        4     21.3333333    5.3333333    .        .
```

```
Level of                 ------------SCORE------------
GROUP          N               Mean          Std Dev


C              6          84.1666667        2.78687400
T              6          89.6666667        6.28225013
```

```
Level of                 ------------SCORE------------
TIME           N               Mean          Std Dev


AFTER          6          90.3333333        5.60951572
BEFORE         6          83.5000000        2.42899156
```

```
Level of     Level of            ------------SCORE------------
GROUP        TIME         N             Mean          Std Dev


C            AFTER        3        85.6666667       2.51661148
C            BEFORE       3        82.6666667       2.51661148
T            AFTER        3        95.0000000       2.64575131
T            BEFORE       3        84.3333333       2.51661148
```

```
Tests of Hypotheses Using the Anova MS for SUBJ(GROUP) as an Error Term
```

Source	DF	Anova SS	Mean Square	F Value	Pr > F
GROUP	1	90.75000000	90.75000000	11.84	0.0263

```
              Tests of Hypotheses Using the Anova MS
                 for SUBJ*TIME(GROUP) as an Error Term
```

Source	DF	Anova SS	Mean Square	F Value	Pr > F
TIME	1	140.0833333	140.0833333	26.27	0.0069
GROUP*TIME	1	44.0833333	44.0833333	8.27	0.0452

Since all sources of variation were included in the MODEL statement, the error sum of squares is zero, and the F value is undefined (it prints as a missing value '.'). The requested tests are shown at the bottom of the listing. Group differences have an F = 11.84 and a p = .0263. TIME and GROUP*TIME have F values of 26.27 and 8.27 and probabilities of .0069 and .0452, respectively.

In this experimental design, it is the interaction of GROUP and TIME that is of primary importance. This interaction term tells us if the pre/post changes were the same for control and treatment subjects. An interaction graph will make this clear. The output from the MEANS request is shown here:

```
Level of                -----------SCORE------------
GROUP        N              Mean            Std Dev

C            6          84.1666667        2.78687400
T            6          89.6666667        6.28225013

Level of                -----------SCORE------------
TIME         N              Mean            Std Dev

AFTER        6          90.3333333        5.60951572
BEFORE       6          83.5000000        2.42899156

Level of     Level of              -----------SCORE------------
GROUP        TIME         N             Mean            Std Dev

C            AFTER        3         85.6666667        2.51661148
C            BEFORE       3         82.6666667        2.51661148
T            AFTER        3         95.0000000        2.64575131
T            BEFORE       3         84.3333333        2.51661148
```

We can use the last set of means (interaction of GROUP and TIME) to plot the interaction graph. We pick one of the independent variables (we will use TIME) to go on the x-axis and then plot means of the dependent variable at each of the levels of the other independent variable (GROUP). A few lines of SAS code will produce the desired interaction plot. Here is the code:

```
PROC MEANS DATA=TWOWAY NOPRINT NWAY;
   CLASS GROUP TIME;
   VAR SCORE;
   OUTPUT OUT=INTER
          MEAN=;
RUN;
OPTIONS LINESIZE=68 PAGESIZE=24;
SYMBOL1 VALUE=CIRCLE COLOR=BLACK INTERPOL=JOIN;
SYMBOL2 VALUE=SQUARE COLOR=BLACK INTERPOL=JOIN;

PROC GPLOT DATA=INTER;
   TITLE "Interaction Plot";
   PLOT SCORE*TIME=GROUP;
RUN;
```

The resulting graph is shown here. (NOTE: The values of PRE and POST were changed to 1-PRE and 2-POST, respectively, so that the PRE values would come before the POST values on the x-axis.)

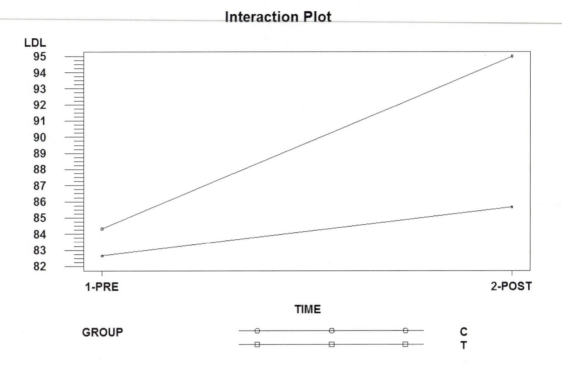

A significant interaction term shows us that the two pre/post lines are not parallel. This tells us that the change from pre to post was different, depending on which GROUP a subject was in, which is precisely what we wanted to know. The treatment group and control group were quite similar in terms of pain relief before the drug was administered (mean = 84.33 and 82.67). After the drug was given (the POST measure), the treatment group showed dramatic gains and the control group only modest gains. The F-statistic for GROUP × TIME (8.27) and its p-value (.045) tell us that this difference in improvement is greater than could be expected by chance alone. The F-statistic for GROUP (F = 11.84, p = .0263) tells us that if we summed over the pre and post tests, the groups were different. This isn't of use to us since it combines the pre measure (where we anticipated them being the same) with the post measure (where we anticipated a difference). The same logic is true for TIME. Here we are summing over the control and treatment groups. Finally, note that the p-value for GROUP × TIME is the same as for the t-test of the difference scores, because we are essentially making the same test in both analyses.

The PROC MIXED statements (without explanation) to analyze this model are:

```
PROC MIXED DATA=TWOWAY;
   TITLE "Mixed Model for Two-way Design";
   CLASS GROUP TIME SUBJ;
   MODEL SCORE = GROUP TIME GROUP*TIME /SOLUTION;
   RANDOM SUBJ(GROUP);
   LSMEANS GROUP TIME;
RUN;
QUIT;
```

If you run this model, you will see that the p-values for the main effects (GROUP and TIME) are the same as the ones in the previous listings. The standard errors of the estimates, however, are different.

Next, we move to a somewhat more complex setting.

F.   TWO-FACTOR EXPERIMENTS WITH REPEATED MEASURES ON BOTH FACTORS

This design is similar to the previous design except that each subject is measured under all levels of both factors. An example follows:

A group of subjects is tested in the morning and afternoon of two different days. On one of the days, the subjects receive a strong sleeping aid the night before the experiment is to be conducted; on the other, a placebo. The subjects' reaction time to a stimulus is measured. A diagram of the experiment is shown here:

Reaction Times by Treatment and Time

	Treatment			
	Control		Drug	
Subj	A.M.	P.M.	A.M.	P.M.
1	65	55	70	60
2	72	64	78	68
3	90	80	97	85

We would like to see if the drug had any effect on the reaction time and if the effect was the same for the whole day. We can use the AM/PM measurements on the control day as a comparison for the AM/PM changes on the drug day.

Since each subject is measured under all levels of treatment (PLACEBO or DRUG) and TIME (AM/PM), we can treat this experiment as a SUBJ by TREATMENT by TIME factorial design. However, we must specify the error terms to test our hypotheses.

To create our SAS data set, we could use the following statements:

```
DATA SLEEP;
   INPUT SUBJ TREAT $ TIME $ REACT;
DATALINES;
1    CONT    AM    65
1    DRUG    AM    70
1    CONT    PM    55
1    DRUG    PM    60
2    CONT    AM    72
2    DRUG    AM    78
2    CONT    PM    64
2    DRUG    PM    68
3    CONT    AM    90
3    DRUG    AM    97
3    CONT    PM    80
3    DRUG    PM    85
;
```

The ANOVA statements can be written:

```
PROC ANOVA DATA=SLEEP;
   TITLE "Two-way ANOVA with a Repeated Measure on Both Factors";
   CLASS SUBJ TREAT TIME;
   MODEL REACT = SUBJ|TREAT|TIME;
   MEANS TREAT|TIME;
   TEST H=TREAT         E=SUBJ*TREAT;
   TEST H=TIME          E=SUBJ*TIME;
   TEST H=TREAT*TIME    E=SUBJ*TREAT*TIME;
RUN;
```

In any design where ALL factors are repeated, such as this one, we can treat the SUBJ variable as being crossed by all other factors (as opposed to nested). The MODEL statement is therefore the same as our factorial design. However, by including the SUBJ term in our model, the error term will be zero (as in our previous example). Thus, our ANOVA table will not show F values or probabilities. These are obtained by specifying TEST statements following the MODEL statement, as previously described.

The error terms to test each hypothesis are simple to remember: For factor X, the error term is SUBJ*X. For example, the error term to test TREAT is SUBJ*TREAT; the error term to test the interaction TREAT*TIME is SUBJ*TREAT*TIME. To specify the correct error term for each main effect and interaction, the three TEST statements following the MODEL statement were added, each specifying a hypothesis to be tested and the error term to be used in calculating the F ratio.

A portion of the output from PROC ANOVA is shown here:

```
Two-way ANOVA with a Repeated Measure on Both Factors

Dependent Variable: REACT

                                 Sum of
Source                   DF      Squares    Mean Square  F Value  Pr > F

Model                    11   1750.666667   159.151515      .       .

Error                     0      0.000000        .

Corrected Total          11   1750.666667

R-Square      Coeff Var      Root MSE     REACT Mean

1.000000          .              .         73.66667

Source                   DF     Anova SS    Mean Square  F Value  Pr > F

SUBJ                      2   1360.666667   680.333333      .       .
TREAT                     1     85.333333    85.333333      .       .
SUBJ*TREAT                2      0.666667     0.333333      .       .
TIME                      1    300.000000   300.000000      .       .
SUBJ*TIME                 2      2.000000     1.000000      .       .
TREAT*TIME                1      1.333333     1.333333      .       .
SUBJ*TREAT*TIME           2      0.666667     0.333333      .       .

Dependent Variable: REACT

 Tests of Hypotheses Using the Anova MS for SUBJ*TREAT as an Error Term

Source                   DF     Anova SS    Mean Square  F Value  Pr > F

TREAT                     1   85.33333333   85.33333333   256.00  0.0039

 Tests of Hypotheses Using the Anova MS for SUBJ*TIME as an Error Term

Source                   DF     Anova SS    Mean Square  F Value  Pr > F

TIME                      1  300.0000000   300.0000000    300.00  0.0033

                Tests of Hypotheses Using the Anova MS
                  for SUBJ*TREAT*TIME as an Error Term

Source                   DF     Anova SS    Mean Square  F Value  Pr > F

TREAT*TIME                1    1.33333333    1.33333333     4.00  0.1835
```

What conclusions can we draw? (1) The drug increases reaction time (F = 256.00, p = .0039); (2) reaction time is longer in the morning compared to the afternoon (F = 300.00, p = .0033); and (3) we cannot conclude that the effect of the drug on reaction time is related to the time of day (the interaction of TREAT and TIME is not significant; F = 4.00, p = 0.1835). Note that this study is not a pre/post study as in the previous example. Even so, had the TREAT by TIME interaction been significant, we would have been more cautious in looking at the TREAT and TIME effects.

If this were a real experiment, we would have to control for the learning effect that might take place. For example, if we measure each subject in the same order, we might find a decrease in reaction time from CONTROL AM to DRUG PM because the subject became more familiar with the apparatus. To avoid this, we would either have to acquaint the subject with the equipment before the experiment begins to assure ourselves that the learning has stabilized, or measure each subject using TREATMENT and TIME in random order.

This design may also be analyzed using the REPEATED statements of PROC ANOVA. If we read in the four reaction times for each subject in the order: AM Control - AM Drug - PM Control - PM Drug, and name our variables REACT1-REACT4, the SAS statements are:

```
DATA REPEAT2;
    INPUT REACT1-REACT4;
DATALINES;
65    70    55    60
72    78    64    68
90    97    80    85
;
PROC ANOVA DATA=REPEAT2;
    MODEL REACT1-REACT4 = / NOUNI;
    REPEATED TIME 2 , TREAT 2 / NOM;
RUN;
```

When there is more than one repeated factor, we need to specify the number of levels of each factor after the factor name. The factor we name first changes the slowest. That is, the first two reaction times are for TIME=AM with REACT1 associated with TREAT=control and REACT2 associated with TREAT=drug. We have now exhausted all levels of TREAT and set TIME=PM; REACT3 and REACT4 are both PM measurements.

G.   THREE-FACTOR EXPERIMENTS WITH A REPEATED MEASURE ON THE LAST FACTOR

For this example, we consider a marketing experiment. Male and female subjects are offered one of three different brands of coffee. Each brand is tasted twice; once immediately after breakfast, the other time after dinner (the order of presentation is randomized for each subject). The preference of each brand is measured on a scale from 1 to 10 (1 = lowest, 10 = highest). The experimental design is shown here:

		BRAND (of coffee)							
	A: subj	Brkfst	Dinner	B: subj	Brkfst	Dinner	C: subj	Brkfst	Dinner
Male	1	7	8	7	4	5	13	8	9
	2	6	7	8	3	5	14	6	9
GENDER	3	6	8	9	3	5	15	5	8
Female	4	5	7	10	3	4	16	6	9
	5	4	7	11	4	4	17	6	9
	6	4	6	12	2	3	18	7	8

In this experiment, the factors BRAND and GENDER are crossed factors while MEAL is a repeated measure factor (each subject tastes coffee after breakfast and dinner). Since a single subject tastes only one brand of coffee and is clearly only one gender, the subject term is said to be nested within BRAND and GENDER (written SUBJ(BRAND GENDER)). We could arrange our data several ways. First, we arrange data so that we can take advantage of the REPEATED statement of ANOVA. To do this, we place all data for each subject on one line. Thus, our program and data will look as follows:

```
DATA COFFEE;
    INPUT SUBJ BRAND $ GENDER $ SCORE_B SCORE_D;
DATALINES;
1    A    M    7    8
2    A    M    6    7
3    A    M    6    8
4    A    F    5    7
5    A    F    4    7
6    A    F    4    6
7    B    M    4    6
8    B    M    3    5
9    B    M    3    5
10   B    F    3    4
11   B    F    4    4
12   B    F    2    3
13   C    M    8    9
14   C    M    6    9
15   C    M    5    8
16   C    F    6    9
17   C    F    6    9
```

```
18  C   F   7   8
;
PROC ANOVA DATA=COFFEE;
   TITLE "Coffee Study";
   CLASS BRAND GENDER;
   MODEL SCORE_B SCORE_D = BRAND|GENDER / NOUNI;
   REPEATED MEAL;
   MEANS BRAND|GENDER;
RUN;
```

Notice that BRAND and GENDER are crossed while MEAL is the repeated measures factor. As before, the option NOUNI on the MODEL statement indicates that we do not want univariate statistics for SCORE_B and SCORE_D.

Selected portions of the output from the previous program are shown here:

```
Coffee Study

   Class Level Information

Class          Levels    Values

BRAND               3    A B C

GENDER              2    F M

Number of Observations Read        18
Number of Observations Used        18

   Repeated Measures Level Information

Dependent Variable    SCORE_B  SCORE_D

    Level of MEAL          1        2

Repeated Measures Analysis of Variance
Tests of Hypotheses for Between Subjects Effects

Source                 DF     Anova SS   Mean Square  F Value  Pr > F

BRAND                   2   83.38888889  41.69444444    51.76  <.0001
GENDER                  1    6.25000000   6.25000000     7.76  0.0165
BRAND*GENDER            2    3.50000000   1.75000000     2.17  0.1566
Error                  12    9.66666667   0.80555556

                                                        [Continued]
```

```
Univariate Tests of Hypotheses for Within Subject Effects

Source                    DF      Anova SS    Mean Square   F Value   Pr > F

MEAL                       1    30.25000000   30.25000000    99.00   <.0001
MEAL*BRAND                 2     1.50000000    0.75000000     2.45   0.1278
MEAL*GENDER                1     0.02777778    0.02777778     0.09   0.7682
MEAL*BRAND*GENDER          2     2.05555556    1.02777778     3.36   0.0692
Error(MEAL)               12     3.66666667    0.30555556
```

```
Level of          ----------SCORE_B----------   ----------SCORE_D----------
BRAND       N        Mean         Std Dev           Mean         Std Dev

A           6     5.33333333     1.21106014      7.16666667     0.75277265
B           6     3.16666667     0.75277265      4.50000000     1.04880885
C           6     6.33333333     1.03279556      8.66666667     0.51639778
```

```
Level of          ----------SCORE_B----------   ----------SCORE_D----------
GENDER      N        Mean         Std Dev           Mean         Std Dev

F           9     4.55555556     1.58989867      6.33333333     2.23606798
M           9     5.33333333     1.73205081      7.22222222     1.56347192
```

```
Level of Level of      ---------SCORE_B---------  ---------SCORE_D---------
BRAND    GENDER   N        Mean       Std Dev        Mean       Std Dev

A        F        3     4.33333333   0.57735027    6.66666667   0.57735027
A        M        3     6.33333333   0.57735027    7.66666667   0.57735027
B        F        3     3.00000000   1.00000000    3.66666667   0.57735027
B        M        3     3.33333333   0.57735027    5.33333333   0.57735027
C        F        3     6.33333333   0.57735027    8.66666667   0.57735027
C        M        3     6.33333333   1.52752523    8.66666667   0.57735027
```

We shall explain the results after the following alternative program:

An alternative program can be written that does not use the REPEATED statement of PROC ANOVA. Although in some models this does not give you the protection of the more conservative multivariate model, you still may want to run such a model. To do so, it is more useful if the data are arranged with two observations per subject and a MEAL variable already in the data set. So, if the data are arranged like this:

SUBJ	BRAND	GENDER	MEAL	SCORE
1	A	M	BRKFST	7
1	A	M	DINNER	8
2	A	M	BRKFST	6
		etc.		

your INPUT statement would look like this:

```
INPUT SUBJ BRAND $ GENDER $ MEAL $ SCORE;
```

The ANOVA statements are written:

```
PROC ANOVA DATA=COFFEE;
   CLASS SUBJ BRAND GENDER MEAL;
   MODEL SCORE = BRAND GENDER BRAND*GENDER SUBJ(BRAND GENDER)
                 MEAL BRAND*MEAL GENDER*MEAL BRAND*GENDER*MEAL
                 MEAL*SUBJ(BRAND GENDER);
   MEANS BRAND|GENDER / SNK E=SUBJ(BRAND GENDER);
   MEANS MEAL BRAND*MEAL GENDER*MEAL BRAND*GENDER*MEAL;
*------------------------------------------------------------*
| The following TEST statements are needed to obtain the     |
| correct F and p-values:                                    |
*------------------------------------------------------------*;
   TEST H=BRAND GENDER BRAND*GENDER
        E=SUBJ(BRAND GENDER);
   TEST H=MEAL BRAND*MEAL GENDER*MEAL BRAND*GENDER*MEAL
        E=MEAL*SUBJ(BRAND GENDER);
RUN;
```

The first TEST statement will test each of the terms (BRAND GENDER and BRAND*GENDER) with the error term SUBJ(BRAND GENDER). The effects MEAL, BRAND*MEAL, GENDER*MEAL, and BRAND *GENDER * MEAL will all be tested with the error term MEAL*SUBJ(BRAND GENDER). We have also made a change in the way the MEANS statements were written. Included after the SNK option is an "E =" specification. This is done because the SNK procedure will use the residual mean square as the error term unless otherwise instructed. Since we have completely defined every source of variation in our model, the residual mean square is zero. The "E =error term" option uses the same error term as the "H =" option of the corresponding TEST statement. Also, since different error terms are used to test different hypotheses, a separate MEANS statement is required each time a different error term is used. Note that we did not need to perform a multiple-comparison test for MEAL since this variable has only two levels.

A portion of the results of running this alternative program are shown here:

```
The ANOVA Procedure

                    Class Level Information

Class         Levels     Values

SUBJ             18       1 2 3 4 5 6 7 8 9 10 11 12 13 14 15 16 17 18

BRAND             3       A B C

GENDER            2       F M

MEAL              2       BRKFST DINNER
```

[Continued]

```
Number of Observations Read        36
Number of Observations Used        36


Dependent Variable: SCORE

                                    Sum of
Source                 DF          Squares    Mean Square   F Value   Pr > F

Model                  35      140.3055556      4.0087302      .         .


Error                   0        0.0000000          .


Corrected Total        35      140.3055556


R-Square      Coeff Var       Root MSE      SCORE Mean

1.000000          .               .          5.861111


Source                 DF       Anova SS    Mean Square   F Value   Pr > F

BRAND                   2     83.38888889   41.69444444      .         .
GENDER                  1      6.25000000    6.25000000      .         .
BRAND*GENDER            2      3.50000000    1.75000000      .         .
SUBJ(BRAND*GENDER)     12      9.66666667    0.80555556      .         .
MEAL                    1     30.25000000   30.25000000      .         .
BRAND*MEAL              2      1.50000000    0.75000000      .         .
GENDER*MEAL             1      0.02777778    0.02777778      .         .
BRAND*GENDER*MEAL       2      2.05555556    1.02777778      .         .
SUBJ*MEAL(BRAN*GEND)   12      3.66666667    0.30555556      .         .

Student-Newman-Keuls Test for SCORE

NOTE: This test controls the Type I experimentwise error rate under the
      complete null hypothesis but not under partial null hypotheses.

Alpha                        0.05
Error Degrees of Freedom       12
Error Mean Square        0.805556

Number of Means              2              3
Critical Range       0.7983504      0.9775036


Means with the same letter are not significantly different.


S
N
K
```

```
G
r
o
u
p
i
n
g          Mean      N    BRAND

A         7.5000    12    C

B         6.2500    12    A

C         3.8333    12    B
```

Level of BRAND	Level of GENDER	N	------------SCORE------------ Mean	Std Dev
A	F	6	5.50000000	1.37840488
A	M	6	7.00000000	0.89442719
B	F	6	3.33333333	0.81649658
B	M	6	4.33333333	1.21106014
C	F	6	7.50000000	1.37840488
C	M	6	7.50000000	1.64316767

Level of MEAL	N	------------SCORE------------ Mean	Std Dev
BRKFST	18	4.94444444	1.66175748
DINNER	18	6.77777778	1.92676369

Level of BRAND	Level of MEAL	N	------------SCORE------------ Mean	Std Dev
A	BRKFST	6	5.33333333	1.21106014
A	DINNER	6	7.16666667	0.75277265
B	BRKFST	6	3.16666667	0.75277265
B	DINNER	6	4.50000000	1.04880885
C	BRKFST	6	6.33333333	1.03279556
C	DINNER	6	8.66666667	0.51639778

Level of GENDER	Level of MEAL	N	------------SCORE------------ Mean	Std Dev
F	BRKFST	9	4.55555556	1.58989867
F	DINNER	9	6.33333333	2.23606798
M	BRKFST	9	5.33333333	1.73205081
M	DINNER	9	7.22222222	1.56347192

[Continued]

```
Level of      Level of      Level of                 ------------SCORE------------
BRAND         GENDER        MEAL          N              Mean              Std Dev

A             F             BRKFST        3          4.33333333          0.57735027
A             F             DINNER        3          6.66666667          0.57735027
A             M             BRKFST        3          6.33333333          0.57735027
A             M             DINNER        3          7.66666667          0.57735027
B             F             BRKFST        3          3.00000000          1.00000000
B             F             DINNER        3          3.66666667          0.57735027
B             M             BRKFST        3          3.33333333          0.57735027
B             M             DINNER        3          5.33333333          0.57735027
C             F             BRKFST        3          6.33333333          0.57735027
C             F             DINNER        3          8.66666667          0.57735027
C             M             BRKFST        3          6.33333333          1.52752523
C             M             DINNER        3          8.66666667          0.57735027


Dependent Variable: SCORE

                 Tests of Hypotheses Using the Anova MS
                 for SUBJ(BRAND*GENDER) as an Error Term


Source                       DF      Anova SS      Mean Square   F Value   Pr > F

BRAND                         2    83.38888889    41.69444444     51.76   <.0001
GENDER                        1     6.25000000     6.25000000      7.76   0.0165
BRAND*GENDER                  2     3.50000000     1.75000000      2.17   0.1566

                 Tests of Hypotheses Using the Anova MS for
                   SUBJ*MEAL(BRAN*GEND) as an Error Term

Source                       DF      Anova SS      Mean Square   F Value   Pr > F

MEAL                          1    30.25000000    30.25000000     99.00   <.0001
BRAND*MEAL                    2     1.50000000     0.75000000      2.45   0.1278
GENDER*MEAL                   1     0.02777778     0.02777778      0.09   0.7682
BRAND*GENDER*MEAL             2     2.05555556     1.02777778      3.36   0.0692
```

What conclusions can we draw from these results? First, we notice that the variables BRAND, MEAL, and GENDER are all significant effects (BRAND and MEAL at p = .0001, GENDER at p = .016). We see from the SNK test that brand C is the preferred brand, followed by A and B. The fact that MEAL (breakfast or dinner) is significant and that BRAND * MEAL is not, tells us that all three brands of coffee are preferred after dinner.

H. THREE-FACTOR EXPERIMENTS WITH REPEATED MEASURES ON TWO FACTORS

As an example of a three-factor experiment with two repeated measures factors, we have designed a hypothetical study involving reading comprehension and a concept called slippage. It is well known that many students will do less well on a reading comprehension test in the early fall compared to the previous spring because of "slippage" during the summer vacation. As children grow older, the slippage should decrease. Also, slippage tends to be smaller with high-SES (socioeconomic status—roughly speaking, "wealthier") children compared to low-SES children, since high-SES children typically do more reading over the summer.

To test these ideas, the following experiment was devised:

A group of high- and low-SES children is selected for the experiment. Their reading comprehension is tested each spring and fall for three consecutive years. A diagram of the design is shown here:

Reading Comprehension Scores

		Years: 1		2		3	
		SPRING	**FALL**	**SPRING**	**FALL**	**SPRING**	**FALL**
HIGH	**subj**						
SES	1	61	50	60	55	59	62
	2	64	55	62	57	63	63
	3	59	49	58	52	60	58
	4	63	59	65	64	67	70
	5	62	51	61	56	60	63
LOW	6	57	42	56	46	54	50
SES	7	61	47	58	48	59	55
	8	55	40	55	46	57	52
	9	59	44	61	50	63	60
	10	58	44	56	49	55	49

Notice that each subject is measured each spring and fall of each year so that the variables SEASON and YEAR are both repeated measures factors. In this design, each subject belongs to either the high-SES or the low-SES group. Therefore, subjects are nested within SES.

We show three ways of writing a SAS program to analyze this experiment. First, using the REPEATED statement of PROC ANOVA:

```
DATA READ_1;
   INPUT SUBJ SES $ READ1-READ6;
   LABEL READ1 = 'SPRING YR 1'
         READ2 = 'FALL YR 1'
         READ3 = 'SPRING YR 2'
         READ4 = 'FALL YR 2'
         READ5 = 'SPRING YR 3'
         READ6 = 'FALL YR 3';
DATALINES;
                                              [Continued]
```

```
 1 HIGH 61 50 60 55 59 62
 2 HIGH 64 55 62 57 63 63
 3 HIGH 59 49 58 52 60 58
 4 HIGH 63 59 65 64 67 70
 5 HIGH 62 51 61 56 60 63
 6 LOW 57 42 56 46 54 50
 7 LOW 61 47 58 48 59 55
 8 LOW 55 40 55 46 57 52
 9 LOW 59 44 61 50 63 60
10 LOW 58 44 56 49 55 49
;
PROC ANOVA DATA=READ_1;
   TITLE "Reading Comprehension Analysis";
   CLASS SES;
   MODEL READ1-READ6 = SES / NOUNI;
   REPEATED YEAR 3, SEASON 2;
   MEANS SES;
RUN;
```

Since the REPEATED statement is confusing when we have more than one repeated factor, we again show you how to determine the order of the factor names. The variables listed on the MODEL statement are in the following order:

Year 1		Year 2		Year 3	
Spring	**Fall**	**Spring**	**Fall**	**Spring**	**Fall**
1	2	3	4	5	6

There are three levels of YEAR and two levels of SEASON. The factors following the keyword REPEATED are placed in order from the one that varies the slowest to the one that varies the fastest. For example, the first number (READ1) is from YEAR 1 in the SPRING. The next number (READ2) is still YEAR 1 but in the FALL. Thus, we say that SEASON is varying faster than YEAR. We must also be sure to indicate the number of levels of each factor following the factor name on the REPEATED statement.

```
REPEATED  YEAR  3,  SEASON  2;
```

This statement instructs the ANOVA procedure to choose the first level of YEAR (1), then loop through two levels of SEASON (SPRING FALL), then return to the next level of YEAR (2), followed by two levels of SEASON, etc. The product of the two levels must equal the number of variables in the dependent variable list of the MODEL statement. To check, $3 \times 2 = 6$ and we have READ1 to READ6 on the MODEL statement.

Results of running this program are shown next (some sections omitted):

```
Reading Comprehension Analysis

The ANOVA Procedure

        Class Level Information

Class          Levels    Values

SES               2       HIGH LOW

Number of Observations Read          10
Number of Observations Used          10

Repeated Measures Analysis of Variance

              Repeated Measures Level Information

Dependent Variable      READ1    READ2    READ3    READ4    READ5    READ6

      Level of YEAR        1        1        2        2        3        3
      Level of SEASON      1        2        1        2        1        2

Repeated Measures Analysis of Variance
Tests of Hypotheses for Between Subjects Effects

Source                  DF      Anova SS    Mean Square  F Value  Pr > F

SES                      1    680.0666667  680.0666667   13.54   0.0062
Error                    8    401.6666667   50.2083333

Univariate Tests of Hypotheses for Within Subject Effects

Source                  DF      Anova SS    Mean Square  F Value  Pr > F

YEAR                     2    252.0333333  126.0166667   26.91   <.0001
YEAR*SES                 2      1.0333333    0.5166667    0.11   0.8962
Error(YEAR)             16     74.9333333    4.6833333

                          Adj Pr > F
Source                  G — G     H — F

YEAR                    0.0002    <.0001
YEAR*SES                0.8186    0.8700
Error(YEAR)

Greenhouse-Geisser Epsilon    0.6757
Huynh-Feldt Epsilon           0.8658

                                                    [Continued]
```

Source	DF	Anova SS	Mean Square	F Value	Pr > F
SEASON	1	680.0666667	680.0666667	224.82	<.0001
SEASON*SES	1	112.0666667	112.0666667	37.05	0.0003
Error(SEASON)	8	24.2000000	3.0250000		

Source	DF	Anova SS	Mean Square	F Value	Pr > F
YEAR*SEASON	2	265.4333333	132.7166667	112.95	<.0001
YEAR*SEASON*SES	2	0.4333333	0.2166667	0.18	0.8333
Error(YEAR*SEASON)	16	18.8000000	1.1750000		

	Adj Pr > F	
Source	G − G	H − F
YEAR*SEASON	<.0001	<.0001
YEAR*SEASON*SES	0.7592	0.8168
Error(YEAR*SEASON)		

```
Greenhouse-Geisser Epsilon     0.7073
Huynh-Feldt Epsilon            0.9221
```

Level of		----------READ1----------		----------READ2----------	
SES	N	Mean	Std Dev	Mean	Std Dev
HIGH	5	61.8000000	1.92353841	52.8000000	4.14728827
LOW	5	58.0000000	2.23606798	43.4000000	2.60768096

Level of		----------READ3----------		----------READ4----------	
SES	N	Mean	Std Dev	Mean	Std Dev
HIGH	5	61.2000000	2.58843582	56.8000000	4.43846820
LOW	5	57.2000000	2.38746728	47.8000000	1.78885438

Level of		----------READ5----------		----------READ6----------	
SES	N	Mean	Std Dev	Mean	Std Dev
HIGH	5	61.8000000	3.27108545	63.2000000	4.32434966
LOW	5	57.6000000	3.57770876	53.2000000	4.43846820

We discuss the statistical results later, after two alternate programs have been presented.

We now present the other two programs that analyze this experiment without use of the REPEATED statement. Here is a second method: We have arranged our data so that each line represents one cell of our design. In practice, this would be tedious, but it will help you understand the last program for this problem in which all data for a subject are read on one line, and the data set is transformed to look like this one.

```
DATA READ_2;
     INPUT SUBJ SES $ YEAR SEASON $ READ;
DATALINES;
1 HIGH 1 SPRING 61
1 HIGH 1 FALL 50
1 HIGH 2 SPRING 60
1 HIGH 2 FALL 55
1 HIGH 3 SPRING 59
1 HIGH 3 FALL 62
2 HIGH 1 SPRING 64
(more data lines)
```

To simplify data entry (with the consequence of making the program more complicated), we can place all the data for each subject on one line. Since there is an alternative easier method (previous), you may skip the following, more elaborate, program and not sacrifice anything in the way of statistical understanding:

```
*---------------------------------------------------------------*
| Alternative Program for reading in the data for the reading   |
| experiment with all the data for one subject on one line.     |
*---------------------------------------------------------------*;
DATA READ_3;
    DO SES = 'HIGH','LOW';    (1)
        SUBJ = 0;    (2)
        DO N = 1 TO 5;    (3)
            SUBJ + 1;    (4)
            DO YEAR = 1 TO 3;    (5)
                DO SEASON = 'SPRING','FALL';    (6)
                    INPUT SCORE @;    (7)
                    OUTPUT;    (8)
                END;
            END;
        END;
    END;
DROP N;    (9)
DATALINES;
```

[Continued]

61	50	60	55	59	62
64	55	62	57	63	63
59	49	58	52	60	58
63	59	65	64	67	70
62	51	61	56	60	63
57	42	56	46	54	50
61	47	58	48	59	55
55	40	55	46	57	52
59	44	61	50	63	60
58	44	56	49	55	49
;					

(NOTE: The indentation is not necessary. It is used as a visual aid to keep the DO loops straight.)

This program is not as complicated as it may seem at first glance. Our data will be arranged in the same order as they appear in the diagram of the experimental design. All high-SES students will be read followed by the low-SES students. Each student will have a spring/fall set of reading comprehension scores for each of the three years.

The data are arranged with all the high-SES students followed by all the low-SES students. The DO loop ① sets the values of SES appropriately. The use of character values in a DO loop may be new to you. This very useful feature of SAS saves you the trouble of using numbers in DO loops and then formatting the numeric values to the character labels. Since there are five students in each SES group, the DO loop ③ goes from 1 to 5. The sum statement ④ will generate a subject number from 1 to 5 (it gets reset after all the data values for the HIGH-SES subjects have been read, shown in statement ②). Since the order of the data for each subject is YEAR, SEASON, the two DO loops ⑤ and ⑥ set the values of YEAR and SEASON before reading in a score in line ⑦. The trailing @ sign in ⑦ is necessary to prevent the pointer from going to a new line until all six scores have been read. The OUTPUT statement ⑧ will output an observation with the variables:

SES SUBJ YEAR SEASON READ

Note that the variable N is not included because of the DROP statement ⑨. It is not necessary to drop N; we simply don't need it. We could leave it in the data set and just not use it.

One final note: Be careful when using character values with DO loops because if the length of the first value in the DO loop is shorter than the other levels, the program will truncate the length of the character variable to the first length it encounters. To avoid this problem, either pad the first value with blanks to be equal to the length of the longest value, or use a LENGTH statement to define the length of the character variable.

This ends the discussion of the alternative program.

Now, regardless of the SAS data statements you used, the ANOVA statements will be the following:

```
PROC ANOVA DATA=READ_3;
    TITLE "Reading Comprehension Analysis";
    CLASS SUBJ SES YEAR SEASON;
    MODEL SCORE = SES SUBJ(SES)
                  YEAR SES*YEAR YEAR*SUBJ(SES)
                  SEASON SES*SEASON SEASON*SUBJ(SES)
                  YEAR*SEASON SES*YEAR*SEASON YEAR*SEASON*SUBJ(SES);
    MEANS YEAR / SNK E=YEAR*SUBJ(SES);
    MEANS SES SEASON SES*YEAR SES*SEASON YEAR*SEASON
          SES*YEAR*SEASON;
    TEST H=SES                      E=SUBJ(SES);
    TEST H=YEAR SES*YEAR            E=YEAR*SUBJ(SES);
    TEST H=SEASON SES*SEASON        E=SEASON*SUBJ(SES);
    TEST H=YEAR*SEASON SES*YEAR*SEASON
          E=YEAR*SEASON*SUBJ(SES);
RUN;
```

Selected portions of the output from this procedure is shown next:

```
Reading Comprehension Analysis

The ANOVA Procedure

Dependent Variable: SCORE

Level of            ------------SCORE------------
SES            N           Mean            Std Dev

HIGH           30      59.6000000       4.92425384
LOW            30      52.8666667       6.23523543

Level of            ------------SCORE------------
SEASON         N           Mean            Std Dev

FALL           30      52.8666667       7.26224689
SPRING         30      59.6000000       3.22276386
```

                                                    [Continued]

```
Level of      Level of                -----------SCORE-----------
SES           YEAR          N              Mean            Std Dev

HIGH          1             10         57.3000000         5.63816361
HIGH          2             10         59.0000000         4.13655788
HIGH          3             10         62.5000000         3.68932394
LOW           1             10         50.7000000         8.02842174
LOW           2             10         52.5000000         5.33853913
LOW           3             10         55.4000000         4.45221543

Level of      Level of                -----------SCORE-----------
SES           SEASON        N              Mean            Std Dev

HIGH          FALL          15         57.6000000         5.96178305
HIGH          SPRING        15         61.6000000         2.47270817
LOW           FALL          15         48.1333333         5.06904706
LOW           SPRING        15         57.6000000         2.61315354

Level of      Level of                -----------SCORE-----------
YEAR          SEASON        N              Mean            Std Dev

1             FALL          10         48.1000000         5.93389510
1             SPRING        10         59.9000000         2.80673792
2             FALL          10         52.3000000         5.71644800
2             SPRING        10         59.2000000         3.15524255
3             FALL          10         58.2000000         6.69659947
3             SPRING        10         59.7000000         3.91719855


Level of      Level of      Level of             -----------SCORE-----------
SES           YEAR          SEASON        N           Mean            Std Dev

HIGH          1             FALL          5        52.8000000        4.14728827
HIGH          1             SPRING        5        61.8000000        1.92353841
HIGH          2             FALL          5        56.8000000        4.43846820
HIGH          2             SPRING        5        61.2000000        2.58843582
HIGH          3             FALL          5        63.2000000        4.32434966
HIGH          3             SPRING        5        61.8000000        3.27108545
LOW           1             FALL          5        43.4000000        2.60768096
LOW           1             SPRING        5        58.0000000        2.23606798
LOW           2             FALL          5        47.8000000        1.78885438
LOW           2             SPRING        5        57.2000000        2.38746728
LOW           3             FALL          5        53.2000000        4.43846820
LOW           3             SPRING        5        57.6000000        3.57770876
```

Tests of Hypotheses Using the Anova MS for SUBJ(SES) as an Error Term

Source	DF	Anova SS	Mean Square	F Value	Pr > F
SES	1	680.0666667	680.0666667	13.54	0.0062

Tests of Hypotheses Using the Anova MS
for SUBJ*YEAR(SES) as an Error Term

Source	DF	Anova SS	Mean Square	F Value	Pr > F
YEAR	2	252.0333333	126.0166667	26.91	<.0001
SES*YEAR	2	1.0333333	0.5166667	0.11	0.8962

Tests of Hypotheses Using the Anova MS
for SUBJ*SEASON(SES) as an Error Term

Source	DF	Anova SS	Mean Square	F Value	Pr > F
SEASON	1	680.0666667	680.0666667	224.82	<.0001
SES*SEASON	1	112.0666667	112.0666667	37.05	0.0003

Tests of Hypotheses Using the Anova MS for
SUBJ*YEAR*SEASO(SES) as an Error Term

Source	DF	Anova SS	Mean Square	F Value	Pr > F
YEAR*SEASON	2	265.4333333	132.7166667	112.95	<.0001
SES*YEAR*SEASON	2	0.4333333	0.2166667	0.18	0.8333

As before, we specify hypotheses and error terms with TEST statements following our MODEL, and we include the appropriate error terms with the SNK requests.

Have our original ideas about "slippage" been confirmed by the data?

First, let us examine each of the main effects and their interactions:

What conclusions can we draw from these results?

1. High-SES students have higher reading comprehension scores than low-SES students (F = 13.54, p = .0062).

2. Reading comprehension increases with each year (F = 26.91, p = .0001). However, this increase is due partly to the smaller "slippage" in the later years [see (5)].

3. Students had higher reading comprehension scores in the spring compared to the following fall (F = 224.82, p = .0001).

4. The "slippage" was greater for the low-SES students (there was a significant SES * SEASON interaction [F = 37.05, p = .0003]).

5. "Slippage" decreases as the students get older (YEAR * SEASON is significant [F = 112.95, p = .0001]).

Repeated measures designs can be a powerful ally for the applied researcher. They can also be a little tricky. For example, in our coffee study, even though we randomized the order of first drinking the coffee with dinner or breakfast, there may be an effect we're overlooking. It may be that one (or all) of the brands take a little "getting used to." This could result in subjects preferring their second drinking of the coffee (whether at breakfast or dinner). We are ignoring this in our study and maybe we shouldn't be. Had we not randomized which drinking came first, we would have confounded drinking order with MEAL. The best way to make sure that you are getting what you want out of a repeated measures design is to consult a text that deals solely with the design and statistical issues involved. (Winer does an excellent job of this.)

PROBLEMS

Remember, you can download all the data sets and programs for these problems from the web site: www.prenhall.com/cody

8.1 A marketing survey is conducted to determine sport shirt preference. A questionnaire is presented to a panel of four judges. Each judge rates three shirts of three brands (X, Y, and Z). The data entry form is shown here:

MARKETING SURVEY FORM

1. Judge ID   ☐   1
2. Brand (1 = X, 2 = Y, 3=Z)   ☐   2
3. Color rating 9 = Best, 1 = Worst   ☐   3
4. Workmanship rating   ☐   4
5. Overall preference   ☐   5

An index is computed as follows:

```
INDEX=(3*OVERALL PREFERENCE + 2*WORKMANSHIP +
          COLOR RATING) / 6.0
```

The collected data follow:

```
11836
21747
31767
41846
12635
22534
32546
42436
13988
23877
33978
43887
```

Using analysis of variance, compare the color rating, workmanship, overall preference, and index among the three brands. (HINT: This is a repeated measures design—each judge rates all three brands.)

8.2 Run the following program to create a SAS data set called STATIN (NOTE: Statins are a class of drugs that lower cholesterol. Some raise the "good" cholesterol [HDL] while lowering the "bad" cholesterol [LDL], as well as the total cholesterol.) Each patient has his or her cholesterol measured at the end of 6 weeks while on drug A, B, or C. Ignoring GENDER and DIET in the data set, compare the total cholesterol (TOTAL), the LDL, and the HDL among the three drugs.

```
DATA STATIN;
   DO SUBJ = 1 TO 20;
      IF RANUNI(1557) LT .5 THEN GENDER = 'FEMALE';
      ELSE GENDER = 'MALE';
      IF RANUNI(0) LT .3 THEN DIET = 'HIGH FAT';
      ELSE DIET = 'LOW FAT';
      DO DRUG = 'A','B','C';
         LDL = ROUND(RANNOR(1557)*20 + 110
                      + 5*(DRUG EQ 'A')
                      - 10*(DRUG EQ 'B')
                      - 5*(GENDER EQ 'FEMALE')
                      + 10*(DIET EQ 'HIGH FAT'));
         HDL = ROUND(RANNOR(1557)*10 + 20
                      + .2*LDL
                      + 12*(DRUG EQ 'B'));
         TOTAL = ROUND(RANNOR(1557)*20 + LDL + HDL + 50
                      -10*(GENDER EQ 'FEMALE')
                      +10*(DIET EQ 'HIGH FAT'));
         OUTPUT;
      END;
   END;
RUN;
```

8.3 A taste test is conducted to determine which city has the best-tasting tap water. A panel of four judges tastes each of the samples from the four cities represented. The rating scale is a Likert scale with 1 = worst to 9 = best. Sample data and the coding scheme are shown here:

COLUMN	DESCRIPTION
1–3	Judge identification number
4	City code:
	1 = New York, 2 = New Orleans,
	3 = Chicago, 4 = Denver
5	Taste rating: 1 = worst 9 = best

Data:

```
00118
00126
00138
00145
```

```
00215
00226
00235
00244
00317
00324
00336
00344
00417
00425
00437
00443
```

Write a SAS program to describe these data and to perform an analysis of variance. Treat the ratings as interval data. Remember that we have a repeated measures design.

*8.4 Using the STATIN data set from problem 8.2, add GENDER to the model and test for GENDER and DRUG effects as well as a GENDER by DRUG interaction. Remember that DRUG is a repeated measure factor and the design is unbalanced. Generate an interaction graph for LDL, HDL, and TOTAL with DRUG on the x-axis and GENDER as the plotting symbol. Use either PROC PLOT or PROC GPLOT.

*8.5 The same data as in problem 8.3 are to be analyzed. However, they are arranged so that the four ratings from each judge are on one line. Thus, columns 1–3 are for the judge ID, column 4 is the rating for New York, column 5 for New Orleans, column 6 for Chicago, and column 7 for Denver. Our reformed data are shown here:

```
0018685
0025654
0037464
0047573
```

Write the DATA statements to analyze this arrangement of the data. You will need to create a variable for CITY and to have one observation per city. Also, run these data using the REPEATED statement of PROC ANOVA. How do the two solutions compare?

*8.6 Again, using the STATIN data set from problem 8.2, add DIET (not a repeated factor) to the model. The independent variables are GENDER, DRUG (repeated), and DIET.

*8.7 A study is conducted to test the area of nerve fibers in NORMAL and DIABETIC rats. A sample from the DISTAL and PROXIMAL ends of each nerve fiber is measured for each rat. Therefore, we have GROUP (NORMAL versus CONTROL) and LOCATION (DISTAL versus PROXIMAL) as independent variables, with location as a repeated measure (each rat nerve is measured at each end of the nerve fiber). The data are shown here:

	RATNO	DISTAL	PROXIMAL
	1	34	38
Normal	2	28	38
	3	38	48
	4	32	38
	5	44	42
Diabetic	6	52	48
	7	46	46
	8	54	50

Write a SAS program to enter these data and run a two-way analysis of variance, treating the location as a repeated measure. Use the REPEATED option for the LOCATION variable. Is there any difficulty in interpreting the main effects? Why?

8.8 Restructure the STATIN data set so that you can use the REPEATED statement of PROC GLM (or PROC MIXED). Do this only for LDL cholesterol. Include in your model GENDER, LDL_A, LDL_B, and LDL_C. Compare the result with the results from problem 8.4.

8.9 What's wrong with this program?

```
1    DATA FINDIT;
2       DO GROUP='CONTROL','DRUG';
3          DO TIME='BEFORE','AFTER';
4             DO SUBJ=1 TO 3;
5                INPUT SCORE
6          END;
7       END;
8       END;
9    DATALINES;
10 13 15 20 (data for subject 1) Order is CONTROL TIME 1,
12 14 16 18 (data for subject 2) CONTROL TIME 2, DRUG TIME 1,
15 18 22 28 (data for subject 3) and DRUG TIME 2
;
10   PROC ANOVA DATA=FINDIT;
11      TITLE "ANALYSIS OF VARIANCE";
12        CLASS SUBJ GROUP TIME;
13        MODEL SCORE = GROUP SUBJ(GROUP)
14              TIME GROUP*TIME TIME*SUBJ(GROUP);
15              TEST H=GROUP E=SUBJ(GROUP);
16        TEST H=TIME GROUP*TIME E=TIME*SUBJ(GROUP);
17        LSMEANS GROUP|TIME;
18   RUN;
```

Note: The comments within parentheses are not part of the program.

8.10 Repeat problem 8.8 except include DIET in the model. Compare your results with the results from problem 8.6.

C H A P T E R   9

Multiple-Regression Analysis

A. Introduction
B. Designed Regression
C. Nonexperimental Regression
D. Stepwise and Other Variable Selection Methods
E. Creating and Using Dummy Variables
F. Using the Variance Inflation Factor to Look for Multicollinearity
G. Logistic Regression
H. Automatic Creation of Dummy Variables with PROC LOGISTIC

A.   INTRODUCTION

Multiple-regression analysis is a method for relating two or more independent variables to a dependent variable. While the dependent variable (the variable you want to predict) must be a continuous variable (except with logistic regression), the independent variables may either be continuous or categorical, such as "gender" or "type of medication." In the case of categorical independent variables, we need to create "dummy" variables rather than use the actual character values (more on this later). If all of your independent variables are categorical (or most of them), you may be better off using analysis-of-variance techniques.

There are two rather distinct uses of multiple regression, and they will be addressed separately. The first use is for studies where the levels of the independent variables have been experimentally controlled (such as amount of medication and number of days between dosages). This use will be referred to as "designed regression." The second use involves settings where a sample of subjects have been observed on a number of naturally occurring variables (i.e., age, income, level of anxiety, etc.), which are then related to some outcome of interest. This use of regression will be referred to as "nonexperimental regression."

It is fairly easy to misuse regression. We will try to note some popular pitfalls, but we cannot list them all. A rule of thumb is to use your common sense. If the results of an analysis don't make any sense, get help. Ultimately, statistics is a tool employed to help us understand life. Although understanding life can be tricky, it is not usually perverse. Before accepting conclusions that seem silly based on statistical analyses, consult with a veteran data analyst. Most truly revolutionary results from data analyses are based on data-entry errors.

B.   DESIGNED REGRESSION

Imagine a researcher interested in the effects of scheduled exercise and the use of a stimulant for weight loss. She constructs an experiment using 24 college sophomores where four levels of stimulant and three levels of exercise are used. There are 24 subjects in the experiment and each is randomly assigned to a level of exercise and stimulant, such that two students are in each of the 12 (3 × 4) possible combinations of exercise and stimulant. After 3 weeks of participation, a measure of weight loss (post − pre weight) is obtained for each subject. The data for the experiment might look as shown here:

Data for Weight-Loss Experiment

Subject	Stimulant (mg/day)	Exercise (hr/week)	Weight Loss (pounds)
1	100	0	−4
2	100	0	0
3	100	5	−7
4	100	5	−6
5	100	10	−2
6	100	10	−14
7	200	0	−5
8	200	0	−2
9	200	5	−5
10	200	5	−8
11	200	10	−9
12	200	10	−9
13	300	0	1
14	300	0	0
15	300	5	−3
16	300	5	−3
17	300	10	−8
18	300	10	−12
19	400	0	−5
20	400	0	−4
21	400	5	−4
22	400	5	−6
23	400	10	−9
24	400	10	−7

These data can be analyzed either as a 3 × 4 analysis of variance, or as a two-variable multiple regression. The regression approach typically assumes that the effects of exercise and medication increase linearly (i.e., in a straight line); the ANOVA model makes no such assumption. If we use the multiple regression approach, the following program will provide the desired results:

```
DATA WEIGHT_LOSS;
    INPUT ID DOSAGE EXERCISE LOSS;
DATALINES;
1   100  0 -4
2   100  0  0
3   100  5 -7
4   100  5 -6
5   100 10 -2
6   100 10 -14
7   200  0 -5
8   200  0 -2
9   200  5 -5
10   200  5 -8
11   200 10 -9
12   200 10 -9
13   300  0 1
14   300  0 0
15   300  5 -3
16   300  5 -3
17   300 10 -8
18   300 10 -12
19   400  0 -5
20   400  0 -4
21   400  5 -4
22   400  5 -6
23   400 10 -9
24   400 10 -7
;
PROC REG DATA=WEIGHT_LOSS;
    TITLE "Weight Loss Experiment - Regression Example";
    MODEL LOSS=DOSAGE EXERCISE / P R;
RUN;
QUIT;
```

The first three lines create the data set. PROC REG performs a wide variety of regression models. The MODEL statement indicates that the dependent variable (the one to the left of the equal sign) is LOSS, and the two independent (or predictor) variables (the ones to the right of the equal sign) are DOSAGE and EXERCISE. The options "P" and "R" specify that we want predicted values and residuals to be computed. Note the use of a QUIT statement at the end of this procedure. PROC REG, as well as PROC ANOVA and PROC GLM, are considered to be "interactive" procedures (also known as "RUN GROUP processing"). That is, after a RUN statement has been encountered and the procedure has executed, you may submit additional statements (new models, for example). The top line of the SAS Display Manager will continue to show the procedure "running" until a new procedure is submitted or until a QUIT statement is submitted. The use of a QUIT statement is therefore optional.

The output from this program is presented here:

```
Weight Loss Experiment - Regression Example

The REG Procedure
Model: MODEL1
Dependent Variable: LOSS

Number of Observations Read        24
Number of Observations Used        24

                        Analysis of Variance

                            Sum of        Mean
Source              DF      Squares       Square   F Value   Pr > F

Model                2    162.97083     81.48542     11.19   0.0005
Error               21    152.98750      7.28512
Corrected Total     23    315.95833


Root MSE              2.69910    R-Square     0.5158
Dependent Mean       -5.45833    Adj R-Sq     0.4697
Coeff Var           -49.44909

                        Parameter Estimates

                    Parameter      Standard
Variable      DF     Estimate         Error    t Value    Pr > |t|

Intercept      1     -2.56250       1.50884      -1.70      0.1042
DOSAGE         1      0.00117       0.00493       0.24      0.8151
EXERCISE       1     -0.63750       0.13495      -4.72      0.0001

                        Output Statistics

         Dependent Predicted   Std Error                Std Error  Student
   Obs    Variable     Value  Mean Predict  Residual  Residual  Residual

    1     -4.0000    -2.4458      1.1425    -1.5542     2.445    -0.636
    2          0     -2.4458      1.1425     2.4458     2.445     1.000
    3     -7.0000    -5.6333      0.9219    -1.3667     2.537    -0.539
    4     -6.0000    -5.6333      0.9219    -0.3667     2.537    -0.145
    5     -2.0000    -8.8208      1.1425     6.8208     2.445     2.789

                                                        [Continued]
```

```
  6   -14.0000   -8.8208    1.1425   -5.1792    2.445   -2.118
  7    -5.0000   -2.3292    0.9053   -2.6708    2.543   -1.050
  8    -2.0000   -2.3292    0.9053    0.3292    2.543    0.129
  9    -5.0000   -5.5167    0.6035    0.5167    2.631    0.196
 10    -8.0000   -5.5167    0.6035   -2.4833    2.631   -0.944
 11    -9.0000   -8.7042    0.9053   -0.2958    2.543   -0.116
 12    -9.0000   -8.7042    0.9053   -0.2958    2.543   -0.116
 13     1.0000   -2.2125    0.9053    3.2125    2.543    1.263
 14          0   -2.2125    0.9053    2.2125    2.543    0.870
 15    -3.0000   -5.4000    0.6035    2.4000    2.631    0.912
 16    -3.0000   -5.4000    0.6035    2.4000    2.631    0.912
 17    -8.0000   -8.5875    0.9053    0.5875    2.543    0.231
 18   -12.0000   -8.5875    0.9053   -3.4125    2.543   -1.342
 19    -5.0000   -2.0958    1.1425   -2.9042    2.445   -1.188
 20    -4.0000   -2.0958    1.1425   -1.9042    2.445   -0.779
 21    -4.0000   -5.2833    0.9219    1.2833    2.537    0.506
 22    -6.0000   -5.2833    0.9219   -0.7167    2.537   -0.283
 23    -9.0000   -8.4708    1.1425   -0.5292    2.445   -0.216
 24    -7.0000   -8.4708    1.1425    1.4708    2.445    0.601
```

Output Statistics

```
                              Cook's
Obs    -2-1 0 1 2               D

  1  |      *|          |       0.029
  2  |       |**        |       0.073
  3  |      *|          |       0.013
  4  |       |          |       0.001
  5  |       |*****     |       0.566
  6  |  ****|           |       0.326
  7  |    **|           |       0.047
  8  |       |          |       0.001
  9  |       |          |       0.001
 10  |     *|           |       0.016
 11  |       |          |       0.001
 12  |       |          |       0.001
 13  |       |**        |       0.067
 14  |       |*         |       0.032
 15  |       |*         |       0.015
 16  |       |*         |       0.015
 17  |       |          |       0.002
 18  |    **|           |       0.076
 19  |    **|           |       0.103
 20  |     *|           |       0.044
 21  |       |*         |       0.011
 22  |       |          |       0.004
 23  |       |          |       0.003
 24  |       |*         |       0.026
```

Sum of Residuals   0
Sum of Squared Residuals   152.98750
Predicted Residual SS (PRESS)   212.03588

The output begins with an analysis-of-variance table, which looks much as it would from a standard ANOVA. We can see that there are two degrees of freedom in this model, one for EXERCISE and one for DOSAGE. There is only one degree of freedom for each since the regression estimates a single straight line for each variable rather than estimating a number of cell means.

The sum of squares for the model (162.971) tells us how much of the variation in weight loss is attributable to EXERCISE and DOSAGE. The mean square for the model (81.485) is the sum of squares (162.971) divided by the degrees of freedom for the model (2). This mean square is then divided by the mean square error (7.285) to produce the F-statistic for the regression (11.185). The p-value for this is reported as .0005. "C TOTAL" means "corrected total" and indicates the total degrees of freedom (23) and sum of squares (315.958) in the dependent variable. In regression, the corrected total degrees of freedom is always one less than the total sample size since one degree of freedom is used to estimate the grand mean. The ROOT MSE (2.699) stands for the square root of the mean square error and represents, in standard deviation units, the variation in the system not attributable to EXERCISE or DOSAGE. Dependent Mean (−5.458) is simply the mean of the dependent variable (LOSS). The R-Square (.5158) is the square of the multiple correlation of EXERCISE and DOSAGE with LOSS. It is the proportion of variance in LOSS explained by (attributable to) the independent variables. Adj R-Sq (.4697) is the adjusted r-square. The adjusted r-square takes into account how many variables were used in the equation and slightly lowers the estimate of explained variance. Coeff Var (−49.449) stands for coefficient of variation and is calculated by dividing the ROOT MSE by the mean and multiplying by 100. The CV is sometimes useful when the mean and standard deviation are related (such as in income data).

The bottom part of the output shows us, observation by observation, the actual LOSS, the predicted value, and the difference between the two (residual). In addition, the column labeled "Student Residuals" expresses the residual as a t-score, and Cook's D is a distance measure that helps us determine how strongly a particular data point affects the overall regression. Large absolute values of D (2 or more) indicate possible problems with your model or data points that require some careful scrutiny.

Having explained the terms in the analysis-of-variance table for the regression, let's summarize what meaning we can infer. Basically, the table indicates that the independent variables were related to the dependent variable (since the F was significant at p = .0005). Furthermore, we find that about 50% of the variation in weight loss is explained by the two experimental treatments. Many researchers are more interested in the r-square statistic than in the p-value, since the r-square represents an estimate of how strongly related the variables were. The bottom half of the printout contains the estimates of the parameters of the regression equation. Three parameters are estimated: (1) the intercept, or constant, term; (2) the coefficient for DOSAGE; and (3) the coefficient for EXERCISE. Each parameter estimate was based on one degree of freedom (always the case in regressions). For each parameter estimate, a standard error was estimated along with a t-statistic and a p-value for the t-statistic. The t-statistic is simply the parameter estimate divided by its standard error, and it is based on the number of degrees of freedom for the error term (21 for this example).

The section of the output labeled "Parameter Estimates" tells us that it was really EXERCISE that caused the weight loss. The regression coefficient for DOSAGE is not statistically significantly different from zero (p = .8151). The fact that the intercept was not significantly different from zero is irrelevant here. The intercept merely tells us where the regression line (or plane, in this case) crosses the y-axis and does not explain any variation.

At this point, many researchers would run a new regression with DOSAGE eliminated, to refine the estimate of EXERCISE. Since this was a designed experiment, we would recommend leaving the regression as is for purposes of reporting. Dropping DOSAGE won't affect the estimated impact of EXERCISE since DOSAGE and EXERCISE are uncorrelated (by design). When the independent variables in a regression are uncorrelated, the estimates of the regression coefficients are unchanged by adding or dropping independent variables. When the independent variables are correlated, dropping or adding variables strongly affects the regression estimates and hypothesis tests.

C.   NONEXPERIMENTAL REGRESSION

Many, if not most, regression analyses are conducted on data sets where the independent variables show some degree of correlation. These data sets, resulting from nonexperimental research, are common in all fields. Studies of factors affecting heart disease or the incidence of cancer, studies relating student characteristics to student achievement, and studies predicting economic trends all utilize nonexperimental data. The potential for a researcher to be misled by a nonexperimental data set is high; for a novice researcher, it is a near certainty. We strongly urge consultation with a good text in this area (Pedhazur's *Multiple Regression in Behavioral Research* is excellent) or with a statistical consultant. Having made this caveat, let's venture into regression analysis for nonexperimental data sets.

The Nature of the Data

There are many surface similarities between experimental and nonexperimental data sets. First, there are one or more outcomes or dependent variables. Second, there are several independent variables (sometimes quite a few). The basic difference here is that the independent variables are correlated with one another. This is because in nonexperimental studies, one defines a population of interest (people who have had heart attacks, sixth grade students, etc.), draws a sample, and measures the variables of interest. The goal of the study is usually to explain variation in the dependent variable by one or more of the independent variables. So far it sounds simple.

The problem is that correlation among the independent variables causes the regression estimates to change depending on which independent variables are being used. That is, the impact of B on A depends on whether C is in the equation or not. With C omitted, B can look very influential. With C included, the impact of B can disappear completely! The reason for this is as follows: A regression coefficient tells us the unique contribution of an independent variable to a dependent variable. That is, the coefficient for B tells us what B contributes all by itself with no overlap with any other variable. If B is the only variable in the equation, this is no problem. But if we add C, and if B and C are correlated, then the unique contribution of B on A will be changed. Let's see how this works in an example.

The subjects are a random sample of sixth grade students from Metropolitan City School District. The following measures have been taken on the subjects:

1. ACH6: Reading achievement at the end of sixth grade.
2. ACH5: Reading achievement at the end of fifth grade.
3. APT: A measure of verbal aptitude taken in the fifth grade.
4. ATT: A measure of attitude toward school taken in fifth grade.
5. INCOME: A measure of parental income (in thousands of dollars per year).

Our data set is listed here. (NOTE: These are not actual data.)

ID	ACH6	ACH5	APT	ATT	INCOME
1	7.5	6.6	104	60	67
2	6.9	6.0	116	58	29
3	7.2	6.0	130	63	36
4	6.8	5.9	110	74	84
5	6.7	6.1	114	55	33
6	6.6	6.3	108	52	21
7	7.1	5.2	103	48	19
8	6.5	4.4	92	42	30
9	7.2	4.9	136	57	32
10	6.2	5.1	105	49	23
11	6.5	4.6	98	54	57
12	5.8	4.3	91	56	29
13	6.7	4.8	100	49	30
14	5.5	4.2	98	43	36
15	5.3	4.3	101	52	31
16	4.7	4.4	84	41	33
17	4.9	3.9	96	50	20
18	4.8	4.1	99	52	34
19	4.7	3.8	106	47	30
20	4.6	3.6	89	58	27

The purpose of the study is to understand what underlies the reading achievement of the students in the district. The following program was written to analyze the data:

```
DATA NONEXP;
    INPUT ID ACH6 ACH5 APT ATT INCOME;
DATALINES;
                                                    [Continued]
```

```
 1   7.5   6.6   104   60   67
 2   6.9   6.0   116   58   29
 3   7.2   6.0   130   63   36
 4   6.8   5.9   110   74   84
 5   6.7   6.1   114   55   33
 6   6.6   6.3   108   52   21
 7   7.1   5.2   103   48   19
 8   6.5   4.4    92   42   30
 9   7.2   4.9   136   57   32
10   6.2   5.1   105   49   23
11   6.5   4.6    98   54   57
12   5.8   4.3    91   56   29
13   6.7   4.8   100   49   30
14   5.5   4.2    98   43   36
15   5.3   4.3   101   52   31
16   4.7   4.4    84   41   33
17   4.9   3.9    96   50   20
18   4.8   4.1    99   52   34
19   4.7   3.8   106   47   30
20   4.6   3.6    89   58   27
;
PROC REG DATA=NONEXP;
    TITLE "Nonexperimental Design Example";
MODEL ACH6 = ACH5 APT ATT INCOME /
    SELECTION = FORWARD;
MODEL ACH6 = ACH5 APT ATT INCOME /
    SELECTION = MAXR;
RUN;
QUIT;
```

By using the SELECTION = FORWARD option of PROC REG, we have specified that a forward regression is to be run with ACH6 as the dependent variable, and ACH5, APT, ATT, and INCOME as independent variables. Each variable will be tested, and the one that produces the largest F value will be entered first (if the p-value for entry is less than the specified or default value). Since we did not specify a criteria for entry into the model, the default value of .50 is used. If we want to change the p-value for entry into the model, we can include the MODEL option SLENTRY= our p-value. Although we did not choose to show it, we could run a stepwise regression (where variables that enter the model can also leave the model later) where we can specify both entry-level p-value (with SLENTRY= our p-value) and a p-value for staying in the model (SLSTAY= our p-value). For selection method STEPWISE, the default entry and staying p-values are both .15.

We also want to run the model using the MAXR technique. Before examining the output, we should discuss briefly stepwise regression and nonexperimental data.

D.   STEPWISE AND OTHER VARIABLE SELECTION METHODS

As mentioned earlier, with nonexperimental data sets, the independent variables are not truly "independent" in that they are usually correlated with one another. If these correlations are moderate to high (say 0.50 and above), then the regression coefficients are greatly affected by that particular subset of independent variables that are in the regression equation. If there are a number of independent variables to consider, coming up with the best subset can be difficult. Variable selection methods, including stepwise, were developed to assist researchers in arriving at this optimal subset. Unfortunately, many of these methods are frequently misused. The problem is that the solution from a purely statistical point of view is often not the best from a substantive perspective. That is, a lot of variance is explained but the regression doesn't make much sense and isn't very useful. We'll discuss this more when we examine the printout.

Stepwise regression examines a number of different regression equations. Basically, the goal of stepwise techniques is to take a set of independent variables and put them into a regression one at a time in a specified manner until all variables have been added or until a specified criterion has been met. The criterion is usually one of statistical significance or the improvement in the explained variance.

SAS software allows for a number of variable selection techniques. Among them are:

1. **FORWARD:** Starts with the best single regressor, then finds the best one to add to what exists, then the next best, etc.
2. **BACKWARD:** Starts with all variables in the equation, then drops the worst one, then the next, etc.
3. **STEPWISE:** Similar to FORWARD except that there is an additional step where all variables in each equation are checked again to see if they remain significant after the new variable has been entered.
4. **MAXR:** A rather complicated procedure, but basically it tries to find the one-variable regression with the highest r-square, then the two-variable regression with the highest r-square, etc.
5. **MINR:** Very similar to the MAXR, except that the selection system is slightly different.

Now, let's examine the printout from the program:

```
Nonexperimental Design Example

The REG Procedure
Model: MODEL1
Dependent Variable: ACH6

Number of Observations Read        20
Number of Observations Used        20

                                              [Continued]
```

```
Forward Selection: Step 1

Variable ACH5 Entered: R-Square = 0.6691 and C(p) = 1.8755

                        Analysis of Variance

                            Sum of        Mean
Source              DF      Squares       Square    F Value   Pr > F

Model                1     12.17625     12.17625     36.40    <.0001
Error               18      6.02175      0.33454
Corrected Total     19     18.19800

                 Parameter     Standard
Variable         Estimate        Error     Type II SS  F Value   Pr > F

Intercept         1.83725      0.71994       2.17866     6.51    0.0200
ACH5              0.86756      0.14380      12.17625    36.40    <.0001

Bounds on condition number: 1, 1
```

```
Forward Selection: Step 2

Variable APT Entered: R-Square = 0.7082 and C(p) = 1.7646

                        Analysis of Variance

                            Sum of        Mean
Source              DF      Squares       Square    F Value   Pr > F

Model                2     12.88735      6.44367     20.63    <.0001
Error               17      5.31065      0.31239
Corrected Total     19     18.19800

                 Parameter     Standard
Variable         Estimate        Error     Type II SS  F Value   Pr > F

Intercept         0.64270      1.05398       0.11616     0.37    0.5501
ACH5              0.72475      0.16814       5.80435    18.58    0.0005
APT               0.01825      0.01210       0.71110     2.28    0.1497

Bounds on condition number: 1.464, 5.8559
```

```
No other variable met the 0.5000 significance level for entry into the
model.

                  Summary of Forward Selection

         Variable    Number   Partial    Model
Step     Entered    Vars In   R-Square   R-Square    C(p)    F Value   Pr > F

  1      ACH5          1       0.6691     0.6691     1.8755   36.40    <.0001
  2      APT           2       0.0391     0.7082     1.7646    2.28    0.1497
```

The REG Procedure
Model: MODEL2
Dependent Variable: ACH6

Number of Observations Read   20
Number of Observations Used   20

Maximum R-Square Improvement: Step 1

Variable ACH5 Entered: R-Square = 0.6691 and C(p) = 1.8755

Analysis of Variance

Source	DF	Sum of Squares	Mean Square	F Value	Pr > F
Model	1	12.17625	12.17625	36.40	<.0001
Error	18	6.02175	0.33454		
Corrected Total	19	18.19800			

Variable	Parameter Estimate	Standard Error	Type II SS	F Value	Pr > F
Intercept	1.83725	0.71994	2.17866	6.51	0.0200
ACH5	0.86756	0.14380	12.17625	36.40	<.0001

Bounds on condition number: 1, 1

The above model is the best 1-variable model found.

Maximum R-Square Improvement: Step 2

Variable APT Entered: R-Square = 0.7082 and C(p) = 1.7646

Analysis of Variance

Source	DF	Sum of Squares	Mean Square	F Value	Pr > F
Model	2	12.88735	6.44367	20.63	<.0001
Error	17	5.31065	0.31239		
Corrected Total	19	18.19800			

Variable	Parameter Estimate	Standard Error	Type II SS	F Value	Pr > F
Intercept	0.64270	1.05398	0.11616	0.37	0.5501
ACH5	0.72475	0.16814	5.80435	18.58	0.0005
APT	0.01825	0.01210	0.71110	2.28	0.1497

Bounds on condition number: 1.464, 5.8559

[Continued]

The above model is the best 2-variable model found.

Maximum R-Square Improvement: Step 3

Variable ATT Entered: R-Square = 0.7109 and C(p) = 3.6194

Analysis of Variance

Source	DF	Sum of Squares	Mean Square	F Value	Pr > F
Model	3	12.93628	4.31209	13.11	0.0001
Error	16	5.26172	0.32886		
Corrected Total	19	18.19800			

Variable	Parameter Estimate	Standard Error	Type II SS	F Value	Pr > F
Intercept	0.80014	1.15586	0.15759	0.48	0.4987
ACH5	0.74740	0.18223	5.53198	16.82	0.0008
APT	0.01973	0.01299	0.75862	2.31	0.1483
ATT	−0.00798	0.02068	0.04893	0.15	0.7048

Bounds on condition number: 1.6336, 14.16

The above model is the best 3-variable model found.

Maximum R-Square Improvement: Step 4

Variable INCOME Entered: R-Square = 0.7223 and C(p) = 5.0000

Analysis of Variance

Source	DF	Sum of Squares	Mean Square	F Value	Pr > F
Model	4	13.14492	3.28623	9.76	0.0004
Error	15	5.05308	0.33687		
Corrected Total	19	18.19800			

Variable	Parameter Estimate	Standard Error	Type II SS	F Value	Pr > F
Intercept	0.91165	1.17841	0.20162	0.60	0.4512
ACH5	0.71374	0.18933	4.78747	14.21	0.0019
APT	0.02394	0.01419	0.95826	2.84	0.1124
ATT	−0.02116	0.02681	0.20983	0.62	0.4423
INCOME	0.00899	0.01142	0.20864	0.62	0.4435

Bounds on condition number: 2.4316, 31.793

The above model is the best 4-variable model found.

No further improvement in R-Square is possible.

Since a forward selection was requested first, that is what was run first. In step 1, the technique picked ACH5 as the first regressor since it had the highest correlation with the dependent variable ACH6. The r-square (variance explained) is 0.669, which is quite high. "C_p" is a statistic used in determining how many variables to use in the regression. You will need to consult one of the references in Chapter 1 or see your friendly statistician for help in interpreting Mallow's C_p statistic. A rule of thumb is to look at the number of independent variables in your equation, plus 1 (for the intercept), and when C_p approaches this value, no further variables need to be entered. The remaining statistics are the same as for the PROC REG program run earlier. On step 2, the technique determined that adding APT would lead to the largest increase in r-square. We notice, however, that r-square has only moved from 0.669 to 0.708, a slight increase. Furthermore, the regression coefficient for APT is nonsignificant (p = .1497). This indicates that APT doesn't tell us much more than we already knew from ACH5. Most researchers would drop it from the model and use the one-variable (ACH5) model. After step 2 has been run, the forward technique indicates that no other variable would come close to being significant. In fact, no other variable would have a p-value less than .50 (we usually require less than .05, although in regression analysis, it is not uncommon to set the inclusion level at .10).

The MAXR approach finds the best one-variable model, then the best two-variable model, etc., until the full model (all variables included) is estimated. As can be seen with these data, both of these techniques lead to the same conclusions: ACH5 is far and away the best predictor; it is a strong predictor, and no other variables would be included, with the possible exception of APT.

There is a problem here, however. Any sixth grade teacher could tell you that the best predictor of sixth grade performance is fifth grade performance. But it doesn't really tell us very much else. It might be more helpful to look at APT, ATT, and INCOME without ACH5 in the regression. Also, it could be useful to make ACH5 the dependent variable and have APT, ATT, and INCOME be regressors. Of course, this is suggesting quite a bit in the way of regressions. There is another regression technique that greatly facilitates looking at a large number of possibilities quickly. This is the RSQUARE selection method of PROC REG. The RSQUARE method will give us the multiple r-square value for every one, two, three, ..., n-way combinations of the variables in the independent variable list. The following lines will generate all of the regressions mentioned so far, as well as the model with ACH5 as the dependent variable (the MODEL option CP was added so that Mallow's C_p statistic would be computed for each of the computed models):

```
PROC REG DATA=NONEXP;
   MODEL ACH6 = INCOME ATT APT ACH5 / SELECTION=RSQUARE CP;
   MODEL ACH5 = INCOME ATT APT / SELECTION=RSQUARE CP;
RUN;
QUIT;
```

The output from PROC REG with RSQUARE the selection option is shown next:

```
Nonexperimental Design Example

The REG Procedure
Model: MODEL1
Dependent Variable: ACH6

R-Square Selection Method

Number of Observations Read          20
Number of Observations Used          20
```

Number in Model	R-Square	C(p)	Variables in Model
1	0.6691	1.8755	ACH5
1	0.3892	16.9947	APT
1	0.1811	28.2357	ATT
1	0.1017	32.5248	INCOME
2	0.7082	1.7646	APT ACH5
2	0.6696	3.8459	INCOME ACH5
2	0.6692	3.8713	ATT ACH5
2	0.4563	15.3711	INCOME APT
2	0.4069	18.0410	ATT APT
2	0.1856	29.9919	INCOME ATT
3	0.7109	3.6194	ATT APT ACH5
3	0.7108	3.6229	INCOME APT ACH5
3	0.6697	5.8446	INCOME ATT ACH5
3	0.4593	17.2116	INCOME ATT APT
4	0.7223	5.0000	INCOME ATT APT ACH5

```
Model: MODEL2
Dependent Variable: ACH5

R-Square Selection Method

Number of Observations Read          20
Number of Observations Used          20
```

Number in Model	R-Square	C(p)	Variables in Model
1	0.3169	2.8134	APT
1	0.2612	4.3496	ATT
1	0.1320	7.9081	INCOME
2	0.4127	2.1748	INCOME APT
2	0.3878	2.8604	ATT APT
2	0.2642	6.2652	INCOME ATT
3	0.4191	4.0000	INCOME ATT APT

The top part contains all of the RSQUAREs for every possible one-, two-, three, and four-variable regression with ACH6 as the outcome variable. It is possible to quickly glean a lot of information from this table.

Let us say that you have just decided that you don't want ACH5 as a regressor. You can quickly see from the one-variable regressions that APT is the next best regressor (r-square = .389). The next question is, "What is the best two-variable regression?" and "Is the improvement large enough to be worthwhile?" Let's look at the two-variable regressions that have APT in them:

```
R-square for APT + ATT = .407
              APT + INCOME = .456
```

(Remember, we're eliminating ACH5 for now.)

APT and INCOME is best, and the gain is .067 (which is equal to 0.456 − 0.389). Is a 6.7% increase in variance explained worth including? Probably it is, although it may not be statistically significant with our small sample size. In explaining the regressions using ACH5 as an outcome variable, we can see that APT and INCOME also look like the best bet there. In interpreting these data, we might conclude that aptitude combined with parental wealth are strong explanatory variables in reading achievement. It is important to remember that statistical analyses must make substantive sense. The question arises here as to how these two variables work to influence reading scores. Some researchers would agree that APT is a psychological variable and INCOME is a sociological variable, and the two shouldn't be mixed in a single regression. It's a bit beyond the scope of this book to speculate on this, but when running nonexperimental regressions, it is best to be guided by these two principles:

1. **Parsimony:** Less is more in terms of regressors. Another regressor will always explain a little bit more, but it often confuses our understanding of life.
2. **Common Sense:** The regressors must bear a logical relationship to the dependent variable in addition to a statistical one. (ACH6 would be a great predictor of ACH5, but it is a logical impossibility.)

Finally, whenever regression analysis is used, the researcher should examine the simple correlations among the variables. The following statements will generate a correlation matrix among all the variables of interest:

```
PROC CORR DATA=NONEXP NOSIMPLE;
   TITLE "Correlations from NONEXP Data Set";
   VAR APT ATTACH5 ACH6 INCOME;
RUN;
```

The output from running this procedure is:

```
Correlations from NONEXP Data Set

The CORR Procedure

  5  Variables:     APT       ATT       ACH5      ACH6       INCOME

              Pearson Correlation Coefficients, N = 20
                     Prob > |r|  under HO: Rho=0

                   APT          ATT          ACH5          ACH6          INCOME

APT            1.00000      0.49741      0.56297       0.62387       0.09811
                            0.0256       0.0098        0.0033        0.6807

ATT            0.49741      1.00000      0.51104       0.42559       0.62638
               0.0256                    0.0213        0.0614        0.0031

ACH5           0.56297      0.51104      1.00000       0.81798       0.36326
               0.0098       0.0213                     <.0001        0.1154

ACH6           0.62387      0.42559      0.81798       1.00000       0.31896
               0.0033       0.0614       <.0001                      0.1705

INCOME         0.09811      0.62638      0.36326       0.31896       1.00000
               0.6807       0.0031       0.1154        0.1705
```

An examination of the simple correlations often leads to a better understanding of the more complex regression analyses. Here we can see why ATT, which shows a fairly good correlation with ACH6, was never included in a final model. It is highly related to INCOME ($r = .626$) and to APT ($r = .497$). Whatever relationship it had with ACH6 was redundant with APT and INCOME. Also notice that INCOME and APT are unrelated, which contributes to their inclusion in the model. The simple correlations also protect against misinterpretation of "suppressor variables." These are a little too complex to discuss here. However, they can be spotted when a variable is not significantly correlated with the dependent variable but in the multiple regression has a significant regression coefficient (usually negative). You should get help (from Pedhazur or from another text or a consultant) with interpreting such a variable.

There is a tendency when using stepwise regression to let the statistical technique take precedence over theory and logic. Many researchers recommend that stepwise techniques should never be used. This might be too strong a statement, but generally speaking, it is wise to use stepwise techniques sparingly, if at all.

E.   CREATING AND USING DUMMY VARIABLES

As we mentioned in the introduction to this chapter, categorical variables, such as gender or race, can be used as independent variables in a multiple regression, provided you first create dummy variables. When a categorical variable has only two levels, such as gender, this is very easy. You code the dummy variable as either FEMALE or NOT FEMALE (or MALE versus NOT MALE). It is traditional to use the values of 0 and 1 for these dummy variables. So, if you have a variable called GENDER with values of 'F' and 'M', you would create a dummy variable with the following statements:

```
IF GENDER = 'F' THEN FEMALE = 1;
ELSE IF GENDER='M' THEN FEMALE = 0;
```

When this dummy variable is used in a regression, the coefficient of FEMALE will show how much to add or subtract (if the coefficient is negative) from the predicted value of the dependent variables if the subject is female. The reason for this is that we have chosen the 0 level for males, which makes the males the "reference" level (anything times 0 is still 0). Notice that we did not code this as:

```
IF GENDER = 'F' THEN FEMALE = 1;
ELSE FEMALE = 0;
```

The reason for this is that any value other than 'F' (missing or a data error) would result in a value of 0 for FEMALE (thus assuming the missing or invalid value was for a male). A good rule of thumb is to be careful when you use an ELSE statement without an IF before it.

A compact SAS statement to create FEMALE would be:

```
IF GENDER IN ('M' 'F') THEN FEMALE = (GENDER EQ 'F');
```

The logical expression, GENDER EQ 'F', returns a 0 or 1. Any missing or invalid data values will result in a missing value for GENDER. This statement is not preferrable to the previous two lines, but it is useful if you want to impress your boss!

What do you do when your independent categorical variable has more than two levels? You choose one of the levels as your "reference" and create k − 1 dummy variables where k is the number of levels for your categorical variable. As an example, suppose a variable RACE has levels of 'WHITE', 'AFRICAN AM', and 'ASIAN'. Arbitrarily choosing 'WHITE' as the reference level, you would create two dummy variables—one representing African American or not; the other representing Asian or not. Here is one way to code this:

```
IF RACE = 'AFRICAN  AM' THEN AF_AM = 1;
ELSE IF RACE NE ' ' THEN AF_AM = 0;
IF RACE = 'ASIAN' THEN ASIAN = 1;
ELSE IF RACE NE ' ' THEN ASIAN = 0;
```

A more compact way would be:

```
IF RACE IN ('AFRICAN AM' 'ASIAN' 'WHITE') THEN DO;
   AF_AM = (RACE EQ 'AFRICAN AM');
   ASIAN = (RACE EQ 'ASIAN');
END;
```

You may find it impractical to create dummy variables for categorical variables with a large number of values.

F. **USING THE VARIANCE INFLATION FACTOR TO LOOK FOR MULTICOLLINEARITY**

In our previous discussion of the models to predict high school achievement based on a number of predictors, we showed how highly correlated independent variables can make it difficult to obtain a good model and how to interpret the results. One of the primary diagnostics to detect the effect of this intercorrelation is the variance inflaction factor (VIF). This diagnostic is a measure of how much the variance of a regression coefficient is inflated by the fact that other independent variables contain the same information as the variable in question. Large values of this diagnostic indicate signs of serious multicollinearity. Values exceeding 10 should be investigated carefully.

To add this diagnostic to our multiple regression output, the MODEL option VIF is needed. So, to include this diagnostic in the model predicting ACH6 from ACH5, APT, ATT, and INCOME you would write:

```
PROC REG DATA=NONEXP;
   TITLE "Adding the Variance Inflation Factor to the Regression";
   MODEL ACH6 = ACH5 APT ATT INCOME / VIF;
RUN;
QUIT;
```

The portion of the output containing the VIF diagnostic is shown here:

Parameter Estimates

Variable	DF	Parameter Estimate	Standard Error	t Value	Pr > |t|	Variance Inflation
Intercept	1	0.91165	1.17841	0.77	0.4512	0
ACH5	1	0.71374	0.18933	3.77	0.0019	1.72141
APT	1	0.02394	0.01419	1.69	0.1124	1.86921
ATT	1	−0.02116	0.02681	−0.79	0.4423	2.43159
INCOME	1	0.00899	0.01142	0.79	0.4435	1.92607

Notice in this output that there are no large VIF values, so you can assume that there are no serious collinearity problems.

G. **LOGISTIC REGRESSION**

When you have a dependent variable with only two levels (such as dead/alive, sick/well), multiple regression techniques are not appropriate. Suppose you coded your dependent variable as a 1 for SICK and a 0 for WELL. You would like the regression equation to predict a number between 0 and 1, which could be interpreted as the probability that the subject was sick or well. However, using the multiple-regression methods described in the previous sections, the prediction equation could result in negative values or values greater than 1. This would be difficult to interpret.

A regression method called logistic regression was developed to handle this problem. Logistic regression uses a transformation (called a logit) that forces the prediction equation to predict values between 0 and 1. A logistic regression equation predicts the natural log of the odds for a subject being in one category or another. In addition, the regression coefficients in a logistic regression equation can be used to estimate odds ratios for each of the independent variables.

Although the details of logistic regression are beyond the scope of this book, we will demonstrate several ways to run logistic regression with the following data set.

We have recorded accident status (did the subject have an accident in the past year?), the age, vision status, driver education status, and gender of a number of individuals. The sample data, stored in a data set called C:\APPLIED\ACCIDENT.DTA is listed here (0 = No, 1 = Yes):

Accident Statistics Based on Age, Vision, and Driver's Education:

Did Subject Have an Accident?	Age	Vision Problem?	Driver Education	Gender
1	17	1	1	F
1	44	0	0	M
1	48	1	0	F
1	55	0	0	M
1	75	1	1	F
0	35	0	1	F
0	42	1	1	F
0	57	0	0	M
0	28	0	1	M
0	20	0	1	F
0	38	1	0	F
0	45	0	1	M
0	47	1	1	F
0	52	0	0	F
0	55	0	1	M
1	68	1	0	M
1	18	1	0	F
1	68	0	0	M
1	48	1	1	F
1	17	0	0	F
1	70	1	1	M
1	72	1	0	F
1	35	0	1	M

(Continued)

Did Subject Have an Accident?	Age	Vision Problem?	Driver Education	Gender
1	19	1	0	F
1	62	1	0	F
0	39	1	1	F
0	40	1	1	F
0	55	0	0	M
0	68	0	1	M
0	25	1	0	M
0	17	0	0	F
0	45	0	1	M
0	44	0	1	F
0	67	0	0	F
0	55	0	1	M
1	61	1	0	M
1	19	1	0	M
1	69	0	0	F
1	23	1	1	M
1	19	0	0	F
1	72	1	1	M
1	74	1	0	M
1	31	0	1	M
1	16	1	0	M
1	61	1	0	F

Our aim is to see if age, vision status, and driver education can be used to predict if the subject had an accident in the past year (we'll use the variable GENDER later on to demonstrate the CLASS statement of PROC LOGISTIC). The following is a program to create a SAS data set called LOGISTIC, which includes the variables just listed plus three new variables (AGEGROUP, YOUNG, and OLD) that will be used in later sections. The statements to run a forward stepwise logistic regression are included:

```
*-----------------------------------------------------*
| Program Name: LOGISTIC.SAS in C:\APPLIED            |
| Purpose: To demonstrate logistic regression         |
| Date: April 13, 2004                                |
*-----------------------------------------------------*;
PROC FORMAT;
   VALUE AGEGROUP   0 = '20 to 65 (inclusive)'
                    1 = '<20 or >65';
   VALUE VISION  0 = 'No Problem'
                 1 = 'Some Problem';
   VALUE YES_NO  0 = 'No'
                 1 = 'Yes';
RUN;
```

```
DATA LOGISTIC;
    INFILE 'C:\APPLIED\ACCIDENT.DTA' MISSOVER;
    INPUT ACCIDENT AGE VISION DRIVE_ED GENDER : $1.;
    IF NOT MISSING(AGE) THEN DO;
        IF AGE GE 20 AND AGE LE 65 THEN AGEGROUP = 0;
        ELSE AGEGROUP = 1;
        IF AGE LT 20 THEN YOUNG = 1;
        ELSE YOUNG = 0;
        IF AGE GT 65 THEN OLD = 1;
        ELSE OLD = 0;
    END;
    LABEL
        ACCIDENT = 'Accident in Last Year?'
        AGE      = 'Age of Driver'
        VISION   = 'Vision Problem?'
        DRIVE_ED = 'Driver Education?';
    FORMAT   ACCIDENT DRIVE_ED YOUNG OLD YES_NO.
             AGEGROUP AGEGROUP.
             VISION VISION.;
RUN;
PROC LOGISTIC DATA=LOGISTIC DESCENDING;
    TITLE "Predicting Accidents Using Logistic Regression";
    MODEL  ACCIDENT = AGE VISION DRIVE_ED /
        SELECTION = FORWARD
        CTABLE PPROB =(0 to 1 by .1)
        LACKFIT
        RISKLIMITS;
RUN;
QUIT;
```

The DATA step statements are straightforward. In case it's new to you, the MISSING function returns a value of TRUE (1) if its argument is missing, FALSE (0) otherwise. Note that this function works with either character or numeric arguments.

Let's explain the PROC LOGISTIC statements. One somewhat peculiar "feature" of PROC LOGISTIC is that the resulting equation predicts the log odds for the LOWER value of the dependent variable. The default for PROC LOGISTIC is to use formatted values to determine the ordering of variables. In this case, 'No' comes before 'Yes' so that PROC LOGISTIC will predict the probability of the lower value ('No') by default. Therefore, the equation would predict the log odds of NOT having an accident. One easy way to reverse this is to use the option DESCENDING. In our example, the use of this option will cause the program to predict the log odds of having an accident given a certain set of predictor or explanatory values. (Another way is to provide a format with values in the correct direction or, if the internal values are in the correct direction, use the PROC LOGISTIC option ORDER=INTERNAL to tell it to use the internal values and not the formatted values to determine which is the lower level.) If your odds ratios all seem to be in the wrong direction, check the top part of the output where the procedure lists which value it is predicting as the outcome.

Next, our MODEL statement looks just like the ones we used with PROC REG. This logistic regression example includes several MODEL options: The selection method is chosen to be FORWARD (the same as for regular regression); a classification table (CTABLE) is requested for all probabilities from 0 to 1 by .1; the Hosmer and Lemeshow Goodness-of-Fit test (LACKFIT) and the odds ratios for each variable in the equation with their 95% confidence limits (RISKLIMITS) are requested. Here are the results:

```
Predicting Accidents Using Logistic Regression

The LOGISTIC Procedure

                       Model Information

Data Set                     WORK.LOGISTIC
Response Variable            ACCIDENT            Accident in Last Year?
Number of Response Levels    2
Model                        binary logit
Optimization Technique       Fisher's scoring


Number of Observations Read         45
Number of Observations Used         45

           Response Profile  ①

   Ordered                   Total
     Value     ACCIDENT    Frequency

       1       Yes             25
       2       No              20


Probability modeled is ACCIDENT='Yes'.


Forward Selection Procedure   ②


Step  0. Intercept entered:

                   Model Convergence Status

        Convergence criterion (GCONV=1E-8) satisfied.


-2 Log L = 61.827

    Residual Chi-Square Test

Chi-Square        DF      Pr > ChiSq

   10.7057         3          0.0134
```

Step 1. Effect VISION entered:

Model Convergence Status

Convergence criterion (GCONV=1E-8) satisfied.

Model Fit Statistics

Criterion	Intercept Only	Intercept and Covariates
AIC	63.827	59.244
SC	65.633	62.857
−2 Log L	61.827	55.244

Testing Global Null Hypothesis: BETA=0

Test	Chi-Square	DF	Pr > ChiSq
Likelihood Ratio	6.5830	1	0.0103
Score	6.4209	1	0.0113
Wald	6.0756	1	0.0137

Residual Chi-Square Test

Chi-Square	DF	Pr > ChiSq
4.9818	2	0.0828

Step 2. Effect DRIVE_ED entered:

Model Convergence Status

Convergence criterion (GCONV=1E −8) satisfied.

Model Fit Statistics

Criterion	Intercept Only	Intercept and Covariates
AIC	63.827	56.287
SC	65.633	61.707
−2 Log L	61.827	50.287

Testing Global Null Hypothesis: BETA=0

Test	Chi-Square	DF	Pr > ChiSq
Likelihood Ratio	11.5391	2	0.0031
Score	10.5976	2	0.0050
Wald	8.5949	2	0.0136

Residual Chi-Square Test

[Continued]

```
Chi-Square        DF     Pr > ChiSq

   0.1293          1         0.7191
```

NOTE: No (additional) effects met the 0.05 significance level for entry
      into the model.

Summary of Forward Selection ③

	Effect		Number	Score		Variable
Step	Entered	DF	In	Chi-Square	Pr > ChiSq	Label
1	VISION	1	1	6.4209	0.0113	Vision Problem?
2	DRIVE_ED	1	2	4.8680	0.0274	Driver Education?

Analysis of Maximum Likelihood Estimates ④

Parameter	DF	Estimate	Standard Error	Wald Chi-Square	Pr > ChiSq
Intercept	1	0.1110	0.5457	0.0414	0.8389
VISION	1	1.7137	0.7049	5.9113	0.0150
DRIVE_ED	1	−1.5000	0.7037	4.5440	0.0330

Odds Ratio Estimates

Effect	Point Estimate	95% Wald Confidence Limits	
VISION	5.550	1.394	22.093
DRIVE_ED	0.223	0.056	0.886

Association of Predicted Probabilities and Observed Responses ⑤

Percent Concordant	67.2	Somers' D	0.532
Percent Discordant	14.0	Gamma	0.655
Percent Tied	18.8	Tau-a	0.269
Pairs	500	c	0.766

Wald Confidence Interval for Adjusted Odds Ratios ⑥

Effect	Unit	Estimate	95% Confidence Limits	
VISION	1.0000	5.550	1.394	22.093
DRIVE_ED	1.0000	0.223	0.056	0.886

Partition for the Hosmer and Lemeshow Test

Group	Total	ACCIDENT = Yes		ACCIDENT = No	
		Observed	Expected	Observed	Expected
1	11	2	2.20	9	8.80
2	11	6	5.80	5	5.20
3	10	6	5.80	4	4.20
4	13	11	11.19	2	1.81

Hosmer and Lemeshow Goodness-of-Fit Test ⑦

Chi-Square	DF	Pr > ChiSq
0.0756	2	0.9629

Classification Table ⑧

	Correct		Incorrect			Percentages			
Prob Level	Event	Non-Event	Event	Non-Event	Correct	Sensi-tivity	Speci-ficity	False POS	False NEG
0.000	25	0	20	0	55.6	100.0	0.0	44.4	.
0.100	25	0	20	0	55.6	100.0	0.0	44.4	.
0.200	23	0	20	2	51.1	92.0	0.0	46.5	100.0
0.300	23	9	11	2	71.1	92.0	45.0	32.4	18.2
0.400	23	9	11	2	71.1	92.0	45.0	32.4	18.2
0.500	17	9	11	8	57.8	68.0	45.0	39.3	47.1
0.600	11	14	6	14	55.6	44.0	70.0	35.3	50.0
0.700	11	18	2	14	64.4	44.0	90.0	15.4	43.8
0.800	11	18	2	14	64.4	44.0	90.0	15.4	43.8
0.900	0	18	2	25	40.0	0.0	90.0	100.0	58.1
1.000	0	20	0	25	44.4	0.0	100.0	.	55.6

Explanation of the Output

Let's examine the salient sections of this output. First is the "Response Profile," ① which lists the number of observations in each category of the outcome variable (AC-CIDENT). Pay careful attention to this, especially the ordered value information. Because we used the DESCENDING option on the PROC LOGISTIC statement, the value of 1 (formatted as 'Yes') is first in the list of ordered values. As we mentioned before, this means that this logistic model will be predicting the odds and probabilities of having an accident based on the explanatory variables.

The next section shows the order that the independent or explanatory variables entered the model ②. We see VISION entered first, with several criteria for assessing the importance of this variable in predicting accidents. The two criteria "−2 LOG L" and "Score" are both used to test whether the independent variable(s) is significant, based on a chi-squared distribution. We see that VISION is a significant explanatory variable using either of these two criteria (p approximately = .01). The other two criteria, "AIC" (Akaike Information Criterion) and "SC" (Schwartz Criterion) serve a similar purpose, except they adjust for the number of explanatory variables and for the number of observations used in the model. These statistics are useful for comparing different models; lower values of these statistics indicate a better-fitting model.

Looking farther down the output, we see DRIVE_ED (driver education) entered next. The overall model improved (based on a lower AIC and SC and a smaller p-value for −2 LOG L). Since no other variables met the default entry-level significance of .05, the model building stopped at this point.

The "Summary" section ③ is printed only for stepwise (FORWARD, BACK-WARD, or STEPWISE) selection methods. It summarizes the order in which the explanatory variables entered the model and shows the chi-square and p-value for each variable.

Section ④ gives us the parameter estimates for the logistic regression model. In this example, the equation is:

$$\text{Log (odds of having an accident)} = .1110 + 1.7137(\text{VISION})$$
$$- 1.5000(\text{DRIVE\_ED})$$

If we substitute values for VISION and DRIVE_ED into this equation, the results are the log (odds) of having an accident. To determine the odds, raise e (the base of common logarithms) to this power. To compute the probability that a person will have an accident, based on vision and driver education, you can use the relationship that:

$$\text{Odds} = \frac{P}{1 - P}$$

where P is the probability.

Solving for P, we get:

$$P = \frac{\text{Odds}}{1 + \text{Odds}}.$$

Let's use this equation to predict the odds and the probability of a person having an accident for given values of VISION and DRIVE_ED. For a person with no vision problem (VISION=0) and who never took a driver education course (DRIVE_ED=0), the calculation would be:

$$\log \text{(odds)} = .1110 + 1.7139 \times 0 - 1.5001 \times 0 = .1110.$$

Therefore, the odds of having an accident for this person are:

$$\text{Odds (of having an accident)} = \exp(.1110) = 1.1174.$$

And the probability of having an accident is:

$$p \text{ (having an accident)} = \frac{1.1174}{1 + 1.1174} = .5277.$$

Taking a similar person, except one with a vision problem (VISION=1), we again compute odds and probabilities:

$$\log \text{(odds)} = .1110 + 1.7139 \times 1 - 1.5001 \times 0 = 1.8249,$$

$$\text{Odds (of having an accident)} = \exp(1.8249) = 6.2022,$$

$$p \text{ (having an accident)} = \frac{6.2022}{1 + 6.2022} = .8612.$$

You can see that the odds of having an accident increase dramatically (from 1.1174 to 6.2022) when a person has a vision problem. We often look at the ratio of these odds,

$$\frac{6.2022}{1.1174} = 5.5506,$$

to describe the effect of an explanatory variable on the odds for an event. This ratio is called the odds ratio and is shown in a later section of the output (see Chapter 3, Section L for more detail on the odds ratio).

The section labeled "Association of Predicted Probabilities" ⑤ is somewhat complicated. It works this way: Take the 990 possible pairs of observations and find those in which the outcomes are not the same. In this example, there are 500 pairs where one unit of the pair had an accident and the other did not. Then compute the probability of each unit of the pair having an accident. If the unit with the higher computed probability is the one that actually experienced the event in question (an accident), this pair is labeled "Concordant." When a pair is in the "wrong" order, it is labeled "Discordant." If both units of the pair have the same probability, the pair is labeled "Tied." It is desirable to have a high concordant percentage and a low discordant percentage.

The "Wald Confidence Interval for Adjusted Odds Ratios" ⑥ is the result of the RISKLIMITS option. For each variable, it lists the odds ratio and the 95% confidence interval for this ratio. Notice that the odds ratio for VISION is 5.550, which is the same as we computed earlier (if we round our result). Since the 95% confidence interval does not contain one, we have additional confirmation that vision is a significant explanatory variable in our model.

The fact that the odds ratio for DRIVE_ED is less than 1 tells us that driver education helps reduce accidents.

The "Hosmer and Lemeshow Goodness-of-Fit" statistics ⑦ is a chi-square-based test to assess goodness of fit. Since you probably do not want to reject the null hypothesis that your data fit the specified model, you would like a high p-value for this test. In this example, the chi-square value of .0756 with two degrees of freedom gives us a p-value of .9629, which means that we do not reject the null hypothesis that these data fit this model.

We finally get to the "Classification Table" ⑧ which gives us the sensitivity, specificity, false positive rate, and false negative rate for several levels of probability. Suppose, for example, that you decide that any predicted probability greater than .3 should be considered a "positive diagnosis" for having an accident. In other words, you want to be somewhat conservative and consider a person an accident risk even though the probability of having an accident is less than .5. Based on the classification table, this cutoff for a "positive diagnosis" would have a high sensitivity (92%) and a relatively low specificity (45%). "Sensitivity," for those not familiar with the term, is the proportion of people who have had the event in question (an accident) and are predicted to have one ($p > .3$ in this case). Specificity would be the proportion of people who did not have an accident and who had a probability less than .3. Looking at this table, you can decide what a convenient cutoff for a "positive diagnosis" might be, depending on your desired sensitivity and specificity. We will discuss a graphical way of looking at this later in this section when we show you how to produce a receiver operator characteristic curve (ROC).

Creating a Categorical Variable from AGE

Either by inspection of the data or by experience, you may be surprised to find that age did not enter into the equation. To investigate this further, let's look at the age distributions for those who had accidents and those who did not. A simple PROC GCHART (or PROC CHART) can quickly do this for us. Here is the code:

```
OPTIONS PS=24;
PATTERN COLOR=BLACK VALUE=EMPTY;
PROC GCHART DATA=LOGISTIC;
    TITLE "Distribution of Ages by Accident Status";
    VBAR AGE / MIDPOINTS=10 TO 90 BY 10
    GROUP=ACCIDENT;
RUN;
```

Here is the resulting output:

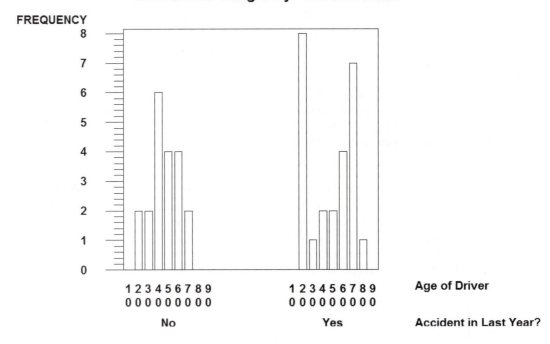

Notice that for the nonaccident group, there are more subjects in the center of the distribution, but in the accident group, there seems to be an excess of very young and older individuals. Based on this finding, we can create a new variable (AGE-GROUP) that will have a value of 0 for subjects between 20 and 65 (inclusive) and a value of 1 otherwise. With foresight, we already created this variable in the original data set. We can therefore use the new variable AGEGROUP instead of AGE and rerun the regression. Here is the modified program with the added option to create an output data set that will contain the sensitivity and 1-specificity (the false positive rate) so we can plot an ROC (receiver operator characteristic curve) later on:

```
PROC LOGISTIC DATA=LOGISTIC DESCENDING;
   TITLE "Predicting Accidents Using Logistic Regression";
   MODEL ACCIDENT = AGEGROUP VISION DRIVE_ED /
         SELECTION=FORWARD
         CTABLE PPROB =(0 to 1 by .1)
         LACKFIT
         RISKLIMITS
         OUTROC=ROC;
RUN;
QUIT;
OPTIONS LS=64 PS=32;
SYMBOL VALUE=DOT COLOR=BLACK INTERPOL=SMS60 WIDTH=2;
PROC GPLOT DATA=ROC;
   TITLE "ROC Curve";
   PLOT _SENSIT_ * _1MSPEC_ ;
   LABEL _SENSIT_ = 'Sensitivity'
         _1MSPEC_ = '1 - Specificity';
RUN;
```

The PROC LOGISTIC statements are basically the same as before except that we substituted AGEGROUP for AGE and included the MODEL option OUTROC= to create an output data set (ROC) that will contain the sensitivity (_SENSIT_) and 1 − specificity (_1MSPEC_), the false positive rate. These are the values necessary to plot an ROC curve. We then used PROC GPLOT to plot the ROC curve. The SYMBOL statement specifies that the interpolation method is SMOOTH 'SM', which uses a spline fit to connect the points, the 'S' following the 'SM' is an instruction to sort the values on the x-axis before computing the line, and the number following the 'SMS' can range from 0 to 99 and controls how much "wiggle" there is in the line (0, the line tries to go through every point, 99, almost a straight line). WIDTH=2 requests a slightly thicker line (the default is WIDTH=1; higher numbers are thicker).

An edited portion of the output is shown next:

```
Predicting Accidents Using Logistic Regression

      Model Fit Statistics

                           Intercept
                Intercept     and
Criterion         Only    Covariates

AIC              63.827      52.434
SC               65.633      57.854
-2 Log L         61.827      46.434


      Analysis of Maximum Likelihood Estimates

                            Standard      Wald
Parameter    DF   Estimate    Error   Chi-Square   Pr > ChiSq

Intercept     1    -1.3334    0.5854     5.1886       0.0227
AGEGROUP      1     2.1611    0.8014     7.2711       0.0070
VISION        1     1.6258    0.7325     4.9265       0.0264


      Odds Ratio Estimates

              Point         95% Wald
Effect      Estimate    Confidence Limits

AGEGROUP      8.680       1.805      41.756
VISION        5.083       1.209      21.359
```

What a difference! Now AGEGROUP enters first, followed by VISION. See how important it is to look at and understand your data before jumping in with both feet and running procedures.

The ROC curve is a traditional method for showing the relationship between sensitivity and the false positive rate. The variables _SENSIT_ and _1MSPEC_ in the output data set (ROC) represent the sensitivity and one minus the specificity (false positive rate), respectively. As mentioned earlier, we can arbitrarily decide what value to call a positive prediction—it doesn't have to be .5. You could declare any value greater than .3 to be a positive prediction. This would increase your sensitivity (anyone who actually had an accident would likely be predicted to be positive) but would also increase your false positive rate (many who did not have accidents would also be predicted to have one). As you can see, one minus the specificity gives us the false positive rate. (Well, maybe you can't see it. Unless you work with these definitions often, it is very easy to get confused.) This is the graph produced by the PROC GCHART procedure:

ROC Curve

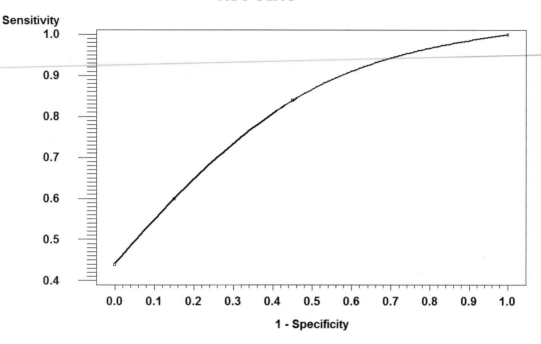

Sensitivity

1 - Specificity

Creating Two Dummy Variables from AGE

For our final trick, let's create two dummy variables from AGE, one called YOUNG, which will be true (1) for anyone less than 20 years old and false (0) otherwise. In a similar manner, the variable OLD will be defined as true for people over 65 and false otherwise. The code to create these two additional variables is included in the original DATA step. The PROC LOGISTIC statements to run an analysis based on the use of these new variables are:

```
PROC LOGISTIC DATA=LOGISTIC DESCENDING;
   TITLE "Predicting Accidents Using Logistic Regression";
   TITLE "Using Two Dummy Variables (YOUNG and OLD) for AGE";
   MODEL ACCIDENT = YOUNG OLD VISION DRIVE_ED /
         SELECTION=FORWARD
         CTABLE PPROB=(0 to 1 by .1)
         LACKFIT
         RISKLIMITS
         OUTROC=ROC;
RUN;
QUIT;
```

Unfortunately, because of the fairly small sample size, there aren't enough subjects in the young and old age groups, so these two variables are not included in the model. However, with a larger data set, this approach may be preferable to the AGE-GROUP approach used earlier since the odds ratios for being young and old can be determined separately.

H.   AUTOMATIC CREATION OF DUMMY VARIABLES WITH PROC LOGISTIC

Suppose we want to add GENDER as a predictor variable. You could create a dummy variable (either MALE or FEMALE, depending on which you want as a reference level), or you can have PROC LOGISTIC do it for you. Let's take the easy way and let PROC LOGISTIC do the work. By adding a CLASS statement, PROC LOGISTIC will automatically create a dummy variable for us. We can also decide which level we want as the reference level. (A word of caution here: An invalid value for a CLASS variable will result in a value of 0 for the dummy variable. Use a CLASS statement only when you are sure you have clean data.) The following program creates a dummy variable for GENDER and uses 'F' (female) as the reference level:

```
PROC LOGISTIC DATA=LOGISTIC DESCENDING;
   TITLE "Predicting Accidents Using Logistic Regression";
   CLASS GENDER (PARAM=REF REF='F');
   MODEL  ACCIDENT = GENDER VISION DRIVE_ED;
RUN;
QUIT;
```

The option PARAM stands for parameterization method, and we have set it equal to REF (for reference) and indicated that a value of 'F' is to be the reference. Using a CLASS statement to handle categorical variables like GENDER or RACE saves a lot of work. Look at a portion of the output from the previous program:

```
Predicting Accidents Using Logistic Regression

The LOGISTIC Procedure

         Response Profile

 Ordered                   Total
   Value    ACCIDENT     Frequency

      1     Yes               25
      2     No                20

Probability modeled is ACCIDENT='Yes'.
```

```
     Class Level Information

                      Design
     Class      Value      Variables

     GENDER     F                0
                M                1

            Analysis of Maximum Likelihood Estimates

                                  Standard        Wald
     Parameter        DF   Estimate     Error   Chi-Square   Pr > ChiSq

     Intercept         1    -0.4078    0.6886      0.3508       0.5537
     GENDER     M      1     0.9841    0.7655      1.6524       0.1986
     VISION            1     2.0074    0.7777      6.6631       0.0098
     DRIVE_ED          1    -1.6362    0.7346      4.9616       0.0259

                   Odds Ratio Estimates

                       Point        95% Wald
     Effect           Estimate   Confidence Limits

     GENDER   M vs F    2.675      0.597    11.995
     VISION             7.444      1.621    34.177
     DRIVE_ED           0.195      0.046     0.822
```

Notice that the odds ratio for GENDER is 2.675 and that this is for males versus females (i.e., the females are the reference level). So in this model, the point estimate for being male is 2.675 and the 95% CI goes from .597 to 11.995 (thus, not significant).

PROBLEMS

Remember, you can download all the data sets and programs for these problems from the web site: www.prenhall.com/cody

9.1 We want to test the effect of light level and amount of water on the yield of tomato plants. Each potted plant receives one of three levels of light ($1 = 5$ hours, $2 = 10$ hours, $3 = 15$ hours) and one of two levels of water ($1 = 1$ quart, $2 = 2$ quarts). The yield, in pounds, is recorded. The results are as follows:

Yield	Light	Water	Yield	Light	Water
12	1	1	20	2	2
9	1	1	16	2	2
8	1	1	16	2	2
13	1	2	18	3	1
15	1	2	25	3	1
14	1	2	20	3	1
16	2	1	25	3	2
14	2	1	27	3	2
12	2	1	29	3	2

Write a SAS program to read these data and perform a multiple regression. Treat the value of light as interval data using values of 5, 10, and 15. Use values of 1 and 2 for the water levels.

9.2 Redo problem 9.1 using dummy variables for light levels and water amounts. For the three light levels, use the 5-hour value as your reference (HINT: you need two dummy variables). For water level, use the one quart level as your reference.

9.3 We would prefer to estimate the number of books in a college library without counting them. Data are collected from colleges across the country of the number of volumes, the student enrollment (in thousands), the highest degree offered (1 = BA, 2 = MA, 3 = PhD), and size of the main campus (in acres). Results of this (hypothetical) study are displayed here:

Books (in millions)	Student Enrollment (in thousands)	Highest Degree	Area (in acres)
4	5	3	20
5	8	3	40
10	40	3	100
1	4	2	50
.5	2	1	300
2	8	1	400
7	30	3	40
4	20	2	200
1	10	2	5
1	12	1	100

Using a forward stepwise regression, show how each of the three factors affects the number of volumes in a college library. Treat DEGREE as a continuous variable (yup, that is a bit silly, but we'll do it the right way in the next problem).

9.4 Repeat problem 9.3, creating two dummy variables for DEGREE using DEGREE = 1 (BA) as your reference level. Also, use the base-e log of AREA instead of AREA in your model (the SAS function LOG computes the base-e log of its argument).

9.5 We want to predict a student's success in college by a battery of tests. Graduating seniors volunteer to take our test battery, and their final grade point average is recorded. Using a MAXR technique, develop a prediction equation for final grade point average using the test battery results. The data are as follows:

GPA	HS GPA	College Board	IQ Test
3.9	3.8	680	130
3.9	3.9	720	110
3.8	3.8	650	120
3.1	3.5	620	125
2.9	2.7	480	110
2.7	2.5	440	100
2.2	2.5	500	115
2.1	1.9	380	105
1.9	2.2	380	110
1.4	2.4	400	110

9.6 Run the following program to create a SAS data set called EXERCISE.

```
DATA EXERCISE;
   DO SUBJ = 1 TO 500;
      IF RANUNI(155) LT .5 THEN GENDER = 'Female';
      ELSE GENDER = 'Male';
      PRESS = INT(RANNOR(0)*20 + 95 + 50*(GENDER EQ 'Male'));
      CURL = RANNOR(0)*10 + 30 + .2*PRESS;
      PUSHUPS = INT(RANNOR(0)*3 + 5 + 5*(GENDER EQ 'Male')
               + .1*CURL);
      SITUPS = INT(RANNOR(0)*10 + 20 + .1*PRESS + PUSHUPS);
      OUTPUT;
   END;
RUN;
```

Generate a correlation matrix among the variables PRESS (weight in bench presses), CURL (curl weight), PUSHUPS (number of push-ups), and SITUPS (number of sit-ups). Then run a stepwise multiple regression predicting PRESS from GENDER, CURL, PUSHUPS, and SITUPS. You will have to make a new data set from EXERCISE with a dummy variable for GENDER (use Female as the reference level).

9.7 Take a sample of 25 people and record their height, waist measurement, length of their right leg, length of their arm, and their weight. Write a SAS program to create a SAS data set of these data, and compute a correlation matrix of these variables. Next, run a stepwise multiple regression using weight as the dependent variable and the other variables as independent.

9.8 Run the following program to create a SAS data set called SMOKING. Outcome values of 1 represent cases (those with lung disease) and outcome values of 0 represent controls (those without lung disease). For this problem, generate a two by two table for outcome (OUTCOME) by smoking status (SMOKING). What is the chi-square and the odds ratio? Now, run a logistic regression with only SMOKING as the predictor variable. HINT: PROC LOGISTIC uses the lower value formatted value of the outcome variable (if there is a format), and since 'Case' comes before 'Control', you don't need to include the DESCENDING option).

```
PROC FORMAT;
   VALUE YESNO 1='YES' 0='NO';
   VALUE OUTCOME 1='Case' 0='Control';
RUN;
DATA SMOKING;
   DO SUBJECT = 1 TO 1000;
      DO OUTCOME = 0,1;
         IF RANUNI(567) LT .1 OR RANUNI(0)*OUTCOME GT .5 THEN
            SMOKING = 1;
         ELSE SMOKING = 0;
         IF RANUNI(0) LT .05 OR
            (RANUNI(0)*OUTCOME + .1*SMOKING) GT .6 THEN ASBESTOS = 1;
         ELSE ASBESTOS = 0;
         IF RANUNI(0) LT .3 OR OUTCOME*RANUNI(0) GT .9 THEN
```

[Continued]

```
            SES = '1-Low    ';
        ELSE IF RANUNI(0) LT .3 OR OUTCOME*RANUNI(0) GT .8 THEN
            SES = '2-Medium';
        ELSE SES = '3-High';
        OUTPUT;
     END;
   END;
   FORMAT SMOKING ASBESTOS YESNO. OUTCOME OUTCOME.;
 RUN;
```

9.9 What's wrong with this program?

```
1    DATA MULTREG;
2       INPUT HEIGHT WAIST LEG ARM WEIGHT;
3    DATALINES;
     (data lines)
     ;
4    PROC CORR DATA=MULTREG;
5       VAR HEIGHT — WEIGHT;
6    RUN;
7    PROC REG DATA=MULTREG;
8       MODEL WEIGHT=HEIGHT WAIST LEG ARM / SELECTION=STEPWISE;
9    RUN;
```

9.10 Using the data set from problem 9.8, conduct a logistic regression predicting outcome (case or control) using SMOKING, ASBESTOS, and SES. Use a CLASS statement to create two dummy variables for SES, setting the reference level to '2-Medium'. (HINT: CLASS SES (PARAM=REF REF='2-Medium'); PROC LOGISTIC uses the lower value formatted value of the outcome variable (if there is a format), and since 'Case' comes before 'Control', you don't need to include the DESCENDING option).

9.11 Add the necessary statements to your logistic regression in problem 9.8 to produce an ROC curve (sensitivity versus $1 -$ specificity). Use the OUTROC $= data$-set-$name$ MODEL option to create a data set so you can plot the curve. Use PROC GPLOT (or PROC PLOT) to generate this curve.

9.12 Repeat problem 9.8, except this time create your own dummy variables (i.e., do not use a CLASS statement). Make the reference level for SES '2-Medium' as before.

9.13 Accident data, similar to Section G, are presented here. This time, we recorded accidents that occurred in the past year, based on the presence of a drinking problem and whether the driver had one or more accidents in the previous year. Run a forward stepwise logistic regression on these data, and write the resulting logistic regression equation. Compute the odds and the probability of an accident for two cases: (1) a person with no drinking problem and no previous accident; (2) a person with a drinking problem but no previous accidents. Take the ratio of the odds for person (2) divided by person (1) and confirm that the odds ratio is the same as that listed in the SAS output (use the MODEL option RL to obtain the risk limits). Here are the data:

Accident Statistics Based on Drinking and Accident History
(1 = Yes, 0 = No)

Accident in Past Year	Drinking Problem	Previous Accident
1	0	1
1	1	1
1	1	1
1	0	0
1	1	1
0	0	1
0	1	0
0	1	0
0	0	1
0	0	0
0	1	0
0	0	1
0	1	0
0	0	0
0	0	0
1	1	0
1	0	1
1	1	1
1	1	1
1	0	1
1	1	1
1	1	0
1	0	1
1	1	0
1	1	1
0	1	1
0	1	1
0	0	1
0	0	1
0	1	0
0	0	0
0	0	1
0	1	0
0	0	0
0	0	0
1	1	1
1	1	0
1	1	1
1	1	0
1	1	1
1	1	1
1	1	1

C H A P T E R   1 0

Factor Analysis

A. Introduction
B. Types of Factor Analysis
C. Principal Components Analysis
D. Oblique Rotations
E. Using Communalities Other Than One
F. How to Reverse Item Scores

A.   INTRODUCTION

Welcome to the wacky world of factor analysis, a family of statistical techniques developed to allow researchers to reduce a large number of variables, such as questions on a questionnaire, to a smaller number of broad concepts called factors. The factors can then be used in subsequent analyses. But beware. Factor analysis can be fairly seductive in that it holds out the promise of taking a large number of baffling variables and turning them into a clear-cut set of just a few factors.

B.   TYPES OF FACTOR ANALYSIS

Before proceeding, a word or two on different types of factor analyses. Generally speaking, there are two types of factor analyses: exploratory and confirmatory. SAS does either, but we are only going to discuss exploratory here, as confirmatory is too complex for the scope of this book. Within exploratory factor analysis, there are also two basic approaches: principal components analysis and factor analysis. Factor analysis purists get a little crazed over the use of principal components analysis, and some even refuse to consider it as a part of factor analysis. We are going to show you a principal components analysis, then a more proper factor analysis of the same data set, we will show you how to rotate the results (doesn't that sound fun), and then talk about interpretation of the results. By the way, Pedhazur and Schmelkin (1991) is a very readable presentation of factor analysis for beginners.

C.   PRINCIPAL COMPONENTS ANALYSIS

We start with principal components analysis since it is conceptually somewhat simpler than factor analysis. This does not mean that we are recommending it; it just allows for a better pedagogical flow. Imagine you are trying to develop a new measure of depression and paranoia (how pleasant). Your measure contains six questions. For each question, the subject is to respond using the following Likert scale:

1 = Very Strongly Disagree
2 = Strongly Disagree
3 = Disagree
4 = No Opinion
5 = Agree
6 = Strongly Agree
7 = Very Strongly Agree

The six questions are:

1. I usually feel blue.
2. People often stare at me.
3. I think that people are following me.
4. I am usually happy.
5. Someone is trying to hurt me.
6. I enjoy going to parties.

As stated, this example was created with two psychological problems in mind: depression and paranoia. Someone who is depressed will likely agree with questions 1 and 6 and disagree with question 3. Someone who is paranoid will probably agree with questions 2, 4, and 5. Therefore, we would expect the factor analysis to come up with two factors. One we can label depression, the other, paranoia.

Some sample data are shown here:

	Question					
SUBJ	**1**	**2**	**3**	**4**	**5**	**6**
1	7	2	3	4	5	6
2	6	3	2	1	3	2
3	3	6	7	3	6	3
4	2	2	2	5	3	4
5	3	4	2	4	2	3
6	6	3	4	2	3	2
7	1	2	3	7	2	2
8	3	3	2	3	4	3
9	2	1	1	6	2	5

(Continued)

	Question					
SUBJ	**1**	**2**	**3**	**4**	**5**	**6**
10	6	2	3	2	2	2
11	3	5	4	2	3	3
12	6	7	6	2	6	2
13	5	1	1	2	6	2
14	2	1	1	6	1	5
15	1	2	1	7	1	7

Assume that we place these data in a file called FACTOR.DTA with the subject number in columns 1 and 2 and the 6 questions in columns 3 through 8 (i.e., no spaces between any values). Some people say that you should have about 10 times the number of subjects as you have variables to be factor analyzed. However, you probably want a minimum of 50 subjects and do not want huge numbers of variables. We violate that rule for this simple example where we have 6 variables and 15 subjects.

Our first step is to create a SAS data set containing the responses to the six questions, as shown here:

```
*----------------------------------------------------------------*
| Program Name: FACTOR.SAS in C:\APPLIED                          |
| Purpose: To perform a factor analysis on psychological Data     |
*----------------------------------------------------------------*;
PROC FORMAT;
   VALUE LIKERT
      1 = 'V. Strong Dis.'
      2 = 'Strongly Dis.'
      3 = 'Disagree'
      4 = 'No Opinion'
      5 = 'Agree'
      6 = 'Strongly Agree'
      7 = 'V. Strong Agree';
RUN;
DATA FACTOR;
   INFILE 'C:\APPLIED\FACTOR.DTA' PAD;
   INPUT @1  SUBJ $2.
         @3 (QUES1-QUES6) (1.);
   LABEL QUES1 = 'Feel Blue'
         QUES2 = 'People Stare at Me'
         QUES3 = 'People Follow Me'
         QUES4 = 'Basically Happy'
         QUES5 = 'People Want to Hurt Me'
         QUES6 = 'Enjoy Going to Parties';
   FORMAT QUES1-QUES6 LIKERT.;
RUN;
```

The INPUT statement in this example uses a list of variables in parentheses (QUES1–QUES6) followed by an informat list. The informat "1." means one column for each of the six responses. If you prefer, you may separate each data item from the next by a space and use the free form or "list" input method. As always, there are several ways to accomplish our goal with SAS. Now, back to our factor analysis example. The SAS statements to perform the factor analysis are:

```
PROC FACTOR DATA=FACTOR PREPLOT PLOT ROTATE=VARIMAX
          NFACTORS=2 OUT=FACT SCREE;
   TITLE "Example of Factor Analysis";
   VAR QUES1-QUES6;
RUN;
```

We have selected several of the more popular PROC FACTOR options in this example. The PREPLOT option will show us a factor plot before rotation; PLOT will produce a factor plot after rotation; ROTATE=VARIMAX requests a VARIMAX rotation; NFACTORS=2 specifies that a maximum of two factors should be extracted; OUT=FACT specifies that you want the factor scores (in this case FACTOR1 and FACTOR2) to be placed in the named data set along with all the variables in the DATA= data set. The SCREE option gives you a scree plot (discussed later). Here is the output from PROC FACTOR:

```
Example of Factor Analysis


The FACTOR Procedure
Initial Factor Method: Principal Components


Prior Communality Estimates: ONE    ①


Eigenvalues of the Correlation Matrix: Total = 6  Average = 1 ②

          Eigenvalue    Difference     Proportion    Cumulative

      1    3.46586271    2.30980407      0.5776        0.5776
      2    1.15605865    0.44820362      0.1927        0.7703
      3    0.70785502    0.31301451      0.1180        0.8883
      4    0.39484051    0.18582810      0.0658        0.9541
      5    0.20901241    0.14264172      0.0348        0.9889
      6    0.06637069                    0.0111        1.0000


2 factors will be retained by the NFACTOR criterion.

                                                  [Continued]
```

Initial Factor Method: Principal Components

Scree Plot of Eigenvalues ③

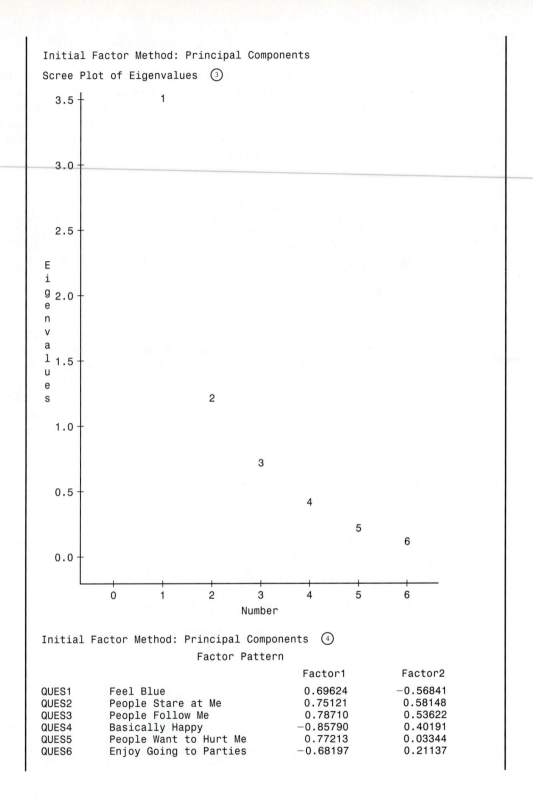

Initial Factor Method: Principal Components ④

Factor Pattern

		Factor1	Factor2
QUES1	Feel Blue	0.69624	−0.56841
QUES2	People Stare at Me	0.75121	0.58148
QUES3	People Follow Me	0.78710	0.53622
QUES4	Basically Happy	−0.85790	0.40191
QUES5	People Want to Hurt Me	0.77213	0.03344
QUES6	Enjoy Going to Parties	−0.68197	0.21137

Variance Explained by Each Factor ⑤

    Factor1   Factor2

  3.4658627   1.1560586

              Final Communality Estimates: Total = 4.621921

        QUES1   QUES2   QUES3   QUES4   QUES5   QUES6

  0.80783204   0.90243480   0.90705927   0.89752979   0.59730074   0.50976472

Initial Factor Method: Principal Components

Plot of Factor Pattern for Factor1 and Factor2 ⑥

                                    Factor1
                                       1
                                      .9
                                      .8   C
                                          E   B
          A   .7
                                      .6
                                      .5
                                      .4
                                      .3
                                      .2
                                      .1   F
                                                                            a
 −1 −.9−.8−.7−.6−.5−.4−.3−.2−.1 0 .1 .2 .3 .4 .5 .6 .7 .8 .9 1.0t
                                     −.1   o
                                                                            r
                                     −.2   2
                                     −.3
                                     −.4
                                     −.5
                                     −.6
                                     −.7   F
                                     −.8
                                                  D
                                     −.9
                                     −1

[Continued]

Initial Factor Method: Principal Components

QUES1=A   QUES2=B   QUES3=C   QUES4=D   QUES5=E
QUES6=F

Rotation Method: Varimax ⑦

Orthogonal Transformation Matrix

	1	2
1	−0.74030	0.67228
2	0.67228	0.74030

Rotated Factor Pattern

		Factor1	Factor2
QUES1	Feel Blue	−0.89755	0.04728
QUES2	People Stare at Me	−0.16521	0.93549
QUES3	People Follow Me	−0.22220	0.92611
QUES4	Basically Happy	0.90530	−0.27922
QUES5	People Want to Hurt Me	−0.54912	0.54384
QUES6	Enjoy Going to Parties	0.64696	−0.30200

Variance Explained by Each Factor

Factor1	Factor2
2.4219214	2.2000000

Final Communality Estimates: Total = 4.621921

QUES1	QUES2	QUES3	QUES4	QUES5	QUES6
0.80783204	0.90243480	0.90705927	0.89752979	0.59730074	0.50976472

Scoring Coefficients Estimated by Regression

Squared Multiple Correlations of the Variables with Each Factor

Factor1	Factor2
1.0000000	1.0000000

Standardized Scoring Coefficients ⑧

		Factor1	Factor2
QUES1	Feel Blue	−0.47926	−0.22893
QUES2	People Stare at Me	0.17769	0.51807
QUES3	People Follow Me	0.14371	0.49605
QUES4	Basically Happy	0.41697	0.09096
QUES5	People Want to Hurt Me	−0.14548	0.17119
QUES6	Enjoy Going to Parties	0.26858	0.00307

```
Plot of Factor Pattern for Factor1 and Factor2  ⑨

                               Factor1
                                 1
               D                .9
                                .8
              F                 .7
                                .6
                                .5
                                .4
                                .3
                                .2
                                .1                                F
                                                                  a
  −1 −.9−.8−.7−.6−.5−.4−.3−.2−.1  0 .1 .2 .3 .4 .5 .6 .7 .8 .9 1.0 t
                               −.1                                o
                                                                  r
                               −.2                             C B 2
                               −.3
                               −.4
                               −.5                    E
                               −.6
                               −.7
                               −.8
                               −.9A
                               −1

  QUES1=A        QUES2=B        QUES3=C        QUES4=D        QUES5=E
  QUES6=F
```

Well, that all looks fairly confusing. Let's walk through what we have. The first thing we learn is that our prior communality estimates were one ①. This means that each variable was given one full unit of variance to be factored in the original correlation matrix. Many factor analysts argue that one should only factor the variance that is shared among the variables. When the prior communalities equal one, the analysis factors all of the variance in all variables and the approach is called principal components analysis. More on this later.

The second thing we encounter are the eigenvalues of the correlation matrix ②. Eigenvalue is a term from matrix algebra. In this analysis, since we let each variable have one unit of variance, and since there are six variables, the total of the eigenvalues equals six. As mentioned earlier, factor analysis tries to reduce variables into a smaller set of factors that "explain" the variance in the original variables. Of the original "6 units" of variance,

3.6683 of these units are explained by the first factor (or more properly, component, since this is a principal components analysis). The second factor explains 1.2400 units of variance, and so on.

It is useful to think about eigenvalues in the following fashion. Imagine the first factor is a new variable, and everyone in the data set had a score on it. We could correlate their score on the factor with their score on each of the six variables. If we then squared those correlations, it would tell us how much of the variance in each original variable was explained by the factor. If we then added those six squared correlations together, the sum would be the eigenvalue for that factor. Thus, the first factor explains 3.6683 units of variance, or $3.6683/6 = .6114$, of the total variance of the original six variables. This .6114 figure is listed in the printout under "Proportion." The "Difference" heading tells us the difference in the proportion explained from the previous total, and the "Cumulative" row gives the cumulative total.

The next thing we see is that two factors will be retained for rotation because that is what we told the program to do.

The next section of printout is the "scree plot" ③. The scree plot is used to help determine how many factors to keep in the analysis. It is simply a plot of the eigenvalue against the number of the factor. One looks for breaks, or "elbows," in the curve. In this graph, it is easy to see that factors 1 and 2 are very different from 3–6, so two factors should be retained. In our example, we knew from a theoretical perspective that we wanted two factors. The results support our notion. In determining the number of factors to keep, it is always best to combine theory and data. If you don't have a strong theory to rely on beforehand, you will have to do one run just to get an idea of how many factors to keep.

The next section ④ presents the initial solution for the analysis. Oversimplified, what the factor analysis tries to do is first find a factor (think of it as a new variable) that will provide the highest set of correlations with the original variables (actually, with the squares of these variables), thus producing the largest eigenvalue. Then it finds a second factor that will correlate as highly as possible with the original variables once the variation from the first factor has been removed. Another way of thinking about this is to say that the factors have to be uncorrelated (or orthogonal). Then a third factor is extracted, which works with the remaining variance, and so on.

What is presented under the heading "Factor Pattern" is the result of this process for factor 1 and factor 2. Had we said we wanted to keep three factors, there would have been a factor 3 here. In the case of principal components analysis, if we don't specify the number of factors to keep, PROC FACTOR will keep as many factors as there are variables. If we specify PRIORS SMC (see Section E), the number of factors retained will be determined by the proportion of the total variance explained and may be less than the total number of variables. For more details on controlling the number of factors to keep, see the *SAS/STAT 9.1 User's Guide* (SAS Institute, Cary NC). The factor pattern displays what are called "factor loadings" for each of the variables. At this point in the analysis, these loadings are the simple correlations of the variables with the factor.

The next part of the printout ⑤ shows the variance explained by each factor (just the first two eigenvalues again) and then the communalities of the variables. The what?

The communalities. Communalities represent how much variance in the original variable is explained by the total of all the factors that are kept in the analysis. We see here that 89% (actually, .890425) of the variance in the first question is explained (or attributable to) the first two factors. Communalities are for original variables what eigenvalues are for factors (more or less).

The next portion of the printout is a plot of the two factors retained ⑥. Note that at the bottom of the plot is a key to what the letters are. There are basically two clusters of variables here: A, B, C, and E; and F and D. One might look at this plot and wonder if the axes could be rotated so that they ran directly through the clusters. What a good idea! The only question would be how to determine just how to do the rotation. SAS provides a number of alternative rotation methods. We look at two here.

The first is called VARIMAX. It maintains the orthogonal (uncorrelated) nature of the factors and tries to get the original variables to load high on one of the factors and low on the rest. When this occurs, it is called simple structure. Other rotation methods also attempt to obtain simple structure. The results of a VARIMAX rotation are presented next ⑦.

There are several issues of note here. First, the factor loadings obtained from a rotation of the axes almost always result in a more readily interpretable solution. Here we see that our original notion of two factors with three variables (questions in our case) on each factor is strongly supported by the data. Second, notice that some of the loadings are negative while others are positive. Although these factor loadings can technically no longer be interpreted as correlation coefficients, it is useful to think of them in the same fashion. We would expect "Feel blue" and "Basically happy" to have opposite signs in their relationship to a depression scale. Next, notice that the variance explained by each factor has changed. Their sum is still the same, but the distribution is more equal now. The communality estimates have not changed since just as much variance is explained in each original variable as before rotation; now, however, more is attributable to the second factor (than before) and less to the first.

There are two more pieces left from this analysis. The first is a standardized scoring coefficient matrix ⑧. You would use this if you actually wanted to calculate factor scores for your subjects (PROC FACTOR will also do this for us automatically with the OUT= procedure option). Although many researchers find this useful, we don't recommend it. Instead, we recommend you simply construct scales using raw data from the variables that load on each factor. There are a number of reasons for this. First, it is much simpler and more straightforward. Second, factor analysis is highly "sample dependent." That is, results tend to be changeable from one data set to the next. Remember, what you are factoring is a matrix of correlations and we all know how sample dependent they are.

The FINAL (yeah!) piece of this analysis is a plot of the rotated factors ⑨. You can see how the new axes really do run right through the clusters. If rotation is confusing you a little, think of it in the three factor case. Here we would have a three-dimensional swarm of points. Now think of looking at any three-dimensional object: a football, a pen, or a diamond ring. It looks different, depending on how you hold it in front of you. Rotating axes is just changing your perspective until you can get the best "look" at something.

D. OBLIQUE ROTATIONS

Well, that was entertaining. But we said there were several rotation methods possible. VARIMAX is one popular method, PROMAX is another. They are similar in some respects, but different in one important aspect: PROMAX does not maintain the orthogonality of the factors. It allows the factors to be correlated. Why should a researcher prefer one over the other? In favor of orthogonal rotation is the argument that it tends to keep things cleaner. If your factors are orthogonal and you then want to use them as independent variables in a regression, you will have fewer problems of collinearity. In favor of a nonorthogonal ("oblique") rotation is the argument that it is usually silly to think that the underlying constructs that are represented by the factors are, in fact, uncorrelated. In our example, it is unreasonable to think that depression and paranoia aren't correlated. Also, it is sometimes easier to obtain simple structure with an oblique rotation. Here is the program to get the oblique rotation called PROMAX from our data set:

```
PROC FACTOR DATA=FACTOR ROTATE=PROMAX NFACTORS=2;
   TITLE "Example of Factor Analysis - Oblique Rotation";
   VAR QUES1-QUES6;
RUN;
```

Now, much of the printout is similar to the orthogonal case, so we will just focus on one section here:

```
Example of Factor Analysis - Oblique Rotation

The FACTOR Procedure
       Inter-Factor Correlations

                 Factor1          Factor2

Factor1          1.00000         -0.41771
Factor2         -0.41771          1.00000


Rotation Method: Promax (power = 3)

  Rotated Factor Pattern (Standardized Regression Coefficients)

                                      Factor1          Factor2

QUES1    Feel Blue                   -0.95376         -0.15988
QUES2    People Stare at Me           0.04303          0.96714
QUES3    People Follow Me            -0.02045          0.94367
QUES4    Basically Happy              0.90739         -0.08770
QUES5    People Want to Hurt Me      -0.46207          0.45587
QUES6    Enjoy Going to Parties       0.62429         -0.17285
```

Here we have the correlation between the factors and the rotated factor loading matrix. The correlation between the factors is −.41771; if we had four factors, this would be a 4 × 4 matrix. Looking at the factor loadings, we see that, in fact, we are closer to simple structure than we were before, even though the results from the orthogonal rotation were quite good.

E.   USING COMMUNALITIES OTHER THAN ONE

The final stop on our journey brings us back to the notion of communalities. When factor analysts get upset over what is really factor analysis, much of the issue has to do with what is placed on the "main diagnonal" of the correlation matrix that is to be factored. When ones (1.0) are used, we are basically factoring what we all know and love as correlation matrices. This approach assumes that each variable is equally as important as the others and has the same amount of interrelatedness with the other variables. In our example, this is a fairly reasonable assumption, but in most cases, some variables are more important than others, have stronger relationships with the variables in the analysis than others, or are measured with less error than others. In this case, it would be better conceptually to have some indication of how much each variable "fits in" with the others. This idea is realized by changing the communalities on the main diagonal to be less than one. Now, there is a whole science to this, but one popular approach is to take each variable and regress all the other variables against it. Then the squared multiple correlation resulting from this regression is used as the communality estimate. "Whoa," you're saying, "that's a lot of work." Indeed, but PROC FACTOR does it all for you. All we have to do is include the statement PRIORS SMC in our PROC FACTOR procedure, and it's done. Here is an example:

```
PROC FACTOR DATA=FACTOR PREPLOT PLOT ROTATE=VARIMAX
        NFACTORS=2 OUT=FACT SCREE;
   TITLE "Example of Factor Analysis";
   VAR QUES1-QUES6;
   PRIORS SMC; ***This is the new line;
RUN;
```

Here is a portion of the output from this modified program:

```
Example of Factor Analysis

The FACTOR Procedure
Initial Factor Method: Principal Factors

            Prior Communality Estimates: SMC

    QUES1        QUES2        QUES3        QUES4        QUES5        QUES6

 0.73873589   0.81127039   0.80947810   0.84388077   0.48543730   0.55088279
                                                                   [Continued]
```

```
         Eigenvalues of the Reduced Correlation Matrix:
           Total = 4.23968523  Average = 0.70661421

        Eigenvalue    Difference    Proportion    Cumulative

   1    3.18866680    2.24478623      0.7521        0.7521
   2    0.94388057    0.65753332      0.2226        0.9747
   3    0.28634725    0.27308801      0.0675        1.0423
   4    0.01325924    0.05986021      0.0031        1.0454
   5   -.04660097     0.09926669     -0.0110        1.0344
   6   -.14586766                    -0.0344        1.0000
```

2 factors will be retained by the NFACTOR criterion.

Initial Factor Method: Principal Factors

```
                    Factor Pattern

                                     Factor1        Factor2

QUES1    Feel Blue                   0.66964       -0.50702
QUES2    People Stare at Me          0.74470        0.52417
QUES3    People Follow Me            0.77759        0.48283
QUES4    Basically Happy            -0.85206        0.39577
QUES5    People Want to Hurt Me      0.68690        0.00487
QUES6    Enjoy Going to Parties     -0.61903        0.14926
```

Variance Explained by Each Factor

```
   Factor1         Factor2

  3.1886668       0.9438806
```

Final Communality Estimates: Total = 4.132547

```
    QUES1         QUES2         QUES3         QUES4         QUES5         QUES6

0.70548390    0.82933259    0.83776353    0.88264013    0.47185185    0.40547537
```

Rotation Method: Varimax

```
    Orthogonal Transformation Matrix


                       1                 2

          1        -0.73430           0.67883
          2         0.67883           0.73430
```

Rotated Factor Pattern

```
                                     Factor1        Factor2

QUES1    Feel Blue                  -0.83589        0.08227
QUES2    People Stare at Me         -0.19101        0.89042
QUES3    People Follow Me           -0.24322        0.88239
QUES4    Basically Happy             0.89433       -0.28779
QUES5    People Want to Hurt Me     -0.50108        0.46986
QUES6    Enjoy Going to Parties      0.55587       -0.31061
```

```
Variance Explained by Each Factor

    Factor1              Factor2
  2.1542516            1.9782958

              Final Communality Estimates: Total = 4.132547

      QUES1          QUES2          QUES3          QUES4          QUES5          QUES6

  0.70548390     0.82933259     0.83776353     0.88264013     0.47185185     0.40547537
```

In all honesty, the results here are not too different than they were before, but remember, this is a simple example. There are important theoretical distinctions among the three analyses we have shown here, but if you'll notice, "Factor Analysis" is only a chapter here; it isn't the whole book. Try Gorsuch (1983) for a thorough discussion of factor analysis or Pedhazur and Schmelkin (1991) for a couple of solid chapters.

F.   HOW TO REVERSE ITEM SCORES

Some researchers prefer to leave the positive and negative forms of questions as they are and interpret the sign of the factor loadings based on that information. Others prefer to modify the scores first so that all scores are in the same direction. For example, a high score (agree) to question 1 implies depression. We can reverse the scoring for questions 4 and 6 so that high scores also imply depression. One way to do this is as follows:

```
DATA FACTOR;
    INFILE 'C:\APPLIED\FACTOR.DTA' PAD;
    INPUT @1   SUBJ $2.
          @3 (QUES1-QUES6) (1.);
    ***Reverse the scores for questions 4 and 6;
    QUES4 = 8 - QUES4;
    QUES6 = 8 - QUES6;
    LABEL QUES1 = 'Feel Blue'
          QUES2 = 'People Stare at Me'
          QUES3 = 'People Follow Me'
          QUES4 = 'Basically Happy'
          QUES5 = 'People Want to Hurt Me'
          QUES6 = 'Enjoy Going to Parties';
    FORMAT QUES1-QUES6 LIKERT.;
RUN;
```

If you had a large number of questions that needed reversing, you could use an ARRAY to do the job (see Chapter 15).

Let's look at the rotated factor loadings when we reverse these two questions:

```
                     Rotated Factor Pattern

                                    Factor1         Factor2

  QUES1    Feel Blue                0.83589         0.08227
  QUES2    People Stare at Me       0.19101         0.89042
  QUES3    People Follow Me         0.24322         0.88239
  QUES4    Basically Happy          0.89433         0.28779
  QUES5    People Want to Hurt Me   0.50108         0.46986
  QUES6    Enjoy Going to Parties   0.55587         0.31061
```

As you can quickly see, the factor loadings are identical to the previous one (with PRIORS=SMC) where the scores were not reversed, except that now none of the loadings are negative.

We originally requested that PROC FACTOR create a new data set for us containing all the original values plus two factor scores. You may choose to use these factor scores in further analyses, but see the caution in Section C. We have reduced the number of variables from six to two. In addition, these two factor scores are uncorrelated to each other, making them particularly useful in regression models. Finally, each of the two factors spans a single psychological dimension (depression or paranoia). Let's run a PROC PRINT on the new data set and see what it contains:

```
PROC PRINT DATA=FACT NOOBS;
    TITLE "Output Data Set (FACT) Created by PROC FACTOR";
    TITLE2 "Questions 4 and 6 Reversed";
RUN;
```

```
Output Data Set (FACT) Created by PROC FACTOR
Questions 4 and 6 Reversed

SUBJ  QUES1  QUES2  QUES3  QUES4  QUES5  QUES6   Factor1    Factor2

  1      7      2      3      4      5      2     0.44069   -0.38174
  2      6      3      2      7      3      6     1.29836   -0.46700
  3      3      6      7      5      6      5    -0.07634    2.07824
  4      2      2      2      3      3      4    -0.57499   -0.32155
  5      3      4      2      4      2      5    -0.44451    0.11733
  6      6      3      4      6      3      6     0.99244    0.12215
  7      1      2      3      1      2      6    -1.44583    0.11565
  8      3      3      2      5      4      5     0.18520   -0.16463
```

9	2	1	1	2	2	3	−0.87359	−0.86028
10	6	2	3	6	2	6	1.10460	−0.45655
11	3	5	4	6	3	5	0.29764	0.85453
12	6	7	6	6	6	6	0.42476	1.87426
13	5	1	1	6	6	6	1.17134	−1.10292
14	2	1	1	2	1	3	−0.89564	−0.88821
15	1	2	1	1	1	1	−1.60415	−0.51928

If factor 1 is depression and factor 2 is paranoia, you can readily spot those subjects who are the most depressed or who are paranoid.

PROBLEMS

Remember, you can download all the data sets and programs for these problems from the web site: www.prenhall.com/cody

10.1 Run a factor analysis on the questionnaire data in Chapter 3, Section B. Use only the variables PRES, ARMS, and CITIES. Request two factors, VARIMAX rotation method, and set the PRIORS estimate of communality to SMC. Include the option to generate a scree plot.

10.2 Run the following program to create the data set PRINCIPAL:

```
DATA PRINCIPAL;
    DO SUBJ = 1 TO 200;
        X1 = ROUND(RANNOR(123)*50 + 500);
        X2 = ROUND(RANNOR(123)*50 + 100 + .8*X1);
        X3 = ROUND(RANNOR(123)*50 + 100 + X1 − .5*X2);
        X4 = ROUND(RANNOR(123)*50 + .3*X1 + .3*X2 + .3*X3);
        OUTPUT;
    END;
RUN;
```

Write the statements necessary to run a factor analysis and produce a scree plot. Looking at this plot, decide how many factors you want to maintain, and add an option to the procedure to extract this many factors and to perform a VARIMAX rotation.

10.3 Using the test scores in Chapter 11, Section B, run a factor analysis on the test, using the scored responses to each of the five questions as the items to factor analyze. Request only one factor and create an output data set containing that one factor score. Do not request any rotations. Use PROC PRINT to list the contents of this output data set. The first unrotated factor from a test is sometimes related to IQ and called factor G (for "general"). Note that there are far too few observations to run a meaningful factor analysis—it is for instructional purposes only.

10.4 Revise your program from problem 10.2 to perform a non-oblique rotation and compare the factor loadings with the results from 10.2. Output a data set containing the factor scores. Use PROC PRINT to list out the first 10 observations from this data set (hint: use the data set option OBS=10). Compare these scores with a simpler method of taking the mean (or sum) of the variables that contribute to the factor. Correlate these two scores.

CHAPTER 11

Psychometrics

A. INTRODUCTION

This chapter contains programs to score a test, to perform item analysis, test reliability (Cronbach's Alpha), and interrater reliability (Coefficient Kappa). In Section E, you will find a complete program for item analysis that you are free to use or incorporate in a larger test-scoring and item-analysis program.

B. USING SAS TO SCORE A TEST

We start with a simple program that will score a five-question multiple-choice test. Later sections enhance this program so that it will be more general and will work with any number of questions. First the program, then the explanation:

```
*----------------------------------------------------------------*
| Program Name: SCORE1.SAS in C:\APPLIED                          |
| Purpose: To score a five-item multiple-choice exam.            |
| Data: The first line is the answer key, remaining lines        |
| contain the student responses                                  |
| Date: April 23, 2004                                           |
*----------------------------------------------------------------*;
```

```
DATA SCORE;
    ARRAY ANS[5] $ 1 ANS1-ANS5; ***Student answers;
    ARRAY KEY[5] $ 1 KEY1-KEY5; ***Answer key;
    ARRAY S[5] 3 S1-S5; ***Score array 1=right,0=wrong;
    RETAIN KEY1-KEY5; ①
    ***Read the answer key;
    IF _N_ = 1 THEN INPUT (KEY1-KEY5)($1.); ②
    ***Read student responses;
    INPUT    @1  ID 1-9 ③
             @11 (ANS1-ANS5)($1.);
    ***Score the test;
    DO I=1 TO 5;  ④
        S[I] = (KEY[I] EQ ANS[I]);  ⑤
    END;
    ***Compute Raw and Percentage scores;
    RAW=SUM (OF S1-S5);  ⑥
    PERCENT=100*RAW / 5;  ⑦
    KEEP ID RAW PERCENT;
    LABEL ID      = 'Social Security Number'
          RAW     = 'Raw Score'
          PERCENT = 'Percent Score';
DATALINES;
ABCDE
123456789 ABCDE
035469871 BBBBB
111222333 ABCBE
212121212 CCCDE
867564733 ABCDA
876543211 DADDE
987876765 ABEEE
;
PROC SORT DATA=SCORE;
    BY ID;
RUN;
PROC PRINT DATA=SCORE LABEL;
    TITLE "Listing of SCORE data set";
    ID ID;
    VAR RAW PERCENT;
    FORMAT ID SSN11.;
RUN;
```

Here is the output from the PROC PRINT:

```
Listing of SCORE data set

      Social
     Security      Raw      Percent
     Number       Score      Score

035-46-9871        1          20
111-22-2333        4          80
123-45-6789        5         100
212-12-1212        3          60
867-56-4733        4          80
876-54-3211        2          40
987-87-6765        3          60
```

Since the answer key for our test is contained in the first line of data, and the student responses in the remaining lines, we can use the automatic SAS DATA step variable _N_, which is incremented by one for each iteration of the DATA step. When the program starts, the variable _N_ will have a value of 1, and the first line of data will be read into the KEY variables ①. By retaining the KEYn variables ②, their value will be available to compare to each of the student responses. Remember that SAS normally sets the value of each variable that is not read from a SAS data set (in a SET or MERGE statement, for instance) to missing before a new data line is read. By retaining the KEYn variables, this initialization does not occur.

The program continues with the next INPUT statement ③ and reads a line of student data. For all subsequent iterations of the data step, _N_ will be greater than 1 and statement ② will not execute again. Thus, the answer key is skipped and the student ID and responses are read.

The scoring DO loop ④ compares each of the student responses with the answer key. Statement ⑤ is somewhat unusual and needs some explanation. The rightmost portion of the statement (KEY[I] EQ ANS[I]) is a logical comparison. If the student answer (ANS[I]) is equal to the answer key (KEY[I]), then the value of S[I] will be 1 (true). Otherwise, it will be a 0 (false). A value of 1 or 0 will then be assigned to the variable S[I]. Instead, you could also score the test with two lines, like this:

```
IF KEY[I] = ANS[I] THEN S[I] = 1;
ELSE ANSf[I] = 0;
```

The SUM statement ⑥ gives us the number of correct answers on the test. A percentage score is computed by dividing the number of correct responses by the number of items on the test and multiplying by 100 ⑦.

We make use of the built-in format SSN11. to print the student social security numbers in standard format (which also ensures that the leading zeros in the number are printed).

C.   GENERALIZING THE PROGRAM FOR A VARIABLE NUMBER OF QUESTIONS

This next program extends the previous program in two ways: First, it can be used to score tests with different numbers of items. Second, several procedures are added to produce class summary reports. Here is the program (to stay with our previous example, the number of questions is set to five):

```
*------------------------------------------------------------------*
| Program Name: SCORE2.SAS in C:\APPLIED                            |
| Purpose: To score a multiple-choice exam with an arbitrary       |
| number of items                                                  |
| Data: The first line is the answer key, remaining lines          |
| contain the student responses                                    |
| Data in file C:\APPLIED\TEST.DTA                                 |
| Date: July 23, 1996                                              |
*------------------------------------------------------------------*;
%LET NUMBER=5; ***The number of items on the test;  ①

DATA SCORE;
    INFILE 'C:\APPLIED\TEST.DTA';  ②
    ARRAY ANS[&NUMBER] $ 1 ANS1-ANS&NUMBER; ***Student answers;
    ARRAY KEY[&NUMBER] $ 1 KEY1-KEY&NUMBER; ***Answer key;
    ARRAY S[&NUMBER] 3 S1-S&NUMBER; ***Score array 1=right,0=wrong;
    RETAIN KEY1-KEY&NUMBER;
    IF _N_=1 THEN INPUT @1 (KEY1-KEY&NUMBER)($1.);
    INPUT @1   ID 1-9
          @11  (ANS1-ANS&NUMBER)($1.);
    DO I=1 TO &NUMBER;
       S[I] = (KEY[I] EQ ANS[I]);
    END;
    RAW=SUM (OF S1-S&NUMBER);
    PERCENT=100*RAW / &NUMBER;
    KEEP ANS1-ANS&NUMBER ID RAW PERCENT;
    LABEL ID      = 'Social Security Number'
          RAW     = 'Raw Score'
          PERCENT = 'Percent Score';
RUN;
PROC SORT DATA=SCORE;  ③
   BY ID;
RUN;
PROC PRINT DATA=SCORE LABEL;  ④
   TITLE "Listing of SCORE data set";
   ID ID;
   VAR RAW PERCENT;
   FORMAT ID SSN11.;
RUN;
                                                        [Continued]
```

```
PROC MEANS DATA=SCORE MAXDEC=2 N MEAN STD RANGE MIN MAX;    ⑤
   TITLE "Class Statistics";
   VAR RAW PERCENT;
RUN;
PATTERN COLOR=BLACK VALUE=EMPTY;
PROC GCHART DATA=SCORE;    ⑥
   TITLE "Histogram of Student Scores";
   VBAR PERCENT / MIDPOINTS=0 TO 100 BY 5;
RUN;
PROC FREQ DATA=SCORE;    ⑦
   TITLE "Frequency Distribution of Student Answers";
   TABLES ANS1-ANS&NUMBER / NOCUM;
RUN;
```

This program uses a macro variable (&NUMBER) that is assigned in the %LET statement ①. Each occurrence of &NUMBER is replaced by this assigned value before the program executes.

One other change from the previous program is that this program reads data from an external file, which is accomplished by the INFILE statement ②. The remainder of the DATA step portion of the program is identical to the previous program.

The first several PROCs are straightforward. We want a student roster in ID order ③ and ④, the class statistics ⑤, a histogram ⑥, and the frequencies of As, Bs, etc., for each of the questions of the test ⑦.

A portion of the output from this program is shown next:

```
Listing of SCORE data set

      Social
    Security      Raw       Percent
     Number      Score       Score

007-34-7889        0           0
010-23-2003        2          40
012-03-4009        3          60
012-12-1122        4          80
012-27-8733        1          20
023-48-1223        2          40
112-34-4987        3          60
               . . .

Class Statistics

The MEANS Procedure
```

Variable	Label	N	Mean	Std Dev	Range
RAW	Raw Score	34	2.91	1.38	5.00
PERCENT	Percent Score	34	58.24	27.58	100.00

Variable	Label	Minimum	Maximum
RAW	Raw Score	0.00	5.00
PERCENT	Percent Score	0.00	100.00

Frequency Distribution of Student Answers

The FREQ Procedure

ANS1	Frequency	Percent
A	19	55.88
B	6	17.65
D	5	14.71
E	4	11.76

ANS2	Frequency	Percent
A	4	11.76
B	17	50.00
C	3	8.82
D	3	8.82
E	7	20.59

. . .

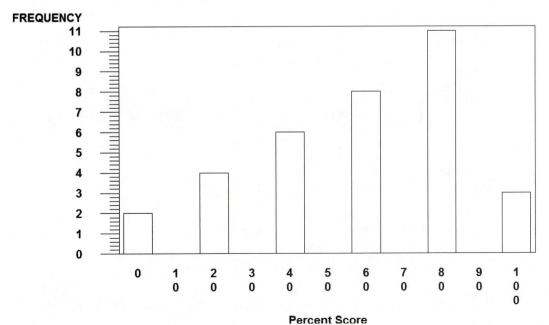

Histogram of Student Scores

FREQUENCY

Percent Score

D.   CREATING A BETTER-LOOKING TABLE USING PROC TABULATE

We can produce a compact table showing answer-choice frequencies using PROC TABU-
LATE. To do this efficiently, we will restructure the data set so that we have a variable
called QUESTION, which is the question number; and CHOICE, which is the answer
choice for that question for each student. We will be fancy and create CHOICE as a character variable that shows the letter choice (A, B, C, D, or E) with an asterisk (*) next to the
correct choice for each question. Again, we offer the program here without much explanation for those who might find the program useful or for those who would like to figure
out how it works. One of the authors (Smith) insists that good item analysis includes the
mean test score for all students choosing each of the multiple-choice items. Therefore, the
code to produce this statistic is included as well. The details of TABULATE are too much
to describe here, and we refer you to the *SAS Guide to Tabulate Processing* or *PROC Tabulate by Example* (by Lauren Haworth), both available from the SAS Institute, Cary, NC.

   The complete program to restructure the data set and produce the statistics previously described is shown next:

```
*----------------------------------------------------------------*
| Program Name: SCORE3.SAS in C:\APPLIED                          |
| Purpose: To score a multiple-choice exam with an arbitrary      |
| number of items and compute item statistics                     |
| Data: The first line is the answer key, remaining lines         |
| contain the student responses. Data is located in               |
| file C:\APPLIED\TEST.DTA                                         |
| Date: April 24, 2004                                            |
*----------------------------------------------------------------*;
OPTIONS LS=64 PS=59 NOCENTER;
PROC FORMAT;    ①
   PICTURE PCT LOW-<0=' ' 0-HIGH='00000%';
RUN;

%LET NUMBER=5; ***The number of items on the test;

DATA SCORE;
   INFILE 'C:\APPLIED\TEST.DTA';
   ARRAY ANS[&NUMBER] $ 2 ANS1-ANS&NUMBER; ***Student answers;
   ARRAY KEY[&NUMBER] $ 1 KEY1-KEY&NUMBER; ***Answer key;
   ARRAY S[&NUMBER] 3 S1-S&NUMBER; ***Score array 1=right,0=wrong;
   RETAIN KEY1-KEY&NUMBER;
   IF _N_=1 THEN INPUT @1 (KEY1-KEY&NUMBER)($1.);
   INPUT @1   ID 1-9
         @11 (ANS1-ANS&NUMBER)($1.);
   DO I=1 TO &NUMBER;
      IF KEY[I] EQ ANS[I] THEN DO;
      S[I] = 1;
      SUBSTR(ANS[I],2,1)='*';   ②
      ***Place an asterisk next to correct answer;
      END;
```

```
        ELSE S[I] = 0;
    END;
    RAW=SUM (OF S1-S&NUMBER);
    PERCENT=100*RAW / &NUMBER;
    KEEP ANS1-ANS&NUMBER ID RAW PERCENT;
    LABEL ID      = 'Social Security Number'
          RAW     = 'Raw Score'
          PERCENT = 'Percent Score';
RUN;

DATA TEMP;    ③
    SET SCORE;
    ARRAY ANS[*] $ 2 ANS1-ANS&NUMBER;
    DO QUESTION=1 TO &NUMBER;
        CHOICE=ANS[QUESTION];
        OUTPUT;
    END;
    KEEP QUESTION CHOICE PERCENT;
RUN;

PROC TABULATE DATA=TEMP;
    TITLE "Item Analysis Using PROC TABULATE";
    CLASS QUESTION CHOICE;
    VAR PERCENT;
    TABLE QUESTION*CHOICE,
    PERCENT=' '*(PCTN<CHOICE>*F=PCT. MEAN*F=PCT.
    STD*F=10.2)  / RTS=20 MISSTEXT=' ';
    KEYLABEL ALL  = 'Total'
             MEAN = 'Mean Score' PCTN='FREQ'
             STD  = 'Standard Deviation';
RUN;
```

A brief explanation of the program follows: PROC FORMAT ① is used to create a picture format so that we can print scores as percentages. (NOTE: There is a PERCENTw. format available as part of the SAS system, but it multiplies by 100 as well as places a percent sign after a number.) The remainder of the SCORE DATA step is the same as the previous program with the exception that the ANS1–ANSn variables are now two bytes in length. We will use this second byte later to place an asterisk next to the correct answer for each item. The SUBSTR function on the left of the equals sign ② is used to place an asterisk in the second position of the correct answer choice. A DATA step ③ is needed to restructure the data set so that it will be in a convenient form for PROC TABULATE. This data set contains n observations per student, where n is the number of items on the test. Selected portions of the output from these procedures are shown in the following table:

Item Analysis Using PROC TABULATE

QUESTION	CHOICE	FREQ	Mean Score	Standard Deviation
1	A*	55%	71%	23.40
	B	17%	26%	24.22
	D	14%	52%	22.80
	E	11%	50%	11.55
2	A	11%	40%	28.28
	B*	50%	75%	16.63
	C	8%	40%	20.00
	D	8%	73%	11.55
	E	20%	28%	22.68
3	A	8%	20%	20.00
	B	17%	46%	24.22
	C*	52%	68%	22.98
	D	8%	53%	46.19
	E	11%	60%	16.33
4	A	14%	32%	30.33
	B	5%	20%	0.00
	C	5%	30%	14.14
	D*	64%	71%	18.16
	E	8%	46%	30.55
5	A	8%	60%	20.00
	B	2%	20%	
	C	14%	16%	16.73
	D	5%	70%	14.14
	E*	67%	67%	21.52

The frequency column shows the percentage of students selecting each item choice. The frequency next to the correct answer (marked by an *) is the item's difficulty (percentage of students answering the item correctly). The column labeled MEAN SCORE shows the mean test score for all students answering each answer choice. For example, for item 1, 55% of the students chose A, which is the correct answer. The students who chose A had a mean score of 71% on the test. Seventeen percent of the students picked choice B, and the mean score of all students who chose B was 26%, and so forth. Looking briefly at item 2, we see that the students who picked choice D did fairly well on the test overall (73% correct). We might want to look at choice D to make sure it is not misleading.

E.   A COMPLETE TEST-SCORING AND ITEM-ANALYSIS PROGRAM

We present here a complete test-scoring and item-analysis program. This is a relatively complex program, and we will not go into any detail about its inner workings. We present it so that you may copy pieces of it, or all of it, and use it to analyze your multiple-choice tests. If you examine the sample output below the program, you will see that a lot of information is presented in a compact table. Each row of the table shows an item number, the answer key, the percentage of the students who chose each of the answer choices, a difficulty (the proportion of students answering the item correctly), a point-biserial correlation coefficient, and the proportion of students (in each quartile of the class) answering the item correctly. Here is the complete program:

```
*----------------------------------------------------------------*
| Program Name: SCORE4.SAS in C:\APPLIED                          |
| Purpose: To score a multiple-choice exam with an arbitrary      |
| number of items                                                 |
| Data: The first line is the answer key, remaining lines         |
| contain the student responses                                   |
| Data in file C:\APPLIED\TEST.DTA                                |
| Date: April 25, 2004                                            |
*----------------------------------------------------------------*;
%LET NUMBER=5; ***The number of items on the test; ①

DATA SCORE;
    INFILE 'C:\APPLIED\TEST.DTA'; ②
    ARRAY ANS [&NUMBER] $ 1 ANS1-ANS&NUMBER; ***Student answers;
    ARRAY KEY [&NUMBER] $ 1 KEY1-KEY&NUMBER; ***Answer key;
    ARRAY S [&NUMBER] S1-S&NUMBER; ***Score array 1=right,0=wrong;
    RETAIN KEY1-KEY&NUMBER;
    IF _N_ = 1 THEN INPUT (KEY1-KEY&NUMBER)($1.);
    INPUT @1  ID 1-9
          @11 (ANS1-ANS&NUMBER)($1)
    DO I=1 TO &NUMBER;
       S[I] = (KEY [I] EQ ANS [I]);
    END;
```
                                                            [Continued]

```
   RAW=SUM (OF S1-S&NUMBER);
   PERCENT=100*RAW / &NUMBER;
   KEEP ANS1-ANS&NUMBER S1-S&NUMBER KEY1-Key&NUMBER
       ID RAW PERCENT;
   ***Note: ANS1-ANSn, S1-Sn, KEY1-KEYnare needed later on;
   LABEL ID      = 'Social Security Number'
         RAW     = 'Raw Score'
         PERCENT = 'Percent Score';
RUN;
*-----------------------------------------------------------*
| You may want to include the procedures in Section C       |
| which print student rosters, histograms, and class        |
| statistics.                                               |
*-----------------------------------------------------------*;
***Section to prepare data sets for PROC TABULATE;
***Write correlation coefficients to a data set;
PROC CORR DATA=SCORE NOSIMPLE NOPRINT
          OUTP=CORROUT(WHERE=(_TYPE_='CORR'));
   VAR S1-S&NUMBER;
   WITH RAW;
RUN;
***Reshape the data set;
DATA CORR;
   SET CORROUT;
   ARRAY S[*] 3 S1-S&NUMBER;
   DO I=1 TO &NUMBER;
      CORR = S[I];
      OUTPUT;
   END;
   KEEP I CORR;
RUN;
***Compute quartiles;
PROC RANK DATA=SCORE GROUPS=4 OUT=QUART(DROP=PERCENT ID);
   RANKS QUARTILE;
   VAR RAW;
RUN;
***Create ITEM variable and reshape again;
DATA TAB;
   SET QUART;
   LENGTH ITEM $ 5 QUARTILE CORRECT I 3 CHOICE $ 1;
   ARRAY S[*] S1-S&NUMBER;
   ARRAY ANS[*] $ 1 ANS1-ANS&NUMBER;
   ARRAY KEY[*] $ 1 KEY1-KEY&NUMBER;
   QUARTILE = QUARTILE+1;
   DO I=1 TO &NUMBER;
      ITEM=RIGHT(PUT(I,3.)) || " " || KEY[I];
      CORRECT=S[I];
```

```
        CHOICE=ANS[I];
        OUTPUT;
     END;
     KEEP I ITEM QUARTILE CORRECT CHOICE;
RUN;
PROC SORT DATA=TAB;
     BY I;
RUN;
***Combine correlations and quartile information;
DATA BOTH;
     MERGE CORR TAB;
     BY I;
RUN;
***Print out a pretty table;
OPTIONS LS=72;
PROC TABULATE FORMAT=7.2 DATA=BOTH ORDER=INTERNAL NOSEPS;
     TITLE "Item Statistics";
     LABEL QUARTILE = 'Quartile'
           CHOICE   = 'Choices';
     CLASS ITEM QUARTILE CHOICE;
     VAR CORRECT CORR;
     TABLE ITEM='# Key'*F=6.,
     CHOICE*(PCTN<CHOICE>)*F=3. CORRECT=' '*MEAN='Diff.'*F=PERCENT5.2
     CORR=' '*MEAN='Corr.'*F=5.2
     CORRECT=' '*QUARTILE*MEAN='Prop. Correct'*F=PERCENT7.2/
        RTS=8;
     KEYLABEL PCTN='%' ;
RUN;
```

Here is a sample of the output from this program:

Item Statistics

# Key	Choices							Quartile			
								1	2	3	4
	A	B	C	D	E			Prop.	Prop.	Prop.	Prop.
	%	%	%	%	%	Diff.	Corr.	Correct	Correct	Correct	Correct
1 A	56	18	.	15	12	56%	0.55	33.3%	28.6%	90.9%	100%
2 B	12	50	9	9	21	50%	0.63	0.00%	42.9%	72.7%	100%
3 C	9	18	53	9	12	53%	0.42	16.7%	50.0%	63.6%	100%
4 D	15	6	6	65	9	65%	0.68	0.00%	64.3%	90.9%	100%
5 E	9	3	15	6	68	68%	0.51	16.7%	71.4%	81.8%	100%

To make this listing clear, let's look at item 1. The correct answer is 'A', which 56% of the students chose. Eighteen percent of the class chose 'B', and so forth. The item difficulty is 56%, and the point-biserial coefficient is .55. Thirty-three percent of the bottom quartile (lowest) answered this item correctly; 29% of the next quartile; 91% of the third quartile; and 100% of the top quartile answered this item correctly. This item is performing very well.

F.   TEST RELIABILITY

PROC CORR has the ability to compute Cronbach's Coefficient Alpha. For test items that are dichotomous, this coefficient is equivalent to the popular Kuder-Richardson formula 20 coefficient. These are the most commonly used estimates of the reliability of a test. They basically assess the degree to which the items on a test are all measuring the same underlying concept. The following lines show how to compute Coefficient Alpha from the data set SCORE in Section C of this chapter:

```
PROC CORR DATA=SCORE NOSIMPLE ALPHA;
   TITLE "Coefficient Alpha from Data Set SCORE";
   VAR S1-S5;
RUN;
```

Since each test item is either right (1) or wrong (0), Coefficient Alpha is equivalent to KR$-$20. Here is a partial listing:

Coefficient Alpha from Data Set SCORE

Cronbach Coefficient Alpha

Variables	Alpha
Raw	0.441866
Standardized	0.444147

Cronbach Coefficient Alpha with Deleted Variable

Deleted Variable	Raw Variables Correlation with Total	Alpha	Standardized Variables Correlation with Total	Alpha
S1	0.219243	0.395887	0.211150	0.404775
S2	0.316668	0.321263	0.317696	0.325870
S3	0.053238	0.511023	0.049289	0.513369
S4	0.404102	0.256819	0.414362	0.248868
S5	0.189630	0.415648	0.196747	0.414978

G. INTERRATER RELIABILITY

In studies where more than one rater rates subjects, you may want to determine how well the two raters agree. Suppose each rater is rating a subject as normal or not normal. By chance alone, the two raters will agree from time to time, even if they are both assigning ratings randomly. To adjust for this, a test statistic called Kappa was developed. If you are running SAS version 6.10 or later, Kappa is requested using the AGREE option on the TABLE statement of PROC FREQ. Suppose each of two raters rated 10 subjects as shown here:

Outcome (N = Normal, X = Not Normal)		
Subject	**Rater 1**	**Rater 2**
1	N	N
2	X	X
3	X	X
4	X	N
5	N	X
6	N	N
7	N	N
8	X	N
9	X	X
10	N	N

The program to compute Kappa is:

```
DATA KAPPA;
   INPUT SUBJECT RATER_1 $ RATER_2 $ @@;
DATALINES;
1 N N 2 X X 3 X X 4 X N 5 N X
6 N N 7 N N 8 X N 9 X X 10 N N
;
PROC FREQ DATA=KAPPA;
   TITLE "Coefficient Kappa Calculation";
   TABLE RATER_1 * RATER_2 / NOCUM NOPERCENT KAPPA;
RUN;
```

The output is shown here:

```
Coefficient Kappa Calculation

The FREQ Procedure

Table of RATER_1 by RATER_2

RATER_1      RATER_2

Frequency│
Row Pct  │
Col Pct  │N        │X        │  Total

N         │    4    │    1    │    5
          │  80.00  │  20.00  │
          │  66.67  │  25.00  │

X         │    2    │    3    │    5
          │  40.00  │  60.00  │
          │  33.33  │  75.00  │

Total          6         4        10

Statistics for Table of RATER_1 by RATER_2

      McNemar's Test

Statistic (S)       0.3333
DF                       1
Pr > S              0.5637

      Simple Kappa Coefficient

Kappa                   0.4000
ASE                     0.2840
95% Lower Conf Limit   -0.1566
95% Upper Conf Limit    0.9566

Sample Size = 10
```

As you can see, Kappa is .4 between these two raters. This may not seem like a terrific reliability, and indeed it isn't. You might say, "But they were in agreement for 7 of 10 cases," and indeed they were. But we would expect 5 out of 10 agreements by flipping a coin (there are only two ratings possible here), so 7 out of 10 is not a wonderful improvement over chance.

PROBLEMS

Remember, you can download all the data sets and programs for these problems from the web site: www.prenhall.com/cody

11.1 Given the following answer key:

Question 1 = 'B'   Question 2 = 'C'   Question 3 = 'D'
Question 4 = 'A'   Question 5 = 'A'

Write a SAS program to grade the six students whose data are shown here. Provide one listing in Social Security number order and another in decreasing test-score order. Compute both a raw score (the number of items correct) and a percentage score. (HINT: Use the SSN11. format for the Social Security number. Be sure to read it as a numeric if you do this.)

Student Data:

Social Security No.	Responses to Five Items
123-45-6789	B C D A A
001-44-5559	A B C D E
012-12-1234	B C C A B
135-63-2837	C B D A A
005-00-9999	E C E C E
789-78-7878	B C D A A

11.2 The results of a 10-item multiple choice test are stored in a text file as displayed here. A three-digit ID is in columns 1–3, and the 10 responses are stored in columns 5–14. (If you wish, you can simply use a DATALINES statement and place the student data in your program.)

```
001 ABCDBECDBE
002 ABCDEABCDE
003 ABCDEABCDD
004 ABCEDABCCE
005 BBCDEBBCDE
006 CABEDACBED
007 DECAACEDAA
008 ABCDEBBBEE
009 DDDDDABCDE
010 ABECDABCDE
```

The answer key is A B C D E A B C D E. Write a program to score this test, either assigning the correct answers in a temporary array (see Chapter 15, Section G) or by simple assignment statements. Compute the raw and percentile score for each student and generate point-biserial correlations for each item.

11.3 Using the test data from problem 11.1, compute the KR − 20 (or Cronbach's Alpha) for the test. Also, compute a point-biserial correlation coefficient for each item. Remember that a point-biserial correlation is equivalent to a Pearson correlation coefficient when one of the scores has values of 0 or 1.

11.4 Two psychiatrists rate each of 19 patients to determine if they are suicidal. The results are as follows (S = suicidal, N = not suicidal):

Subject	1	2	3	4	5	6	7	8	9	10	11	12	13	14	15	16	17	18	19
Rater 1	S	N	S	N	S	S	N	N	N	S	S	N	N	S	N	S	N	S	S
Rater 2	S	N	S	N	N	S	S	N	N	N	S	N	N	S	S	S	N	S	S

Compute Kappa to determine the strength of agreement between these two psychiatrists. The easier way to solve this problem is to enter the data in subject order. Your data would look like this (remember the trailing double @; see Chapter 12):

```
S S   N N   S S   N N   S N   S S   N S   N N   N N   S N
S S   N N   N N   S S   N S   S S   N N   S S   S S
```

If you would like a challenge, you can try to enter the data by rater (as displayed in the previous table).

11.5 Two pathologists viewed 14 slides and made a diagnosis of cancer or not cancer. Using the following data, compute Kappa, an index of interrater reliability (C = cancer, X = not cancer):

Rater 1	Rater 2	Rater 1	Rater 2
C	C	C	X
X	X	C	C
X	X	X	X
C	X	C	C
X	C	C	C
X	X	X	X
X	X	C	C

CHAPTER 12

The SAS INPUT Statement

A. INTRODUCTION

Throughout the examples in the statistics section of this book, we have seen some of the power of the SAS INPUT statement. In this chapter, the first in a section on SAS programming, we explore the power of the INPUT statement. (NOTE: To learn the basics of the INPUT statement, return to Chapter 1.)

B. LIST INPUT: DATA VALUES SEPARATED BY SPACES

SAS can read data values separated by one or more spaces. This form of input is sometimes referred to as list input. The rules here are that we must read every variable on a line, the data values must be separated by one or more spaces, and missing values are represented by periods. A simple example is shown here:

```
DATA QUEST;
    INPUT ID GENDER $ AGE HEIGHT WEIGHT;
DATALINES;
1   M    23    68    155
2   F    .     61    102
3      M      55    70    202
;
```

Notice that character variable names are followed by a $. The multiple spaces between the data values in the third line of data will not cause a problem.

C.   READING COMMA-DELIMITED DATA

Sometimes we are given data that are comma-delimited (i.e., separated by commas) instead of the spaces that SAS is expecting. We have two choices here: We can use an editor and replace all the commas with blanks, or we can leave the commas in the data and use the DLM = option of the INFILE statement to make the comma the data delimiter. (See Chapter 13 for details on the INFILE statement and its options.) As an example, suppose you were given a file on a diskette called SURVEY.DTA. All the data values are separated by commas. The first three lines are shown next:

```
1,M,23,68,155
2,F,.,61,102
3,  M, 55, 70,   202
```

To read this file, we code:

```
DATA HTWT;
    INFILE 'A:SURVEY.DTA' DLM=',';
    INPUT ID GENDER $ AGE HEIGHT WEIGHT;
RUN;
```

The INFILE statement directs the INPUT statement to read an external file called SURVEY on the diskette in drive A and to use commas as the data delimiter.

Another useful INFILE option for reading comma-delimited files is DSD (delimiter-sensitive data). Besides treating commas as delimiters, this option performs several other functions. First, if it finds two adjacent commas, it will assign a missing value to the corresponding variable in the INPUT list. Second, it will allow text strings surrounded by quotes to be read into a character variable and will strip off the quotes in the process. This option is ideal for reading comma-separated variables (CSV) files. To illustrate the DSD option, suppose the following three lines of data are stored in the file A:SURVEY.DTA.

```
1,"M",23,68,155
2,F,,61,102
3, M, 55, 70, 202
```

A SAS DATA step to read these three lines of data is shown next:

```
DATA HTWT;
    INFILE 'A:SURVEY.DTA' DSD;
    INPUT ID GENDER $ AGE HEIGHT WEIGHT;
RUN;
```

The resulting data set will be identical to the earlier example using the DLM option. Notice that the GENDER for the first person will be the value 'M' without the quotes, and the AGE for the second person will be assigned a missing value.

D.   USING INFORMATS WITH LIST INPUT

We may have data, such as date values, that we want to read with a date informat but still want to use list input. We have two choices here. One is to precede the INPUT statement with an INFORMAT statement, assigning an informat to each variable. An INFORMAT statement uses the same syntax as a FORMAT statement but is used to supply an informat instead of a format for a variable. An example, using an INFORMAT statement, is shown here:

```
DATA INFORM;
    INFORMAT DOB VISIT MMDDYY10.;
    INPUT ID DOB VISIT DX;
DATALINES;
1 10/21/1946 6/5/1989 256.20
2 9/15/1944 4/23/1989 232.0
etc.
```

An alternative to the INFORMAT statement is to supply the informats directly in the INPUT statement. We do this by following the variable name with a colon, followed by the appropriate informat. A program using this method with the same data set is:

```
DATA FORM;
    INPUT ID DOB : MMDDYY10. VISIT : MMDDYY10. DX;
DATALINES;
1 10/21/1946 6/5/1989 256.20
2 9/15/1944 4/23/1989 232.0
etc.
```

Either method can also be used to override the default eight-character limit for character variables. So, to read a file containing last names (some longer than eight characters) we could use either of the next two programs:

```
*Example with an INFORMAT statement;
DATA LONGNAME;
    INFORMAT LAST $20.;
    INPUT ID LAST SCORE;
DATALINES;
1 STEVENSON 89
2 CODY 100
3 SMITH 55
4 GETTLEFINGER 92
etc.

*Example with INPUT informats;
DATA LONGNAME;
    INPUT ID LAST : $20. SCORE;
DATALINES;
1 STEVENSON 89
2 CODY 100
3 SMITH 55
4 GETTLEFINGER 92
etc.
```

Before we leave list input, there is one more "trick" you should know about. Suppose you wanted to read a first and last name into a single variable. If we used spaces as data delimiters, we could not have a blank between the first and last name. However, the very clever people at SAS have thought about this problem and have come up with the ampersand (&) informat modifier. An ampersand modifier following a variable name changes the default delimiter of one space to two or more spaces. To see how this works, look at the next program:

```
DATA FIRSTLST;
    INPUT ID NAME & $30. SCORE1 SCORE2;
DATALINES;
```

```
1 RON CODY   97 98
2 JEFF SMITH   57 58
etc.
```

Notice that there are at least two spaces between the name and the first score.

E.   COLUMN INPUT

Many of the examples in this book use INPUT statements that specify which columns to read for each data value. The syntax is to list the variable names, followed by the column or columns to read. In addition, we follow the variable name by a $ sign if we are reading character values. A simple example is:

```
DATA COL;
    INPUT ID        1-3
          GENDER $ 4
          HEIGHT   5-6
          WEIGHT   7-11;
DATALINES;
001M68155.5
2  F61 99.0
 3 M  233.5
(more data lines)
```

Notice that the ID number for subject number 2 is not right-adjusted. In some programming languages, this would cause a problem; not so for SAS. We could have placed the '2' in any of the first three columns, and it would have been read properly. Notice also that we can include decimal points in numeric fields. Just remember to leave the extra columns for them. Finally, remember that we can simply leave columns blank when we have missing values. The downside of using column input is that we can only read character and standard numeric data. To read dates and nonstandard numeric data (such as numbers with commas), you need pointers and informats, described next.

F.   FORMATTED INPUT

An alternative to specifying the starting and ending columns (column input) is to specify a starting column and an INFORMAT (which also specifies how many columns to read).

This form of input is referred to as "formatted input." It is specially useful when we are given a coding layout like the following:

Variable	Starting Column	Length	Type	Description
ID	1	3	NUM	SUBJECT ID
GENDER	4	1	CHAR	GENDER M=MALE F=FEMALE
AGE	9	2	NUM	AGE IN YEARS
HEIGHT	11	2	NUM	HEIGHT IN INCHES
V_DATE	15	8	DATE	VISIT DATE IN MMDDYYYY form

Rather than doing all the high-level arithmetic to compute ending columns for each of these variables, we can use a pointer (@ sign) to specify the starting column and an informat, which will not only tell SAS how to read the data value, but also how many columns to read. Here is the program to read the previous data layout:

```
DATA POINT;
    INPUT @1   ID       3.
          @4   GENDER  $1.
          @9   AGE      2.
          @11  HEIGHT   2.
          @15  V_DATE   MMDDYY8.;
```

The @ symbol, called an absolute column pointer, indicates the starting column for each variable. In the previous INPUT statement, some of the pointers are redundant, such as the @4 before GENDER. As data are read, an internal pointer moves along the data line. Since ID started in column 1 and was three columns in length, this internal pointer was ready to read data in column 4. We recommend using an absolute column pointer before every variable as in this example; it makes for a neater program and reduces the possibility of reading the wrong column. The informat W. is used for a numeric variable of length W; $W. is the informat used for a character variable of length W. The general form for a SAS numeric informat is W.d, where W is the number of columns to read and d is the number of digits to the right of an implied decimal point (if a decimal point is not included in the data value). For example, the number 123 read with an informat 3.2 would be interpreted as 1.23 (we are reading three columns, and there are two digits to the right of the decimal point). Using this notation, we can read numbers with an "implied" decimal point. By the way, we can also read numbers with decimal points. The number 1.23 read with the informat 4. (remember the decimal point takes up a column) would be interpreted as 1.23. That is, any decimal point in the data value itself overrides the implied decimal point in the informat. The informat MMDDYY8. was one of the date formats we used in Chapter 4 to read date values. We used a separate line for each variable simply as a matter of programming style.

G.   READING MORE THAN ONE LINE PER SUBJECT

When we have more than one line of data for each subject, we can use the row pointer (#) to specify which row of data we want to read. Just as with the column pointer, we can move anywhere within the multiple rows of data per subject. Keep in mind that we must have the same number of rows of data for each subject. Here is an example where two lines of data were recorded for each subject:

```
DATA COLUMN;
INPUT #1 ID       1-3
         AGE      5-6
         HEIGHT 10-11
         WEIGHT 15-17
      #2 SBP      5-7
         DBP      8-10;
DATALINES;
001   56   72   202
    140080
002   45   70   170
    130070
;
```

If you have N lines of data per subject but don't want to read data from all N lines, make sure to end your input statement with #N, where N is the number of data lines per subject. For example, if you have six lines of data per subject but only want to read two lines (as in the previous example), you would write:

```
INPUT #1 ID 1-3 AGE 5-6 HEIGHT 10-11 WEIGHT 15-17
#2 SBP 5-7 DBP 8-10 #6;
```

H.   CHANGING THE ORDER AND READING A COLUMN MORE THAN ONCE

It is possible to move the absolute column pointer to any starting column, in any order. Thus, we can read variables in any order, and we can read columns more than once. Suppose we had a six-digit ID where the last two digits represented a county code. We could do this:

```
INPUT @1 ID       6.
         @5 COUNTY $2.
      etc.;
```
or
```
INPUT ID       1-6
         COUNTY $ 5-6
      etc.;
```

We can also read variables in any order. The following INPUT statements are valid:

```
INPUT ID 1-3 HEIGHT 5-6 GENDER $ 4 WEIGHT 7-9;
INPUT @1 ID 3. @5 HEIGHT 2. @4 GENDER $1. @7 WEIGHT 3.;
```

I. INFORMAT LISTS

You may place a group of variables together within parentheses in an INPUT statement and follow this list by one or more INFORMATS, also in parentheses. Here is a simple example:

```
INPUT (X Y Z C1-C3)(1. 2. 1. $3. $3. $3.);
```

Now, this isn't very useful as shown. Where you save time is when several variables in a row all use the same INFORMAT. You can have fewer informats in the IN-FORMAT list than there are variables in the variable list. When this happens, the INFORMAT list is recycled as many times as necessary to provide an INFORMAT for each of the variables in the variable list. The following INPUT statement illustrates this:

```
INPUT (X1-X50)(1.);
```

This indicates to read X1, X2, X3, ..., X50 all with the 1. INFORMAT. Here is an example showing how you can shorten a fairly long INPUT statement using variable lists and INFORMAT lists. First, the INPUT statement without using variable and INFORMAT lists:

```
DATA NOVICE;
   INPUT @1   ID      $3.
         @4   QUES1   1.
         @5   QUES2   1.
         @6   QUES3   1.
         @7   QUES4   1.
         @8   QUES4B  1.
         @9   QUES4C  1.
         @10  QUES5   1.
         @11  QUES6   1.
         @12  QUES7   1.
         @13  QUES8   1.
         @14  QUES9   1.
         @20  DOB        MMDDYY8.
         @28  ST_DATE  MMDDYY8.
         @36  END_DATE MMDDYY8.;
```

The previous program, rewritten to use informat lists is:

```
DATA ADVANCED;
   INPUT @1 ID $3.
         @4   (QUES1-QUES4 QUES4B QUES4C QUES5-QUES9)(1.)
         @20  (DOB ST_DATE END_DATE)(MMDDYY8.);
```

Well, it didn't really save much typing, but if you had hundreds or thousands of variables, the savings would be considerable.

The informat list can also contain something called relative column pointers. Using a + sign followed by positive or negative value, we can move the pointer forward or backward in the record. Next, we show you a novel INPUT statement where relative pointers saved a lot of coding.

A researcher coded 12 systolic and diastolic blood pressures for each subject (the number was reduced for illustrative purposes). They were entered in pairs. A straightforward INPUT statement would be:

```
INPUT @1  ID $3.
      @4   SBP1 3.
      @7   DBP1 3.
      @10  SBP2 3.
      @13  DBP2 3.
   etc. ;
```

A more compact method, using relative pointers is:

```
INPUT @1  ID $3.
      @4  (SBP1-SBP12)(3. +3)
      @7  (DBP1-DBP12)(3. +3);
```

The INFORMAT list (3. + 3) says to read a variable using the 3. INFORMAT and then move the pointer over three spaces. Thus, we "skip over" the diastolic pressures the first time, set the pointer back to column 7, and repeat the same trick with the diastolic pressures.

J.   "HOLDING THE LINE"—SINGLE- AND DOUBLE-TRAILING @S

There are times when we want to read one variable and, based on its value, decide how to read other variables. To do this, we need more than one INPUT statement in a DATA step. Normally, when SAS finishes an INPUT statement, the pointer is moved to the next line of data. So, if we had more than one INPUT statement, the second INPUT statement would be reading data from the next line or record. We have two ways to prevent this: the single- and double-trailing @ symbols. A single @ sign, placed at the end of an INPUT statement, means to "hold the line." That is, do not move the pointer to the next record until an INPUT statement without a trailing @ is reached or until the next iteration of the DATA step begins. The double-trailing @ symbol "holds the line strongly." That is, the pointer will not be moved to the next record, even if a DATALINES or RUN statement (end of the DATA step) is encountered. It will move to the next record only if there are no more data values to be read on a line or if an INPUT statement without a double-trailing @ sign is executed.

Here are some examples to help make this clear. In the first example, we ran a survey for 1989 and 1990. Unfortunately, in 1990, an extra question was added—in the middle of the questionnaire (where else!). We want to be able to combine these data in a single file and read each record according to the data layout for that year. The year of the survey was coded in columns 79 and 80 (a value of 89 for 1989 and a value of 90 for 1990). In 1989, we had 10 questions in columns 1–10. In 1990, there was an extra question (let's call it 5B) placed in column 6, and questions 6–10 wound up in columns 7–11. We will use a trailing @ to read these data:

```
DATA QUEST;
   INPUT YEAR 79-80 @; *** HOLD THE LINE;
   IF YEAR = 89 THEN INPUT @1 (QUES1-QUES10)(1.);
   ELSE IF YEAR = 90 THEN INPUT @1 (QUES1-QUES5)(1.)
                            @6 QUES5B 1.
                            @7 (QUES6-QUES10)(1.);
DATALINES;
```

A simple example where a double-trailing @ is needed is shown next. Suppose we want to read in pairs of Xs and Ys and want to place several X,Y pairs per line. We could read these data with a double-trailing @:

```
DATA XYDATA;
   INPUT X Y @@;
DATALINES;
1 2 7 9 3 4 10 12
15 18 23 67
;
```

The data set XYDATA would contain six X,Y pairs, namely (1,2), (7,9), (3,4) (10,12), (15,18), and (23,67). Without the double-trailing @, the data set would contain only two X,Y pairs—(1,2) and (15,18).

K. SUPPRESSING THE ERROR MESSAGES FOR INVALID DATA

If invalid data values are read by the SAS system (such as character data in a numeric field or two decimal points in a number), an error message is placed in the SAS LOG, the offending record is listed, and a missing value is assigned to the variable. Below is an example of a SAS LOG where a character value ('a') was read into a numeric field:

```
1   data example;

2   input x y z;

3   datalines;

NOTE: Invalid data for X in line 5 1-1.

RULE:----+----1----+----2----+----3----+----4----+----5----+----6

  5 a 3 4

X=. Y=3 Z=4 _ERROR_=1 _N_=2

  6 run;

NOTE: The data set WORK.EXAMPLE has 2 observations and 3 variables.

NOTE: The DATA statement used 2.00 seconds.
```

Although this information can be very useful, there are times where we know that certain fields contain invalid data values, and we want missing values to be substituted for the invalid data. For large files, the SAS LOG may become quite large, and the processing time will be longer when these error messages are processed. There are two ways to reduce the error message handling. First, a single question mark placed after the variable name will suppress the invalid data message but will still print the offending line of data. Two question marks following the variable name will suppress all error messages and prevent the automatic _ERROR_ variable from being incremented. Here is the INPUT statement to suppress error messages when an invalid value is encountered for X:

    INPUT X ?? Y Z;

If you are using column input, you would write:

    INPUT X ?? 1-2 Y 3-4 Z 5-6;

To allow invalid values for X, Y, and Z, we could write:

    INPUT @1 (X Y Z)(?? 2.);

L.   READING "UNSTRUCTURED" DATA

Almost all the examples in this text have been either small data sets or balanced data sets that were relatively easy to read using standard INPUT statements. However, in the real world, we often encounter data sets that are not so clean. For example, we might have a varying number of records for each subject in a study. Another example would be an unbalanced design where there were different numbers of subjects in each treatment. As these data sets become large, reading them without error sometimes becomes the most difficult part of the data-processing problem. The techniques shown in this section will allow you to read almost any type of unstructured data easily.

The key to all the examples that follow is to embed "tags" in the data to indicate to the program what type of data to read. A t-test example with unequal n's and an unbalanced ANOVA will serve to illustrate the use of tags and stream data input.

Example 1. Unbalanced T-test

The amount and complexity of the data have been reduced to make the examples short and easy to follow. The strength of the techniques is their use with larger, more complicated data sets.

We want to analyze an experiment where we had five control and three treatment subjects, and we recorded a single variable per subject. The data are shown here:

GROUP	
Control	Treatment
20	40
25	42
23	35
27	
30	

The simplest, most straightforward method to read these data is shown next:

Example 1-A

```
***Traditional INPUT Method;
DATA EX1A;
    INPUT GROUP $ X @@;
DATALINES;
C 20 C 25 C 23 C 27 C 30
T 40 T 42 T 35
;
PROC TTEST DATA=EX1A;
    CLASS GROUP;
    VAR X;
RUN;
```

For larger amounts of data, this program contains some problems. It is tedious and time-consuming to repeat the group identification before each variable to be read. This can be corrected in two ways: We can put the information concerning the number of observations per group in the program (Examples 1-B and 1-C), or we can put this information in the data itself (Example 1-D). As previously mentioned, if the number of observations was large (several hundred or more), a single mistake in counting would have disastrous consequences.

Example 1-B

```
DATA EX1B;
   GROUP='C';
   DO I=1 TO 5;
      INPUT X @;
      OUTPUT;
   END;
   GROUP='T';
   DO I=1 TO 3;
      INPUT X @;
      OUTPUT;
   END;
   DROP I;
DATALINES;
20 25 23 27 30
40 42 35
;
PROC TTEST DATA=EX1B;
   CLASS GROUP;
   VAR X;
RUN;
```

Example 1-C

```
DATA EX1C;
   DO GROUP = 'C','T';
      DO I=1 TO 5*(GROUP EQ 'C') + 3*(GROUP EQ 'T');
         INPUT X @;
         OUTPUT;
      END;
   END;
   DROP I;
DATALINES;
20 25 23 27 30
40 42 35
;
PROC TTEST DATA=EX1B;
   CLASS GROUP;
   VAR X;
RUN;
```

Example 1-D

```
DATA EX1D;
   DO GROUP='C','T';
      INPUT N;
      DO I=1 TO N;
         INPUT X @;
         OUTPUT;
      END;
   END;
   DROP N I;
DATALINES;
5
20 25 23 27 30
3
40 42 35
;
PROC TTEST DATA=EX1C;
   CLASS GROUP;
   VAR X;
RUN;
```

The method we are suggesting for large data sets is shown in Example 1-E:

Example 1-E

```
***Reading the Data with Tags;
DATA EX1E;
   RETAIN GROUP;
   INPUT DUMMY $ @@;
   IF DUMMY='C' OR DUMMY='T' THEN GROUP=DUMMY;
   ELSE DO;
      X = INPUT(DUMMY,5.0);
      OUTPUT;
   END;
   DROP DUMMY;
DATALINES;
C 20 25 23 27 30
T 40 42 35
;
PROC TTEST DATA=EX1D;
   CLASS GROUP;
   VAR X;
RUN;
```

With this program we can add or delete data without making any changes to our program. The three important points in the program are:

1. All data items are read as a character and are interpreted. If a 'C' or 'T' is found, the variable GROUP is set equal to DUMMY, and the DATA step returns to read the next number. The RETAIN statement prevents the variable GROUP from being reinitialized to missing each time the DATA step iterates—it will keep its value of 'C' or 'T' until reset.

2. The INPUT function is used to "read" a character variable with a numeric informat. The INPUT function takes two arguments. The first is the variable to "reread," the second is the informat with which to read that value. Thus, although DUMMY is a character variable, X will be stored as a numeric. We chose the informat 5. since we knew it would be larger than any of our data values.

3. Because there is an OUTPUT statement in the ELSE DO block, the program will not output an observation when DUMMY is equal to a 'C' or a 'T'. Whenever an explicit OUTPUT statement is used in a DATA step, the automatic implied OUTPUT does not occur at the end of the DATA step.

This same program can read data that are not as ordered as Example 1-D. For instance, the data set

```
C 20 25 23 T 40 42
C 30 T 35
```

will also be read correctly. For large data sets, this structure is less prone to error than Examples 1-A through 1-D. (Of course, we pay additional processing costs for the alternative program, but the ease of data entry and the elimination of counting errors is probably worth the extra cost.)

Example 2. Unbalanced Two-way ANOVA

The next example is an unbalanced design for which we want to perform an analysis of variance. Our design is as follows:

		GROUP		
		A	**B**	**C**
		20	70	90
		30	80	90
	M	40	90	80
		20		90
		50		
Gender				
		25	70	20
		30	90	20
	F	45	90	30
		30	80	
		65	85	
		72		

The straightforward method of entering these data would be:

```
DATA EX2A;
    INPUT GROUP $ GENDER $ SCORE;
DATALINES;
A M 20
A M 30
etc.
```

This is a lengthy and wasteful data-entry method. For small data sets of this type, we could follow the example of the unbalanced t-test problem and enter the number of observations per cell, either in the program or embedded in the data. A preferable method, especially for a large number of observations per cell where counting would be inconvenient, is shown in Example 2-A:

Example 2-A

```
***First Method of Reading ANOVA Data with Tags;
DATA EX2A;
    DO GENDER='M','F';
        DO GROUP='A','B','C';
            INPUT DUMMY $ @;
            DO WHILE (DUMMY NE '#');
                SCORE=INPUT(DUMMY,6.0);
                OUTPUT;
                INPUT DUMMY $ @;
            END;
        END;
    END;
    DROP DUMMY;
DATALINES;
20 30 40 20 50 # 70 80 90
# 90 90 80 90 # 25 30 45 30
65 72 # 70 90 90 80 85 # 20 20 30 #
;
PROC GLM DATA=EX2A;
etc.
```

This program reads and assigns observations to a cell until a "#" is read in the data stream. The program then finishes the innermost loop, and the next cell is selected. We can read as many lines as necessary for the observations in a given cell.

An improved version of this program is shown next (Example 2-B). With this program, we can read the cells in any order and do not have to supply the program with the cell identification since it is incorporated right in the tags. Let's look over the program first, and then we will discuss the salient features:

Example 2-B

```
***More Elegant Method for Unbalanced ANOVA Design;
DATA EX2B;
    RETAIN GROUP GENDER;
    LENGTH GROUP GENDER $ 1;
    INPUT DUMMY $ @@;
    IF VERIFY (DUMMY,'ABCMF ') = 0 THEN DO;
        GROUP = SUBSTR (DUMMY,1,1);
        GENDER = SUBSTR (DUMMY,2,1);
        DELETE;
    END;
    ELSE SCORE = INPUT (DUMMY,6.);
    DROP DUMMY;
DATALINES;
AM   20   30   40   20   50
BM   70   80   90
CM   90   80   80   90
AF   25   30   45   30   65   72
BF   70   90   90   80   85
CF   20   20   30
;
PROC GLM DATA=EX2B;
etc.
```

This program allows us to enter the cells in any order and even use as many lines as necessary for the observations from a cell. This form of data entry is also convenient when we will be adding more data at a later time. The analysis can be rerun without any changes to the program. Additional observations can even be added at the end of the original data.

Special features of this program are the use of the VERIFY and SUBSTR functions. The VERIFY function returns 0 if all the characters in the variable DUMMY can be found as one of the characters in the second argument of the function. Note that a blank is included in argument 2 of the VERIFY function, since the length of DUMMY is, by default, equal to eight bytes, which means that it will contain two letters and six blanks. The SUBSTR function picks off the GROUP and GENDER values from the DUMMY string, and the INPUT function converts all character values back to numeric. (See Chapters 17 and 18 for a more detailed discussion of SAS functions.) A CONCLUDING NOTE: One of the authors of this book writes these programs with relative ease. The other author calls him when in need of help. You have to rely on this chapter. So, be careful about how you structure your data sets.

PROBLEMS

Remember, you can download all the data sets and programs for these problems from the web site: www.prenhall.com/cody

12.1 You have five subjects in a placebo group and five in a drug group. One way to structure your data is like this:

GROUP	SCORE
P	77
P	76
P	74
P	72
P	78
D	80
D	84
D	88
D	87
D	90

(a) Write an INPUT statement to read these data, assuming you have one or more spaces between the group designation and the score.

(b) Suppose you prefer to arrange your values on two lines like this:

```
P 77 P 76 P 74 P 72 P 78
D 80 D 84 D 88 D 87 D 90
```

Write an INPUT statement for this arrangement.

(c) This time, the five scores for the placebo group are on the first line, and the five scores for the drug group are on another, like this:

```
77 76 74 72 78
80 84 88 87 90
```

Write a DATA step to read these data. Be sure the data set contains a GROUP as well as a SCORE variable.

(d) Modify the program in part (c) so that each of the 10 subjects has a subject number from 1 to 10.

12.2 You have raw data consisting of a date of birth, a visit date, a gender, and a last name. The last name may be up to 10 characters in length. Write a SAS DATA step to read the following values. Note that the data values are separated by commas, and you need to read the two dates with the appropriate informat. Make the length of GENDER one byte in length and make NAME 10 bytes. You may use the following sample data lines to test your program. (Hint: Use INFILE DATALINES to apply the appropriate option.)

```
10/21/1950,03MAY2004,M,Schneider
11/12/1944,05DEC2004,F,Strawderman
01/01/1960,25APR2004,M,Smith
```

12.3 Given the three lines of data:

```
1,3,5,7
2,4,6,8
9,8,7,6
```

Write a SAS DATA step to read these data, assigning the four data values to the variables X1 to X4.

12.4 Write a SAS DATA step to read the following lines of data. Notice that two commas in a row indicate a missing value, and you want to strip off the quotes from quoted strings (remember there is an INFILE option to do this). The variables are NAME (up to 20 characters), AGE, HEIGHT, and WEIGHT.

```
Bradley,35,68,155
"Bill Johnson",,70,200
"Smith,Jeff",27,70,188
```

*12.5 Given the five lines of data:

```
1,,"HELLO",7
2,4,TEXT,8
9,,,6
21,31,SHORT
100,200,"LAST LINE",999
```

Write a SAS DATA step to read these data, assigning the four data values to the variables X, Y, C, and Z. Variable C should be a character variable. (Note that the value of variable C in the last line is nine characters.) The double quotes should be stripped off the text strings, and two adjacent commas should be interpreted as containing a missing value. Careful, line four has only three values on it. Remember the INFILE option to handle short lines.

12.6 Using column input, read in NAME, GENDER, and DOB, where NAME is in columns 1–10, GENDER is in column 12, and DOB is in columns 13–22. Note that with column input, you must read the date of birth as a character. You may use the following sample lines to test your program:

```
          1         2
1234567890123456789012
----------------------
Cody       M05/11/1981
McMaster   F11/11/1967
Bill Smith M12/25/1999
```

*12.7 You are given data values, separated by one or more spaces, representing a patient ID, date of visit, DX code, and cost. Create a SAS data set called OFFICE from these data:

```
1 10/01/1996    V075    $102.45
2   02/05/1997    X123456789 $3,123
3   07/07/1996    V4568
4   11/11/1996    A123    $777.
```

NOTES: (1) There is no cost value for patient 3, and the line is not padded with blanks. (2) The maximum length for a DX code is 10. (3) The largest cost is $99,999.

12.8 You have the data lines listed here. The first line is a header line and should not be read (use the INFILE option FIRSTOBS=2). The remaining lines contain an ID number (read this as character), a GENDER, a date of birth (DOB), and two scores (SCORE1 and SCORE2). Note that there are some missing values for the scores, and you want to be sure that SAS does not go to a new line to read these values. Write a SAS DATA step to read these lines of data and create a temporary SAS data set (PROB12_8). Format the DOB with DATE9. Here are the lines of data:

```
***Header line: ID GENDER DOB SCORE1 SCORE2
001 M 10/10/1976 100 99
002 F 01/01/1960 89
003 M 05/07/2001 90 98
```

Make sure your data set has three observations.

12.9 Given the data layout:

Variable	Starting Column	Ending Column	Type
SUBJECT	1	3	Char
A1	5	5	Char
X	7	8	Num
Y	9	10	Num
Z	11	12	Num

Write a SAS DATA step to read these data using starting and ending column specifications in your INPUT statement. Here are some sample lines of data for you to test your program:

```
         1         2
12345678901234567890
--------------------
A12 X 111213
A13 W 102030
```

*12.10 This problem tests your knowledge of the relative column pointer (plus sign). Rewrite the following program using a variable list followed by a single informat list (using the relative column pointer) for the GROUP variables and using a variable list and an informat list for the SCORE variables.

```
DATA PAIRS;
    INPUT @1  GROUP1 $1.
          @2  SCORE1  3.
          @6  GROUP2 $1.
          @7  SCORE2  3.
          @11 GROUP3 $1.
          @12 SCORE3  3.;
DATALINES;
A100 B 90 C 76
C 87 A 86 B 88
C 93 B 92 A 90
;
```

12.11 Create a SAS data set using the same data as problem 12.9, except use column pointers and INFORMATS instead of starting and ending columns.

*12.12 You have data from a clinical trial where the first number on each line tells you how many scores to read for that subject (all the lines contain more values than you need). Using these data, create a SAS data set (CLINICAL) with variables SUBJ (you have to create this variable in the DATA step), SCORE1–SCORE5, and AGE. Some of the SCORE variables will have missing values. Use the following three lines of data to test your program. (HINT: You may want to use ARRAYS to solve this problem.)

```
5 90 80 70 77 88 23
2 100 99 25
3 87 85 88 35
```

To make this clear, a listing of this data set should look like this:

```
Listing of PROB12_12

Obs    SCORE1    SCORE2    SCORE3    SCORE4    SCORE5    NUMBER    SUBJ    AGE

 1        90        80        70        77        88         5        1      23
 2       100        99         .         .         .         2        2      25
 3        87        85        88         .         .         3        3      35
```

*12.13 Someone gives you the following data layout:

Variable	Start Column	Length	Description
ID	1	3	Num
GENDER	4	1	Char
DOB	10	6	MMDDYY
VISIT	16	6	MMDDYY
DISCHRG	22	6	MMDDYY
SBP1	30	3	Num
DBP1	33	3	Num
HR1	36	2	Num
SBP2	38	3	Num
DBP2	41	3	Num
HR2	44	2	Num
SBP3	46	3	Num
DBP3	49	3	Num
HR3	52	2	Num

Write a SAS DATA step to read the following two lines of sample data. Use variable lists and informat lists to read these data. See if you can find a way to read the SBPs and DBPs other than the straightforward `@30 SBP1 3. @33 DBP1 3. @36 HR1 2.`, etc. That is, try to read all the SBPs together (SBP1–SBP3)(your informat) and so on.

```
          1         2         3         4         5
1234567890123456789012345678901234567890123456789012345
-------------------------------------------------------
123M    102146111196111396  130 8668134 8872136 8870
456F    010150122596020597  220110822101028424012084
```

*12.14 Write a SAS DATA step to read the following lines of data. Each line contains a three-digit
number, a two-digit number, and a four-digit number, with four sets of numbers for each
subject. Name your variables THREE1–THREE4, TWO1–TWO4, and FOUR1–FOUR4,
where the THREE variables represent the three-digit numbers, etc. Use variable lists and
informat lists to solve this problem. Here are some sample data lines:

```
12312123421787444412387234587323 5432
19283746574839291928374737281818 2838
```

```
Listing of PROB12_14

       T    T    T    T
       H    H    H    H                          F    F    F    F
       R    R    R    R    T    T    T    T       O    O    O    O
   O   E    E    E    E    W    W    W    W       U    U    U    U
   b   E    E    E    E    O    O    O    O       R    R    R    R
   s   1    2    3    4    1    2    3    4       1    2    3    4

   1  123  217  123  873  12   87   87   23    1234 4444 2345 5432
   2  192  748  283  818  83   39   74   18    7465 2919 7372 2838
```

12.15 Write a SAS DATA step to read the following data:

	Variable	Starting Column	Length	Type
Line 1	ID	1	2	Num
	X	4	2	Num
	Y	6	2	Num
Line 2	A1	3	3	Char
	A2	6	1	Char

Here are some sample data lines on which to test your program:

```
          1         2
12345678901234567890
--------------------
01 2345
 AAAX
02 9876
 BBBY
```

12.16 Each subject in a study has two lines of data. Line 1 contains an ID, GENDER (M or F), and
DOB; line 2 contains HEIGHT and WEIGHT. All data values are separated by commas.
Write a SAS DATA step to read the following sample lines of data:

```
001,M,06/14/1944
68,155
002,F,12/25/1967
52,99
003,M,07/04/1983
72,128
```

12.17 Write a SAS DATA step to read a series of X,Y pairs where there are several X,Y pairs per line. Each X,Y pair forms one observation. Here are some sample lines of data on which to test your program:

```
1 2   3 4   5 6   7 8
11 12   13 14
21 22   23 24   25 26   27 28
```

12.18 You have lines of data containing information on males (GENDER = M) and females (GENDER = F). Create a SAS data set consisting of data for females only. Do this by first testing the value of GENDER and only reading the remaining values if the line contains female data. The data layout is DATE in columns 1–10 (in MMDDYYYY form), GENDER in column 11, AGE in columns 12–13, and SCORE in columns 14–16. You may use the following sample data lines to test your program:

```
04/04/2004M15 90
05/12/2004F16 95
07/23/2004M18 88
01/20/2004F17100
```

*12.19 You conducted two surveys.

Survey ONE format is:

Variable	Starting Column	Length	Type
ID	1	3	Char
HEIGHT	4	2	Num
WEIGHT	6	3	Num

Survey TWO format is:

Variable	Starting Column	Length	Type
ID	1	3	Char
AGE	4	2	Num
HEIGHT	6	2	Num
WEIGHT	8	3	Num

All lines of data using the ONE format have a '1' in column 12. Lines of data using the TWO format have a '2' in column 12. Create a data set using the following sample data:

```
          1         2
12345678901234567890
--------------------
00168155   1
00272201   1
0034570170 2
0045562 90 2
```

CHAPTER 13

External Files: Reading and Writing Raw and System Files

A.   INTRODUCTION

New SAS users are often confused by the different ways SAS can read and write data to external files. This is due to the fact that SAS programs can read and write many different types of data files. For example, simple ASCII files (or EBCDIC text files on most mainframes) are read with INFILE and INPUT statements, whereas SAS data sets use two-level SAS data set names and do not require INPUT statements. This chapter discusses several ways that SAS software can read and write a variety of data types. The use of temporary and permanent SAS data sets and the advantages and disadvantages of each type is discussed.

B. DATA IN THE PROGRAM ITSELF

Before discussing how to read data from external files, let's review how SAS reads data lines that are part of the program itself, following a DATALINES statement. For example:

```
DATA EX1;
    INPUT GROUP $ X Y Z;
DATALINES;
CONTROL 12 17 19
TREAT 23 25 29
CONTROL 19 18 16
TREAT 22 22 29
;
```

The INPUT method used, whether column specification, informats, pointers, etc., will not change any of our examples so, for the most part, simple list input is used. The DATALINES statement tells the program that the data lines will follow. The program reads data lines until it encounters a line that ends in a semicolon. The examples in this book use a semicolon on a line by itself (called a null statement) to end instream data. The word DATALINES replaces an older term, CARDS, obviously a throwback to the past when actual punched cards were read into a card reader. However, since your children have probably never even heard of computer cards (maybe you haven't either), SAS Institute decided that the statement DATALINES was more appropriate than CARDS.

Before we leave this topic, here is one more (and rare) possibility you may encounter. What happens when your data contains semicolons? For example, suppose you had:

```
DATA TEST;
    INPUT AUTHOR $10. TITLE $40.;
DATALINES;
SMITH The Use of the ; in Writing
FIELD Commentary on Smith's Book
;
```

The program, recognizing the semicolon in the first line of data, treats the line as a SAS statement and generates more error messages than you can shake a stick at. The solution to this rare problem is to use the special SAS statement DATALINES4, which requires four semicolons in a row (;;;;) to indicate the end of your data. The corrected example would look like this:

```
DATA TEST;
   INPUT AUTHOR $10. TITLE $40.;
DATALINES4;
SMITH The Use of the ; in Writing
FIELD Commentary on Smith's Book
;;;;
```

C.   READING DATA FROM AN EXTERNAL TEXT FILE (ASCII OR EBCDIC)

It's not unusual to receive data in an external file to be analyzed with SAS. Whether on a diskette, a CD-ROM on a microcomputer, or on a tape used with a mainframe computer, we want a way to have our SAS program read data from an external source. For this example, we assume that the data file is either an ASCII (American Standard Code for Information Interchange) file or a "card image" file on tape (also called raw data). To read this file is surprisingly easy.

The only changes to be made to a program that reads "instream" data with a DATALINES statement are: (1) Precede the INPUT statement with an INFILE statement. (2) Omit the DATALINES statement, and, of course, the lines of data. An INFILE statement is the way we tell a SAS program where to find external raw data.

If you are running a batch version of SAS software on a platform where JCL (Job Control Language) is needed, the INFILE name will correspond to a DDname (DD stands for Data Definition), which gives information on where to find the file. On MVS (Multiple Virtual Systems) batch systems, the DDname is included in the JCL. On systems such as VM (Virtual Memory), the DDname is defined with a FILEDEF statement. For any of these mainframe systems, a FILENAME statement can also be used (and is preferable). For a microcomputer or UNIX system, the INFILE statement can either name a file directly (placed within quotes), or it can be a fileref defined with a FILENAME statement. We show examples of all of these variations.

```
*--------------------------------------------------*
| Personal Computer or UNIX Example                |
| Reading ASCII data from an External Data File    |
*--------------------------------------------------*;
DATA EX2A;
   INFILE 'B:MYDATA';
   ***This INFILE statement tells the program that
      our INPUT data is located in the file MYDATA
      on a diskette in the B drive;
   INPUT GROUP $ X Y Z;
RUN;
```

```
PROC MEANS DATA=EX2A N MEAN STD STDERR MAXDEC=2 ;
   VAR X Y Z;
RUN;
```

File MYDATA (located on the diskette in drive B) looks like this:

```
CONTROL 12 17 19
TREAT 23 25 29
CONTROL 19 18 16
TREAT 22 22 29
```

An alternate way of writing the INFILE statement in the previous example is to use a FILENAME statement to create an alias, or "fileref," for the file (a short "nickname" that we associate with the file). This is shown here:

```
DATA EX2B;
   FILENAME GEORGE 'B:MYDATA';
   INFILE GEORGE;
   ***This INFILE statement tells the program that
      our INPUT data is located in the file MYDATA
      on a diskette in the B drive;
   INPUT GROUP $ X Y Z;
RUN;
PROC MEANS DATA=EX2B N MEAN STD STDERR MAXDEC=2 ;
   VAR X Y Z;
RUN;
```

Note the difference between these two INFILE statements. The first INFILE statement refers to the external file directly, and the filename is placed within quotes. The second INFILE example defines an alias first with a FILENAME statement and then uses the alias with the INFILE statement. Notice that when we use a fileref, it is not in quotes. This point is important since it is the only way that the program can distinguish between an actual file name and a fileref.

A Mainframe Example Using JCL

The mainframe example shown next is basically the same as the previous microcomputer example. The only difference is in the way we create the fileref. On an MVS batch system, we would create the fileref with a DD statement in the JCL like this:

```
//JOBNAME JOB (ACCT,BIN),'RON CODY'
//         EXEC SAS
//SAS.GEORGE DD DSN=ABC.MYDATA,DISP=SHR
```
                                                              [Continued]

```
//SAS.SYSIN DD *
DATA EX2C;
   INFILE GEORGE;
   ***This INFILE statement tells the program that the
      file ABC.MYDATA contains our external data file
      (Assume it is catalogued);
   INPUT GROUP $ X Y Z;
RUN;
PROC MEANS DATA=EX2C N MEAN STD STDERR MAXDEC=2 ;
   VAR X Y Z;
RUN;
```

This example on a VM system would be the same except that a FILEDEF statement would be used to associate the DDname with the file instead of the DD statement in the JCL. Here it is:

```
***CMS FILEDEF GEORGE DISK MYDATA DATA B;
***The file MYDATA DATA is on the B minidisk;
DATA EX2D;
   INFILE GEORGE;
   ***This INFILE statement tells the program that the data are
      located in the file with FILENAME MYDATA, FILETYPE DATA, and
      FILEMODE B.;
   INPUT GROUP $ X Y Z;
RUN;
PROC MEANS DATA=EX2D N MEAN STD STDERR MAXDEC=2 ;
   VAR X Y Z;
RUN;
```

These last two examples could also use a FILENAME statement to point to our data source. Whatever method we use, once we know how to create a DDname or a fileref on our particular platform, the SAS statements for reading the files are the same. You will need to refer to your manual on how to create a fileref with TSO (Time Sharing Option) or CMS. Again, the SAS statements will not change.

D. INFILE OPTIONS

There are a variety of options that can be used with an INFILE statement to control how data are read and to allow the SAS program more control over the input operation. These options are placed after the word INFILE and before the semicolon. We now demonstrate several useful options.

Option END=variable name

This option will automatically set the value of "variable name" to 0 (false) until the INPUT statement starts reading the last data record. This option can be used when you want to read several different files and combine their data, or when you want to do special processing after you've read the last record in a file. (An alternative is to use the EOF=label option, which branches to "label" when the end of file is reached.)

The next example first reads data from a file called OSCAR on a diskette in the B: drive and then from a file called BIGBIRD.TXT located in a subdirectory C: \DATA.

```
DATA EX2E;
   IF TESTEND NE 1 THEN INFILE 'B:OSCAR' END=TESTEND;
   ELSE INFILE 'C:\DATA\BIGBIRD.TXT';
   INPUT GROUP $ X Y Z;
RUN;
PROC MEANS DATA=EX2E N MEAN STD STDERR MAXDEC=2;
   VAR X Y Z;
RUN;
```

Notice here that we can conditionally execute an INFILE statement, thus giving us complete control over the file-reading operation.

Another use of the END= INFILE option is to output summary information after all the data lines have been read. Look at the following program:

```
***Program to count missing values;
TITLE;
DATA _NULL_;
   INFILE 'C:\SASDATA\CLINIC.DAT' PAD END=LAST;
   INPUT @1  ID      $3.
         @4  GENDER $1.
         @5  DOB MMDDYY10.
         @15 HR       3.
         @18 SBP      3.
         @21 DBP      3.;
   IF MISSING(GENDER)  THEN N_GENDER + 1;
   IF MISSING(DOB)     THEN N_DOB + 1;
   IF MISSING(HR)      THEN N_HR + 1;
   IF MISSING(SBP)     THEN N_SBP + 1;
   IF MISSING(DBP)     THEN N_DBP + 1;
                                          [Continued]
```

```
        FILE PRINT;
        IF LAST THEN
            PUT "Summary Report of Missing Values" /
                40*'-' /
                "Number of Missing Values for GENDER: " N_GENDER /
                "Number of Missing Values for DOB: " N_DOB /
                "Number of Missing Values for HR: " N_HR /
                "Number of Missing Values for SBP: " N_SBP /
                "Number of Missing Values for DBP: " N_DBP;
    RUN;
```

There are several points about this program, but the one we want to emphasize here is the use of the END=LAST option on the INFILE statement. The MISSING function is used to detect missing values. Remember that the MISSING function works for both character and numeric variables and returns values of "true" and "false" (1 or 0). When a missing value is found, the appropriate counter is incremented. The special data set name _NULL_ is used when you don't need to create a SAS data set (a good efficiency tool). The FILE PRINT statement directs the results of the PUT statement to the output device (the output window of the Display Manager on a PC or UNIX platform). Finally, the 40*'-' is shorthand for printing 40 dashes, and the forward slashes in the PUT statement cause each missing value to be printed on a separate line. Since LAST is a logical variable, you can either write IF LAST = 1 or just IF LAST, the way we did here. So, the summary report is written after all the lines of data have been read. By the way, END= is also an option on a SET statement, and it is true when you are reading the last observation from a SAS data set.

The listing here shows what the output of this program looks like:

```
Summary Report of Missing Values

Number of Missing Values for GENDER: 2
Number of Missing Values for DOB: 2
Number of Missing Values for HR: 3
Number of Missing Values for SBP: 3
Number of Missing Values for DBP: 2
```

Option MISSOVER

The MISSOVER option is very useful when you have records of different lengths and have missing values at the end of a record. This is frequently the case when a text file is created with a word processor and the records are not padded on the right with blanks. Suppose our file called MYDATA has a short record and looks like:

```
CONTROL 1 2 3
TREAT 4 5
CONTROL 6 7 8
TREAT 8 9 10
```

The program EX2A or EX2B would have a problem reading the second record of this file. Instead of assigning a missing value to the variable Z, it would go to the next record and read "CONTROL" as the value for Z and print an error message (since CONTROL is not a numeric value). The SAS LOG would also contain a NOTE telling us that SAS went to a new line when the INPUT statement reached past the end of a line. The remainder of the third record would not be read, and the next observation in our data set would be GROUP=TREAT, X=8, Y=9, and Z=10. To avoid this problem, use the MISSOVER option on the INFILE statement. This will set all variables to missing if any record is short. The entire program would look like:

```
DATA EX2F;
    INFILE 'B:MYDATA' MISSOVER;
    INPUT GROUP $ X Y Z;
RUN;
PROC MEANS DATA=EX2F N MEAN STD STDERR MAXDEC=2 ;
    VAR X Y Z;
RUN;
```

Option PAD

When you read data in fixed columns, using either column or formatted input, the option PAD can be used to prevent problems with short records. For example, to read data values from fixed columns from a file called C:\DATA MYDATA.TXT, you could write:

```
DATA EX2G;
    INFILE 'C:\DATA\MYDATA.TXT' PAD;
    INPUT GROUP $ 1
          X       2-3
          Y       4-5
          Z       6-7;
RUN;
PROC MEANS DATA=EX2G N MEAN STD STDERR MAXDEC=2 ;
    VAR X Y Z;
RUN;
```

Option LRECL=record-length

You may need to specify your logical record length (the number of columns on a line, roughly speaking) if it exceeds the default value for your system. When in doubt, add the LRECL (this stands for logical record length and is pronounced El-Rec-el) option to the INFILE statement. It will not cause a problem if you specify an LRECL larger than your actual record length. For example, suppose you have an ASCII file on a diskette with 310 characters per line and your system default LRECL is 256. To read this file, you would write the INFILE statement thus:

```
INFILE fileref LRECL=310;
```

There are many other INFILE options that allow you more control over how data are read from external files. They can be found in the SAS Language Reference, version 8 or in the Online Doc™ version 9 (see Chapter 1 for complete references).

There will be times where you have data within the SAS program itself (following a DATALINES; statement) and not in an external file, yet you want to use one or more of the INFILE options to control the input data. We can still use most of these options (END= and EOF= is not allowed with DATALINES) by specifying a special fileref called DATALINES, followed by any options you wish. Suppose you want to use MISSOVER and have included the data within the program. You would proceed as follows:

```
DATA EX2H;
   INFILE DATALINES MISSOVER;
   INPUT X Y Z;
DATALINES;
1 2 3
4 5
6 7 8
;
PROC MEANS DATA=EX2H;
etc.
```

E.   READING DATA FROM MULTIPLE FILES (USING WILDCARDS)

Suppose you have data in several files with similar file names and you want to read raw data from all the files. You can use "wildcards" in naming your input data sets (* and ? in a PC environment) and SAS will read data from each of the files successively. For example, suppose you have three files in your C:\SASDATA folder: FILE001.DAT, FILE002.DAT, AND FILE003.DAT. You could read data from all three (in this example, there are three numbers on each line of the files, and they are separated by commas):

```
DATA ALL_THREE;
   INFILE 'C:\BOOKS\APPLIED_5TH\FILE*.DAT' MISSOVER DLM=',';
   INPUT X Y Z;
RUN;
```

Another useful (and more flexible) way to read data from multiple raw files is to use a FILENAME statement with multiple file names. For example, suppose you have three raw data files, AAA.DAT, BBB.DAT and CCC.DAT, and you want to read data from all three. The following program will work nicely:

```
DATA READ_MANY;
    FILENAME MIKE ('AAA.DAT' 'BBB.DAT' 'CCC.DAT');
    INFILE MIKE MISSOVER;
    INPUT X Y Z;
RUN;
```

This program will read all the records from file AAA.DAT, followed by all the records from BBB.DAT, followed by all the records from CCC.DAT. The reason we use MIKE as a file name is to acknowledge Mike Zdeb, one of our reviewers, who brought this method of reading multiple files to our attention. Thanks Mike. The list of files on the FILENAME statement can also include the '?' and '*' DOS-style wildcards.

F.   WRITING ASCII OR RAW DATA TO AN EXTERNAL FILE

We may have reason to have our SAS program write data to an external text file in ASCII or EBCDIC format. Writing raw data to a file would have the advantage of being somewhat "universal" in that most software packages would be able to read it. On most microcomputer systems, an ASCII file could be read by a word-processing program, a spreadsheet program, or a database management package. Writing raw data to a file is very much like reading data from an external file. We use one statement, FILE, to tell the program where to send the data and another, PUT, to indicate which variables and in what format to write them. Thus, to read raw data files, we use the statements INFILE and INPUT; to write raw data files, we use the statements FILE and PUT. Here is a simple example of a program that reads a raw file (MYDATA), creates new variables, and writes out the new file (NEWDATA) to a diskette:

```
DATA EX3A;
    ***This program reads a raw data file, creates a new
       variable, and writes the new data set to another file;
    INFILE 'C:MYDATA';***Input file;
    FILE 'A:NEWDATA';***Output file;
    INPUT GROUP $ X Y Z;
    TOTAL = SUM (OF X Y Z);
    PUT GROUP $ 1-7 @11 (X Y Z TOTAL) (5.);
RUN;
```

Running this program produces a new file called NEWDATA, which looks like this:

```
              1         2         3
123456789012345678901234567890123456789012

-------------------------------------------------
CONTROL   12   17   19   48
TREAT     23   25   29   77
CONTROL   19   18   16   53
TREAT     22   22   29   73
```

Notice that with the PUT statement we can employ any of the methods of specifying columns or formats that are permissible with an INPUT statement. In the previous example, we specified columns for the first variable (GROUP) and a format (in a format list) for the remaining four variables. This gives us complete control over the structure of the file to be created. It goes without saying that this example will work just the same on a mainframe under MVS, VM, or Z/OS, provided that the correct filename statements are issued. Note that on a mainframe system, if we are creating a new file, we have to provide all the parameters (such as RECFM, DISP, UNIT, DSN, DCB, etc.) necessary for your system.

G.   WRITING CSV (COMMA-SEPARATED VARIABLES) FILES USING SAS

On PCs, comma-separated variable (CSV) files can be written or read by many applications such as spreadsheets or databases. CSV files contain lines of data where data values are separated by commas. In addition, character strings are surrounded by double quotes, and missing values are indicated by two commas together. The first line of a CSV file may or may not contain a header line with the variable names (column headings in the case of a spreadsheet) separated by commas. Don't forget that on PC platforms, the Access™ product from SAS allows you to import and export SAS data sets between a variety of forms, including Excel™ spreadsheets and various database formats.

One easy way to create such a file is with the Output Delivery System (ODS) of SAS. The following is a program that creates a CSV file. Note that you can open this file directly in Excel™ or in any application that can read CSV files. Here is the program:

```
OTIONS MISSING=" ";
DATA COMMA_DELIMITED;
    INPUT NAME $ X Y Z;
DATALINES;
CODY 1 2 3
SMITH 4 5 6
MISS . 8 9
;
ODS LISTING CLOSE;   ***Turn of listing;
ODS CSV FILE='C:\MYFOLDER\COMMA_ODS.CSV';
PROC PRINT DATA=COMMA_DELIMITED NOOBS;
    TITLE;
RUN;
ODS CSV CLOSE;
ODS LISTING;   ***Turn listing back on;
```

A listing of the CSV file is:

```
"NAME","X","Y","Z"
"CODY","1","2","3"
"SMITH","4","5","6"
"MISS"," ","8","9"
```

If all you need is a comma-delimited file, you can have the FILE option DSD precede your PUT statement. This will insert commas between your data values but will not surround all the data values in quotes.

H.   CREATING A PERMANENT SAS DATA SET

Thus far, we have seen how to read raw "card image" data and to write the same type of data to an external file. We now demonstrate how to create a permanent SAS data set.

A SAS data set, unlike a raw data file, is not usable by software other than SAS (although people who do not have SAS installed on their computer can use the freely distributed SAS Viewer to view and print SAS data sets). It contains not only the data, but also such information as the variable names, labels, and formats (if any). Before we show you an example, let's first discuss the pros and cons of using permanent SAS data sets.

First, some cons: As we mentioned, non-SAS programs cannot read SAS data sets. You cannot use a typical editor to read or modify a SAS data set. To read, update, or modify a SAS data set requires using SAS and either writing a program (for example, to change a data value) or using the SAS Viewtable to display and/or update an observation. When you write a SAS program or SAS Viewtables to modify a SAS data set, you must keep in mind that the original raw data are not modified, and you can no longer re-create the SAS data set from the raw data without making the modification again. Finally, in the minus column, SAS data sets typically use more storage than the original data set (although they can be compressed) and are usually kept in addition to the original raw data, thus more than doubling the system storage requirements.

With all these negatives, why create permanent SAS data sets? Probably the most compelling reason is speed. Considerable computer resources are required to read raw data and create a SAS data set. If you plan to be running many different analyses on a data set that will not be changing often, it is a good idea to make the data set permanent for the duration of the analyses. SAS data sets are also a good way to transfer data to other users provided they have SAS available. Knowing the data structure is no longer necessary since all the variables, labels, and formats have already been defined. We will see shortly how to use PROC CONTENTS to determine what is contained in a SAS data set.

Our first example in this section is to write a SAS program that has the data in the program itself and to create a permanent SAS data set.

```
    *-------------------------------------------------------------*
    | This program reads data following the datalines statement   |
    | and creates a permanent SAS data set in a subdirectory      |
    | called C:\SASDATA                                           |
    *-------------------------------------------------------------*;
    LIBNAME FELIX 'C:\SASDATA';

    DATA FELIX.EX4A;
       INPUT GROUP $ X Y Z;
    DATALINES;
    CONTROL 12 17 19
    TREAT 23 25 29
    CONTROL 19 18 16
    TREAT 22 22 29
    ;
```

SAS data set names consist of two parts. The first part of the two-level name (the part before the period) names a location, called a library by SAS, where the SAS data set will be written. In a PC or UNIX environment, this location is usually a folder (also called a subdirectory, for those of us who remember DOS). For this library, SAS uses a LIBNAME statement to assign an alias or nickname, referred to as a libref by SAS. We can have many SAS data sets contained within a single library. The second part (after the period) is the actual data set name. When you start a SAS session, SAS automatically creates two library locations for you, WORK and SASUSER. WORK is a special library since data sets placed in this library are temporary and will disappear when you close your SAS session. We recommend that you do not use the permanent SASUSER library but create your own permanent libraries. If you use a single SAS data set name (without a period), SAS assumes that the library is WORK and the data set will be temporary. For example, DATA ONE and DATA WORK.ONE are equivalent.

When the previous program executes, the data set EX4A will be a permanent SAS data set located in the C:\SASDATA folder. If we look at a list of files in C:\SAS-DATA, there will be a file called EX4A.SAS7BDAT. The extension SAS7BDAT (SAS version 7 compatible binary data) may be different for different platforms. Note that on any SAS system, the first-level name does not remain with the data set; it is only used to point to a SAS library. The only requirement is that the first-level name match the LIBNAME within a program.

Now that we have created a permanent SAS data set, let's see how to read it and determine its contents.

I.   READING PERMANENT SAS DATA SETS

Once we have created a permanent SAS data set, we can use it directly in a procedure after we have defined a libref. We now show you a SAS program that uses this permanent data set.

```
LIBNAME ABC 'C:\SASDATA';
PROC MEANS DATA=ABC.EX4A N MEAN STD STDERR MAXDEC=3;
   VAR X Y Z;
RUN;
```

You can see right away how useful it is to save SAS data sets. Notice that there is no DATA step in the previous program. All that is needed is to define a SAS library (where the SAS data set is located) and to use a DATA= option with PROC MEANS to indicate on which data set to operate. First, observe that the libref ABC is not the same name we used when we created the data set. The libref ABC is defined with the LIBNAME statement and indicates that we are using the folder C:\SASDATA. Therefore, the first part of the two-level SAS data set name is ABC. The second part of the two-level name tells the system which of the SAS data sets located in C:\SASDATA is to be used. It is important to remember that we must use the DATA= option with any procedure where we are accessing previously stored SAS data sets, because the program will not know which data set to use (when we create a SAS data set in a DATA step, the system keeps track of the "most recently created data set" and uses that data set with any procedure where you do not explicitly indicate which data set to use with a DATA= option). Just so that we don't shortchange the mainframe users, the same program, written on an MVS system, would look something like this:

```
//GROUCH JOB (1234567,BIN),'OSCAR THE'
//    EXEC SAS
//SAS.ABC DD DSN=OLS.A123.S456.CODY,DISP=SHR
//SAS.SYSIN DD *
PROC MEANS DATA=ABC.EX4A N MEAN STD STDERR MAXDEC=3;
   VAR X Y Z;
/*
//
```

The DDname was defined in the JCL, indicating the SAS data set was stored in the MVS data set called OLS.A123.S456.CODY that was catalogued. On a VM system, the DDname, or first part of a two-level name, is what VM calls the filetype in the general, filename filetype filemode method of defining a file. The filename corresponds to the SAS second-level name. Thus, without even issuing a FILEDEF command, we could write:

```
PROC MEANS DATA=ABC.EX4A N MEAN STD STDERR MAXDEC=3;
```

which is valid as long as we have a filetype of ABC and a filename of EX4A.

J. HOW TO DETERMINE THE CONTENTS OF A SAS DATA SET

As mentioned earlier, we cannot use our system editor to list the contents of a SAS data set. How can we "see" what is contained in a SAS data set? On a PC or UNIX platform, we can use the SAS Explorer to show us SAS data set information. On any SAS platform, we use PROC CONTENTS. This very useful procedure will tell us important information about our data set: the number of observations, the number of variables, the record length, and an alphabetical listing of variables (which includes labels, length, and formats). Note that the order of the variables in the alphabetical listing are case sensitive—uppercase names come before lowercase names. The statement: OPTIONS VALIDVARNAME=UPCASE will force the listing to ignore case. If you use the VARNUM option, you can obtain a list of variables in the order of their position in the data set. Here are the statements to display the contents of the permanent SAS data set EX4A previously created:

```
LIBNAME SUGI 'C:\SASDATA';
PROC CONTENTS DATA=SUGI.EX4A VARNUM;
   TITLE "Demonstrating PROC CONTENTS";
RUN;
```

Output from this procedure is shown here:

```
Demonstrating PROC CONTENTS


The CONTENTS Procedure


Data Set Name          SUGI.EX4A              Observations          4
Member Type            DATA                   Variables             4
Engine                 V9                     Indexes               0
Created                Tuesday, May 04,       Observation Length    32
                       2004 11:21:34 AM
Last Modified          Tuesday, May 04,       Deleted Observations  0
                       2004 11:21:34 AM
Protection                                    Compressed            NO
Data Set Type                                 Sorted                NO
Label
Data Representation    WINDOWS_32
Encoding               wlatin1  Western (Windows)
```

```
              Engine/Host Dependent Information

  Data Set Page Size             4096
  Number of Data Set Pages       1
  First Data Page                1
  Max Obs per Page               126
  Obs in First Data Page         4
  Number of Data Set Repairs     0
  File Name                      C:\sasdata\ex4a.sas7bdat
  Release Created                9.0101B0
  Host Created                   XP_PRO


   Variables in Creation Order

  #    Variable    Type    Len

  1    GROUP       Char    8
  2    X           Num     8
  3    Y           Num     8
  4    Z           Num     8
```

One final point of information: The DATA= option of PROC CONTENTS can be used to list all the SAS data sets contained in a SAS library instead of a single data set. Use the form libref._ALL_ instead of libref.data_set_name. This will display all the SAS data sets stored in the library referred to by the libref.

If you are working in an interactive, Windows-type environment, you can also use the two commands DIR and VAR (or pointing and clicking appropriately), which will show you any permanent SAS data set and a variable list.

K.   PERMANENT SAS DATA SETS WITH FORMATS

One special note is needed to caution you about saving permanent SAS data sets in which you have assigned user-created formats to one or more of the variables in the DATA step. If you try to use that data set (in a procedure, for example), you will get an error saying that formats are missing. The important thing to remember is this: If you create a permanent SAS data set that assigns user-created formats to variables, you must make the format library permanent as well. Also, if you give someone else the data set, make sure you give him or her the format library.

To make your format library permanent, add the LIBRARY=libref option to PROC FORMAT. You may either use a special libref called LIBRARY or one of your own choosing. If you choose your own libref, you need to supply a system option called FMTSEARCH=(libref) to tell the program where to look for user-defined formats. Since this sounds rather complicated, we show the code needed to create a permanent format library and the code to access a permanent data set where formats were used.

Code to Create a Permanent Format Library and Assign the Format to a Variable:

```
LIBNAME FELIX 'C:\SASDATA';
OPTIONS FMTSEARCH=(FELIX);
***We will place the permanent SAS data sets and the
   formats in C:\SASDATA;
PROC FORMAT LIBRARY=FELIX;
   VALUE $XGROUP 'TREAT'   = 'TREATMENT GRP'
                 'CONTROL' = 'CONTROL GRP';
RUN;

DATA FELIX.EX4A;
   INPUT GROUP $ X Y Z;
   FORMAT GROUP $XGROUP.;
DATALINES;
CONTROL 12 17 19
TREAT 23 25 29
CONTROL 19 18 16
TREAT 22 22 29
;
```

Program to Read a Permanent SAS Data Set with Formats

```
LIBNAME C 'C:\SASDATA';
OPTIONS FMTSEARCH=(C) ;
***Tell the program to look in C:\SASDATA for user
   defined formats;
PROC PRINT DATA=C.EX4A;
RUN;
```

In this example, the libref FELIX was used when the SAS data set and the permanent SAS format was created. A different libref, C, was used in the subsequent program when the data set was accessed. This was for illustrative purposes only. In practice, most SAS programmers usually use the same libref to point to a specific library.

If someone gives you a SAS data set that has user formats defined and does not give you the format library, don't despair! You can at least get the procedures to work if you use the system option NOFMTERR. This option will allow you to process the data set that contains missing formats by supplying SAS system defaults to the character and numeric variables.

L. WORKING WITH LARGE DATA SETS

Special consideration is necessary when we process large data sets. Of course, "large" is a relative term. On a PC, 500,000 observations with 50 variables might be considered large. On a mainframe, users frequently process data sets with millions of observations. A few simple techniques described here can reduce the processing time and memory usage (and cost if you're paying for it) for processing a large file.

1. Don't read a file unnecessarily. For example:
   Inefficient Way:

```
LIBNAME INDATA 'C:\MYDATA';
DATA TEMP;
   SET INDATA.STATS;
RUN;
PROC PRINT DATA=TEMP;
   VAR X Y Z;
RUN;
```

Efficient Way:

```
LIBNAME INDATA 'C:\MYDATA';
PROC PRINT DATA=INDATA.STATS;
   VAR X Y Z;
RUN;
```

The DATA step in the "Inefficient" example is unnecessary. It simply copies one data set into another so that the PROC PRINT can use the temporary data set. Surprisingly, this is a common error.

2. Drop all unnecessary variables. Not only do more variables take up more space, they slow down DATA step processing as well.
   Inefficient Way (if all you want is the quiz average):

```
DATA QUIZ;
   INPUT @1 (QUIZ1-QUIZ10) (3.);
   QUIZAVE = MEAN (OF QUIZ1-QUIZ10);
DATALINES;
```

Efficient Way:

```
DATA QUIZ;
   INPUT @1 (QUIZ1-QUIZ10) (3.);
   QUIZAVE = MEAN (OF QUIZ1-QUIZ10);
   DROP (QUIZ1-QUIZ10);
DATALINES;
```

3. Use a DROP (or KEEP) option on the SET statement rather than a DROP (or KEEP) statement in the DATA step. When a DROP or KEEP option is used, only the variables not dropped or kept will be brought into the program data vector. This can result in a significant decrease in processing time.

Inefficient Way:

```
DATA NEW;
    SET OLD;
    DROP X1-X20 A B;
etc.
```

Efficient Way:

```
DATA NEW
    SET OLD (DROP=X1-X20 A B);
etc.
```

4. Do not sort data sets more than necessary. For example, if you need your data in DAY order and know that later in the program you need it in DAY-HOUR order, do the two-level sort first.

Inefficient Way:

```
PROC SORT DATA=MYDATA;
    BY DAY;
RUN;
etc.
PROC SORT DATA=MYDATA
    BY DAY HOUR;
RUN;
etc.
```

Efficient Way:

```
PROC SORT DATA=MYDATA;
    BY DAY HOUR;
RUN;
etc.
```

5. Think about using a CLASS statement instead of a BY statement with PROC MEANS. This will eliminate the need to sort the data but will require more memory to run.

Inefficient Way:

```
PROC SORT DATA=MYDATA;
   BY DAY;
RUN;
PROC MEANS DATA=MYDATA N MEAN NWAY;
   BY DAY;
   VAR variable-list;
RUN;
```

Efficient Way:

```
PROC MEANS DATA=MYDATA N MEAN NWAY;
   CLASS DAY;
VAR variable-list;
RUN;
```

6. When a small subset is selected from a large file, use the WHERE statement instead of a subsetting IF statement. (Note: While this was especially true in version 6, on a PC platform under version 9, there seems to be little difference.)

Inefficient Way:

```
DATA ALPHA;
   SET BETA;
   IF X GE 20;
RUN;
```

Efficient Way:

```
DATA ALPHA;
   SET BETA;
   WHERE X GE 20;
RUN;
```

or

```
DATA ALPHA;
   SET BETA(WHERE=(X GE 20));
RUN;
```

7. Use a WHERE statement in a PROC when you only need to run a single procedure on a subset of the data.
Inefficient Way:

```
DATA TEMP;
   SET MYDATA;
   WHERE AGE GE 65;
RUN;
PROC MEANS DATA=TEMP N MEAN STD;
   VAR variable-list;
RUN;
```

Efficient Way:

```
PROC MEANS DATA=MYDATA N MEAN STD;
   WHERE AGE GE 65;
   VAR variable-list;
RUN;
```

or

```
PROC MEANS DATA=MYDATA(WHERE=(AGE GE 65)) N MEAN STD;
   VAR variable-list;
RUN;
```

8. Use ELSE IF instead of multiple IF statements.
Inefficient Way:

```
DATA SURVY;
   INPUT ID AGE HEIGHT WEIGHT;
   IF 0 LE AGE LT 20 THEN AGEGRP=1;
   IF 20 LE AGE LT 30 THEN AGEGRP=2;
   IF 30 LE AGE LT 40 THEN AGEGRP=3;
   IF AGE GE 40 THEN AGEGRP=4;
RUN;
```

Efficient Way:

```
DATA SURVY;
   INPUT ID AGE HEIGHT WEIGHT;
   IF 0 LE AGE LT 20 THEN AGEGRP=1;
   ELSE IF 20 LE AGE LT 30 THEN AGEGRP=2;
   ELSE IF 30 LE AGE LT 40 THEN AGEGRP=3;
   ELSE IF AGE GE 40 THEN AGEGRP=4;
RUN;
```

9. When using multiple IF statements, place first the one most likely to be true. Inefficient Way (most of the subjects are over 65):

```
DATA SURVEY;
   SET OLD;
   IF 0 LE AGE LT 20 THEN AGEGRP=1;
   ELSE IF 20 LE AGE LT 30 THEN AGEGRP=2;
   ELSE IF 30 LE AGE LT 40 THEN AGEGRP=3;
   ELSE IF AGE GE 40 THEN AGEGRP=4;
RUN;
```

Efficient Way (most of the subjects are over 65):

```
DATA SURVEY;
   SET OLD;
   IF AGE GE 40 THEN AGEGRP=4;
   ELSE IF 30 LE AGE LT 40 THEN AGEGRP=3;
   ELSE IF 20 LE AGE LT 30 THEN AGEGRP=2;
   ELSE IF 0 LE AGE LT 20 THEN AGEGRP=1;
RUN;
```

10. Save summary statistics in a permanent SAS file if you plan to do further computations with it.
Inefficient Way:

```
LIBNAME C 'C:\MYDATA';
PROC MEANS DATA=C.INDATA;
   CLASS RACE GENDER;
   VAR variable-list;
RUN;
```

Efficient Way:

```
LIBNAME C 'C:\MYDATA';
PROC MEANS DATA=C.INDATA NWAY;
   CLASS RACE GENDER;
   VAR variable-list;
   OUTPUT OUT=C.SUMMARY MEAN=variable-list;
RUN;
```

11. Use _NULL_ as a data set name when you only want to process records from a file (such as data cleaning) but do not want to keep the resulting data set.
Inefficient Way:

```
DATA TEMP;
   SET OLD;
   FILE 'ERRORS';
   IF AGE GT 110 THEN PUT ID= AGE= ;
RUN;
```

Efficient Way:

```
DATA _NULL_;
   SET OLD;
   FILE 'ERRORS';
   IF AGE GT 110 THEN PUT ID= AGE= ;
RUN;
```

12. Save your data as a SAS data set if you plan to do further processing of the data. Reading a SAS data set is much more efficient than reading raw data.

13. Use the data set option "OBS=n", where n is either zero or a small number when testing your code. For example:

```
DATA TESTIT;
   SET VERY_BIG(OBS=10);
   (More SAS code)
RUN;
```

```
or

PROC PRINT DATA=VERY_BIG(OBS=10);
   TITLE "First 10 Obs from data set VERY_BIG";
RUN;
```

14. Use PROC DATASETS instead of a DATA step to rename variables or change variable labels.

Inefficient Way:

```
DATA XXX.TEST;
   SET XXX.TEST;
   LABEL QUES1 = 'Are you over 21?'
         QUES2 = "Do you have a driver's license?";
   RANAME HT = HEIGHT
          WT = WEIGHT;
RUN;
```

Efficient Way:

```
PROC DATASETS LIBRARY=XXX;
   MODIFY TEST;
   LABEL QUES1 = 'Are you over 21?'
         QUES2 = "Do you have a driver's license?";
   RANAME HT = HEIGHT
          WT = WEIGHT;
RUN;
```

15. Use PROC APPEND to add new data to a large file. (Note: if you are adding just a few observations to a very large file, this can result in enormous time savings.)

Inefficient Way:

```
DATA COMBINE;
   SET BIGFILE NEWFILE;
RUN;
```

Efficient Way:

```
PROC APPEND BASE=BIGFILE DATA=NEWFILE;
RUN;
```

PROBLEMS

Remember, you can download all the data sets and programs for these problems from the web site: www.prenhall.com/cody

13.1 You receive a text (ASCII) file called FRODO on a diskette. The data layout is as follows:

Variable	Col(s)
ID	1–3
AGE	5–6
HR	8–10
SBP	12–14
DBP	16–18

A few sample records are shown here:

```
123456789012345678          (Columns listed here,
------------------          not on the diskette)
001 56  64 130 80
002 44  72 180              Note: No DBP recorded for
003 64  78 140 88           this ID (short record)
```

You place this diskette in the A: drive of your computer. Write a SAS program that will read this file and do the following:

(a) Create a permanent SAS data set called BILBO on the diskette in drive A. This data set should contain AGE, HR, SBP, and AVEBP, where AVEBP is defined as two-thirds of the diastolic blood pressure (DBP) plus one-third of the systolic blood pressure (SBP). (This is actually a weighted average of the DBP and SBP with weights of 2/3 and 1/3, since the heart spends more time in diastole than systole.)

(b) Create another SAS data set called HIBP, which contains only records of subjects with AVEBP greater than or equal to 100.

13.2 You are given a raw data file (INPUT.DAT) containing three numbers per line, separated by commas. Some sample lines are as follows:

```
1,2,3
4,9999,5
9999,6,7
1,2
6,4,3
```

Notice that one of the lines only contains two numbers. Read this data file, using a DATA _NULL_ DATA step, and create another raw data file (OUTPUT.DAT) containing the same data as INPUT.DAT except that values of 9999 are replaced by a SAS missing value. The output file should look like this:

```
1,2,3
4,.,5
.,6,7
1,2,.
6,4,3
```

13.3 You are given a diskette with two files: SURVEY.SAS7BDAT and FORMATS.SAS7BCAT. The former is a SAS data set (compatible with your version of SAS software), and FOR-MATS.SAS7BCAT is a user-defined format library. You copy these two files to a subdirectory called C: \SASDATA on your hard disk. Two variables of special interest in this data set are ICD_9 and AGE. The format library contains a format for ICD_9, and that format has been assigned in the data set to ICD_9. Write a SAS program for a computer system you use (mainframe MVS, VM, Windows XP, Windows2000, UNIX, etc.) that will read that data set, recognize the format library, and produce a frequency distribution of the ICD_9 codes in decreasing order of frequency (PROC FREQ option ORDER=FREQ). Also, compute descriptive statistics for AGE (n, mean, standard deviation, standard error, minimum, and maximum).

13.4 You have three CSV (comma-separated variable) files called FILE1.CSV, FILE2.CSV, and FILE3.CSV. Each line in the file contains values for ID (character), X, Y, and Z. Create a SAS data set from these three raw data files. Each of these three files is listed here. (HINT: Option DSD will be useful.)

```
File FILE1.CSV
"001",1,2,3
"002",,4,5
"003",,6,7

File FILE2.CSV
"004",10,20,30
"005",40,50,60

File FILE3.CSV
"006",100,200,300
"007",400,,800
```

13.5 You have collected demographic data for the years 1996 and 1997. The data for 1996 is placed in a file called DEM_1996, and the data for 1997 is placed in a file called DEM_1997. These two files use the same data layout (see following). Both files are located on a diskette that you place in the A: drive of your computer. Write a program to read all the data from both files, and create a single, permanent SAS data set to be located in C:\MYDATA. Use the END = INFILE option, not wildcards, to do this. Call the SAS data set DEM_9697. The data layout for both files is:

Variable	Starting Column	Length	Type
ID	1	3	Char
AGE	4	2	Num
JOB_CODE	6	1	Char
SALARY	7	6	Num

13.6 Run the following SAS program to create two files: BIG and SMALL. First, add the contents of SMALL to the end of BIG, using a SET statement. Next, do the same thing using PROC APPEND. Compare the time used (from the LOG). You may want to increase the size of BIG if you have a really fast computer and want to compare times.

```
DATA BIG;
    DO I = 1 TO 100000;
        X = RANUNI(0);
        OUTPUT;
    END;
    DROP I;
RUN;
DATA SMALL;
    X = .5;
    OUTPUT;
    X = .6;
    OUTPUT;
RUN;
```

13.7 You have a raw data file called SAMPLE.DTA on a diskette. The file contains 100 numbers per line, and the values are separated by one or more spaces. The length of the longest line is 320 bytes, and some lines contain fewer than 100 numbers. Write a SAS DATA step that will read this file from the diskette in drive B: and will assign the 100 values to the variables X1–X100. Assume that the default logical record length for your system is less than 320 bytes.

13.8 Read the raw data file SURVEY.DAT (listed here) and create a permanent SAS data set called SURVEY in a folder of your choice. The values in the file SURVEY.DAT are GENDER and Q1–Q5, and the values are separated by blanks. The first line of the file is a header line and should be skipped. Make formats for Q1–Q5 (1 = Yes, 0 = No) and GENDER ('M' = 'Male' and 'F' = 'Female'). Make these formats permanent as well. Look in your folder and check that the data set (SURVEY.SAS7BDAT) and format library (FORMATS.SAS7BCAT) are there. Remember to use the system option FMTSEARCH= as well as the PROC FORMAT option LIBRARY=.

Listing of SURVEY.DAT

```
***2004 Survey***
M 1 0 1 0 1
F 0 0 1 1 0
M 1 1 1 1 1
F 0 0 0 1 0
```

13.9 Run the following program to create a SAS data set called MILTON. Next, write a SAS DATA step that will read the values from MILTON and write the data for variables A, B, and C to a raw data file to the subdirectory C:\MYDATA. Call the output data file OUT-DATA, and write the values for A, B, and C to columns 1–3, 4–6, and 7–9, respectively.

```
***DATA step to create MILTON;
DATA MILTON;
    INPUT X Y A B C Z;
DATALINES;
1 2 3 4 5 6
11 22 33 44 55 66
;
```

Data Set Subsetting, Concatenating, Merging, and Updating

A.   INTRODUCTION

This chapter covers some basic data set operations. Subsetting is an operation where we select a subset from one data set to form another. We may also want to combine data from several data sets into a single SAS data set; this chapter explores several ways of doing this. Let's take these topics one at a time.

B.   SUBSETTING

We have already seen some examples of data subsetting. The key here is the SET statement, which "reads" observations from a SAS data set to form a new SAS data set. As we process the observations from the original data set, we can modify any of the values, create new variables, or make a decision whether to include the observation in the new data set. A simple example:

```
DATA WOMEN;
   SET ALL;
   IF GENDER = 'F';
RUN;
```

In this example, data set ALL contains a variable, GENDER, which has values of 'M' and 'F'. The IF statement, used in this context, is called a subsetting IF. To those of you familiar with other programming languages, this is a funny-looking IF statement—there is no THEN clause. If the IF statement is NOT true, the DATA step will return to the top (and no automatic output will occur). If the IF statement is true, the DATA step will continue in the normal fashion. You can think of a subsetting IF statement as a valve. If the statement following the IF is true, the valve opens and processing continues; if the statement following the IF is not true, the logic flow returns to the top of the DATA step. Therefore, any observation with GENDER equal to 'F' will be written to the new data set WOMEN. You could also write (assuming you have values of 'M' and 'F' in your data set):

```
IF GENDER NE 'M';
```

Be careful here. If there are any observations with missing or miscoded values (i.e., any value that is not an 'M') for GENDER, the previous statement would add those observations to data set WOMEN. It is usually better to indicate what you want, rather than what you do not want.

We can use any logical expression in the IF statement to subset the data set. For example, we could have:

```
DATA OLDWOMEN;
   SET ALL;
   IF GENDER = 'F' AND AGE > 65;
RUN;
```

An alternative to the subsetting IF statement, when data is coming from SAS (i.e., a SET, MERGE, or UPDATE statement), is called the WHERE statement. Just to be sure this is clear, you cannot include a variable in a WHERE statement that is being read from raw data or is created by an assignment statement in the DATA step. Although there are several subtle differences between using IF and WHERE statements, we can subset a data set just as easily by substituting WHERE for IF in the previous programs. When the data set we are creating is a small subset of the original data set, the WHERE statement may be more efficient. The previous program, using a WHERE statement, would be:

```
DATA OLDWOMEN;
   SET ALL;
   WHERE GENDER = 'F' AND AGE > 65;
RUN;
```

In addition, we also have the option of using the WHERE statement in a SAS PROCEDURE. So, to compute frequencies of RACE and INCOME only for females, we could write:

```
PROC FREQ DATA=ALL;
   WHERE GENDER = 'F';
   TABLES RACE INCOME;
RUN;
```

If we want to run a procedure for all levels of a variable, we should use a BY statement instead. We have found the WHERE statement particularly useful when we run t-tests or ANOVAs, and we want to eliminate one or more groups from the analysis. Suppose the variable GROUP has three levels (A, B, and C), and we want to run a t-test between groups A and B. Using the WHERE statement greatly simplifies the job:

```
PROC TTEST DATA=data_set_name;
   WHERE GROUP='A' OR GROUP='B';
   CLASS GROUP;
   VAR variable-list;
RUN;
```

By the way, a nice alternative to multiple ORs is the IN operator. For example, the previous WHERE statement could be written:

```
WHERE GROUP IN ('A' 'B');
```

You can separate the values in the list following the IN with either blanks or commas. You can also use the IN operator with numeric data. So, if DOSE is a numeric variable and you want only doses of 10, 20, and 30, you could write:

```
WHERE DOSE IN (10 20 30);
```

Again, you have the choice of blanks or commas as the separators.

C.   COMBINING SIMILAR DATA FROM MULTIPLE SAS DATA SETS

Assume that we have several SAS data sets, each containing the same variables. To create a SAS data set from multiple SAS data sets, use the SET statement to name each of the data sets to be combined. For example, to combine the data from two SAS data sets

MEN and WOMEN into a single data set called ALLDATA, the following program could be used:

```
DATA ALLDATA;
   SET MEN WOMEN;
RUN;
```

Data set ALLDATA will contain all the observations from the data set MEN, followed by all the observations from the data set WOMEN.

D. COMBINING DIFFERENT DATA FROM MULTIPLE SAS DATA SETS

Important note: The use of Social Security numbers in this chapter is for illustrative purposes only. In real life, it is not advisable (or may not be legal) to use Social Security numbers in any listing that could be accessed by the public.

In this section, we demonstrate how to combine different information from multiple SAS data sets. Suppose we have a master student data set that contains Social Security (SS) numbers and student names (NAME). We then give a test and create a data set that contains student SS numbers and test scores. We now want to print out a list of student numbers, names, and scores. Here are sample master and test listings:

SS	Name	
123-45-6789	CODY	
987-65-4321	SMITH	
111-22-3333	GREGORY	Master Data
222-33-4444	HAMER	
777-66-5555	CHAMBLISS	

SS	Score	
123-45-6789	100	
987-65-4321	67	Test Data
222-33-4444	92	

To merge the student names in the MASTER data set with the SS numbers in the TEST data set, we first sort both data sets by SS:

```
PROC SORT DATA=MASTER;
   BY SS;
RUN;
PROC SORT DATA=TEST;
   BY SS;
RUN;
```

To merge these two data sets into a new data set, we use a MERGE statement:

```
DATA BOTH;
   MERGE MASTER TEST;
   BY SS;
   FORMAT SS SSN11.;
RUN;
```

Since the MASTER and TEST data sets are now sorted by SS, the MERGE operation will attempt to combine observations from each of the two data sets when the SS has the same value in each observation. Let's look at the observations from these two data sets, in sorted order, side by side:

Data Set MASTER

SS	Name
111-22-3333	GREGORY
123-45-6789	CODY
222-33-4444	HAMER
777-66-5555	CHAMBLISS
987-65-4321	SMITH

Data Set TEST

SS	Score
123-45-6789	100
222-33-4444	92
987-65-4321	67

The first SS number in data set MASTER is not found in data set TEST. Therefore, when the MERGE takes place, the first observation in data set BOTH will have a missing value for SCORE. The next observation in MASTER has a SS number of 123-45-6789. This number is found in both data sets, so the second observation in data set BOTH will contain a value of 123-45-6789 for SS, the NAME 'CODY', and a SCORE of 100. This process continues until all the observations in data sets MASTER and TEST have been processed. The resulting data set BOTH will have the following observations:

Data Set BOTH

SS	Name	Score
111-22-3333	GREGORY	.
123-45-6789	CODY	100
222-33-4444	HAMER	92
777-66-5555	CHAMBLISS	.
987-65-4321	SMITH	67

Most likely, we would like the merged data set to contain only those observations for which there was a test score. The data set option IN=*contributor_variable* following either (or both) of the data set names listed in the MERGE statement gives us control over which observations are added to the merged data set. The value of *contributor_variable* will be true (1) if that data set has made a contribution to (has a nonmissing value for the

BY variable) the current observation being created. If not, it has a value of false (0). The *contributor_variable* created by the IN= data set option is a temporary variable that can be used anywhere in the DATA step but is not added to the new data set. Let's see how we can use the IN= data set option to omit those observations in data set MASTER who did not take the test. We modify the previous program as follows:

```
DATA BOTH;
   MERGE MASTER TEST (IN=FRODO);
   BY SS;
   IF FRODO;
   FORMAT SS SSN11.;
RUN;
```

The IN= option following the data set name TEST creates the logical variable FRODO. To limit the merged data set to only those students in the TEST data set, we use a subsetting IF statement to make sure that the student had an observation in the TEST data set. The resulting merged data set (BOTH) is:

Data Set BOTH

SS	Name	Score
123-45-6789	CODY	100
222-33-4444	HAMER	92
987-65-4321	SMITH	67

The general syntax for the MERGE statement is:

```
MERGE data_set_one   (IN=var_name1)
      data_set_two   (IN=var_name2)
      data_set_n     (IN=var_namen);
   BY match_vars;
```

In this syntax example, data_set_one, data_set_two, to data_set_n are to be merged; the IN= option that follows each data set name, if used, can control which observations will be included in the merged data set. The BY statement will tell the program how to select observations from the merged data sets. Now, it is possible to merge without a BY statement. This is **very** dangerous and should usually be avoided. If you omit a BY statement, you will be merging observation one from each of the data sets, observation two, etc., regardless if those observations belong together or not. A very useful system option, MERGENOBY, allows you to select whether or not a merge without a BY statement should be considered an error. The syntax is:

```
OPTIONS MERGENOBY = ERROR | WARN | NOWARN;
```

where you choose one of the three keywords.

The default is NOWARN. With that option in effect, a merge without a BY statement will occur, and no warnings or errors will be produced. We recommend that you set this option to ERROR (in which case the program stops and you see an error message in the LOG) or, at least, to WARN (where the merge proceeds but you see a warning message in the LOG, if you are careful enough to read the LOG).

Suppose there were observations in the TEST data set without a corresponding one in the MASTER data set. How would we use the IN= data set option to create a merged data set that only contained observations where there was a contribution from both of the data sets? We could use an IN= option for each of the data sets and test that both logical variables were true, like this:

```
DATA BOTH;
   MERGE MASTER(IN=BILBO)
         TEST  (IN=FRODO);
   BY SS;
   IF BILBO AND FRODO;
   FORMAT SS SSN11.;
RUN;
```

If the BY variable has a different variable name in one of the data sets, you can use a RENAME option in the MERGE statement. For example, if SS were called ID in the TEST data set, we would write:

```
MERGE MASTER TEST (IN=INTEST RENAME=(ID=SS));
```

This brings in an observation from data set TEST and renames ID to SS for purposes of the MERGE. Note that the variable name in the data set TEST remains ID.

E.   "TABLE LOOK UP"

This section explores some other ways that merging can be used to perform a "table look up." By table look up we mean that one can pull information from a data set based on one or more criteria and add that information to the current data set. Some simple examples will make this clear. We have one data set that contains ID numbers, YEAR, and white blood cell count (WBC). Some sample observations are shown here:

ID	Year	WBC	
1	1940	6000	
2	1940	8000	
3	1940	9000	Data set WORKER
1	1941	6500	
2	1941	8500	
3	1941	8900	

Next, we have a data set that tells us the benzene exposure for these subjects for each year.

Year	Exposure
1940	200
1941	150
1942	100
1943	80

Data set EXP

What we want is to add the correct exposure to each observation in the WORKER data set. The SAS statements to perform the merge are:

```
PROC SORT DATA=WORKER;
   BY YEAR;
RUN;
PROC SORT DATA=EXP;
   BY YEAR;
RUN;
DATA COMBINE;
   MERGE WORKER (IN=INWORK) EXP;
   BY YEAR;
   IF INWORK;
RUN;
```

The resulting data set, COMBINE is shown next:

ID	Year	WBC	Exposure
1	1940	6000	200
2	1940	8000	200
3	1940	9000	200
1	1941	6500	150
2	1941	8500	150
3	1941	8900	150

Data set COMBINE

We now extend this problem to include two BY variables. We want to assign an exposure based on the YEAR and the WORK assignment. Our look up table consists of years, work codes, and exposures. Here is the look up table:

Year	Work	Exposure
1940	MIXER	190
1940	SPREADER	200
1941	MIXER	140
1941	SPREADER	150
1942	MIXER	90
1942	SPREADER	100
1943	MIXER	70
1943	SPREADER	80

Data set EXP

The WORKER data set now contains the YEAR, WORK code, and WBC counts:

ID	Year	Work	WBC
1	1940	MIXER	6000
2	1940	SPREADER	8000
3	1940	MIXER	9000
1	1941	MIXER	6500
2	1941	MIXER	8500
3	1941	SPREADER	8900

Data set WORKER

To add the correct exposure to each observation in the WORKER data set, we have to "look up" the exposure for the correct YEAR and WORK code. A MERGE statement with two BY variables will accomplish this for us:

```
PROC SORT DATA=WORKER;
   BY YEAR WORK;
RUN;
PROC SORT DATA=EXP;
   BY YEAR WORK;
RUN;
DATA COMBINE;
   MERGE WORKER (IN=INWORK) EXP;
   BY YEAR WORK;
   IF INWORK;
RUN;
```

The merged data set (COMBINE) is shown next:

ID	Year	Work	WBC	Exposure
1	1940	MIXER	6000	190
3	1940	MIXER	9000	190
2	1940	SPREADER	8000	200
1	1941	MIXER	6500	140
2	1941	MIXER	8500	140
3	1941	SPREADER	8900	150

F. UPDATING A MASTER DATA SET FROM AN UPDATE DATA SET

Many business applications contain a master data set that needs to be updated with new information. For example, we might have part numbers and prices in the master data set. The update data set would contain part numbers and new, updated prices. Typically, the update data set would be smaller than the master data set and contain observations only for those part numbers with new prices. In addition, the update data set may contain part numbers that are not present in the master data set. Here is an example:

PART_NO	Price
1	19
4	23
6	22
7	45

MASTER data set

PART_NO	Price
4	24
5	37
7	.

UPDATE data set

We sort both data sets by PART_NO and then perform the UPDATE:

```
DATA NEWMASTR;
   UPDATE MASTER UPDATE;
   BY PART_NO;
RUN;
```

The result is:

PART_NO	Price
1	19
4	24
5	37
6	22
7	45

NEWMASTR data set

Note that the missing value in UPDATE does NOT replace the corresponding observation in MASTER.

PROBLEMS

Remember, you can download all the data sets and programs for these problems from the web site: www.prenhall.com/cody

14.1 You have a file containing gymnastic scores for boys and girls, as follows:

ID	Gender	Age	Vault	Floor	P_BAR
3	M	8	7.5	7.2	6.5
5	F	14	7.9	8.2	6.8
2	F	10	5.6	5.7	5.8
7	M	9	5.4	5.9	6.1
6	F	15	8.2	8.2	7.9

(a) Create a SAS data set called GYM from these data.
(b) Create a subset of these data for males only. Call it MALE_GYM.
(c) Create another subset of GYM for all females greater than or equal to 10 years of age. Call it OLDER_F.

14.2 Run the program from problem 4.9 to create the data set BLOOD.

(a) Create a data set called TIME_ONE containing all the observations from BLOOD, where TIME is equal to 1. Omit the variable TIME from this data set. Use PROC PRINT to list the observations.
(b) Create a data set called HIGH containing all the observations from BLOOD, where GROUP is equal to "A" and the WBC is greater than or equal to 8000 or the RBC is greater than or equal to 5.

14.3 You have two data files, one from the year 1996 and the other from the year 1997, as follows:

File for 1996 (DATA96)

ID	Height	Weight
2	68	155
1	63	102
4	61	111

File for 1997 (DATA97)

ID	Height	Weight
7	72	202
5	78	220
3	66	105

Create a SAS data set from each file (call them YEAR1996 and YEAR1997, respectively). Combine the data from each data set into a single file (call it BOTH).

14.4 Run the following program to create three data sets, ONE, TWO, and THREE. List the contents of all three using PROC PRINT. Now, combine the three data sets into a single data set called TOGETHER.

```
DATA ONE TWO THREE;
    DO I = 1 TO 5;
        DO TIME = 1 TO 3;
            DOB = INT(10000 + RANUNI(0)*365);
            WEIGHT = RANNOR(0)*50 + 150;
            IF TIME = 1 THEN OUTPUT ONE;
            ELSE IF TIME = 2 THEN OUTPUT TWO;
            ELSE IF TIME = 3 THEN OUTPUT THREE;
        END;
    END;
    DROP I;
    FORMAT DOB DATE9.;
RUN;
```

14.5 You have a separate file on the children in problem 14.1. This file contains ID numbers, income ranges, and the parents' last name as follows:

ID	Income	L_NAME
3	A	Klein
7	B	Cesar
8	A	Solanchick
1	B	Warlock
5	A	Cassidy
2	B	Volick

Note that there are IDs for which there is no GYM data and vice versa. First, create a SAS data set called MONEY from the previous data. Next, merge the two data sets (call the merged data set GYMMONEY), including only those IDs that are present in the GYM data set. Next, print out a list showing ID, last name, gender, and age. Have this list in ID order.

14.6 You have temperature data for each day from Jan 1, 2004, to Jan 15, 2004. Run the following program to create a data set TEMP:

```
DATA TEMP;
    IF _N_ = 1 THEN DATE = '01JAN2004'D;
    ELSE DATE + 1;
    FORMAT DATE MMDDYY10.;
    INPUT T @@;
DATALINES;
30 32 28 26 25 12 18 20 22 24 36 38 38 39 44
;
```

The number of heart attack patients arriving at a hospital for each of the 15 days is also recorded. Run the following program to create a data set called MI (myocardial infarction, also known as a heart attack).

```
DATA MI;
    IF _N_ = 1 THEN DATE = '01JAN2004'D;
    ELSE DATE + 1;
    FORMAT DATE MMDDYY10.;
    INPUT NUMBER @@;
DATALINES;
9 7 11 12 15 23 20 18 8 9 13 12 14 13 14
;
```

Create a SAS data set called BOTH from the temperature data and the heart attack incidence data, using DATE as the BY variable.

14.7 Combine the GYMMONEY data set from problem 14.5 with the data set BOTH from problem 14.3. Call the resulting data set FREDDY. Include only those IDs with data in both data sets. List the contents of this data set.

*14.8 Ten students took a test. The tests were scored by an optical mark sense reader (scanner), and the ID and test score (SCORE) were recorded for each student and stored in a SAS data set called TEST. The school maintains a SAS data set called STUDENTS with the ID and NAME of every student. You want to produce a roster with the student ID, NAME, and test score (SCORE). There are IDs and names in the STUDENT data set for students who did not take the test. Also, there was one student who took the test who was not in the STUDENT file. So, in the roster, include this student but assign a value of "Not in Student File" for the NAME. Run the following program to create the two data sets TEST and STUDENT (note that neither data set is sorted):

```
DATA STUDENT;
    INPUT ID NAME & $30. @@;
DATALINES;
1 John Torres  5 Alex Antoniou  3 Thomas Friend
2 Sheldon Goldstein  11 Joanne Dipietro  12 Bill Murray
21 Janet Reno  4 Deborah Smith  6 Don Dubin  7 Alice Ford
8 Diane Farley  9 Laura Easton  10 Brian Fishmann
13 Eric Garrett  14 James Galt  15 Toni Gilman
;
DATA TEST;
    INPUT ID SCORE @@;
DATALINES;
15 95  1 80  3 98  21 75  4 87  14 67  13 91  11 85  12 57
29 93
;
```

As a bonus, see if you can produce the roster in alphabetical order by last name (with the Not in Student File name last). Hint: For the student who is not in the student file, set the value of last name (LAST) equal to "ZZZ." Then, for the other names, use the SCAN function to select the last name (a minus value for the "word" argument scans from the right). Sort by the last name but don't include that variable in the VAR list of PROC PRINT.

14.9 You have a financial plan based on income range and gender. Using the GYMMONEY data set from problem 14.3, create a new data set called FINAL, which contains all the data from GYMMONEY along with the correct financial plan based on the table here:

Income Range	Gender	Financial Plan
A	M	W
A	F	X
B	M	Y
B	F	Z

Produce a listing of this data set.

14.10 Your company price list is stored in a SAS data set called PRICES. Run the following program to create this data set. You want to update some prices and quantities in this data set as follows:

New price for Part #103 is $18.99

New price and quantity for Part #111 is $29.95 and the quantity is 45

New quantity for part #113 is 35

New price for part #123 is $129.96

The program to create data set PRICES is:

```
DATA PRICES;
    INPUT PART_NUMBER QUANTITY PRICE @@;
DATALINES;
100 23 29.95  102 12 9.95  103 21 15.99  123 9 119.95  113 40 56.66
111 55 39.95  105 500 .59
;
```

14.11 You have some new information on the gymnasts in problem 14.1. Subject 3 now has a 6.7 on P_BAR; subject 5 is now 15 and has scores of 8.1 and 7.2 on VAULT and P_BAR, respectively; subject 7 was incorrectly entered as a male and should be female. Create an update data set of this new information, and update the GYM file from problem 14.1. Call the updated data set GYM_2 and provide a listing.

CHAPTER 15

Working with Arrays

A.   INTRODUCTION

SAS arrays are a facility that can reduce the amount of coding in a SAS DATA step. Although often thought of as an advanced programming tool, there are many applications of arrays that can be easily mastered. This chapter demonstrates some of the more common uses of SAS arrays.

One of the most common uses of arrays is to shorten a program that repeats one or more lines of code with the only change being the variable names referenced in each of the lines. You will see that you can write "model" lines, replacing the variable names with array names, and by placing these model lines in a looping structure, you can effectively replace hundreds or thousands or millions or … lines of code with just a few lines. OK, we sometimes get carried away with how useful arrays can be!

B.   SUBSTITUTING ONE VALUE FOR ANOTHER FOR A SERIES OF VARIABLES

One of the best ways to learn how to use arrays is to first write a few lines of SAS code without them. Once you see the pattern, you can write your array statement, your "model" lines of code, and decide how to place those model lines in a DO loop. So, for this example and for most of the examples in this chapter, you will see some lines of SAS code without using arrays and the corresponding coding using arrays.

For this first example, imagine that you have been given a SAS data set where a value of 999 was entered whenever there was a missing numeric value. (Yes, you guessed it, probably a converted SPSS data set!) Here is a DATA step that converts the values of 999 to missing, without using SAS arrays:

```
*------------------------------------------------------------*
| Example 1: Converting 999 to missing without using an array |
*------------------------------------------------------------*;
DATA MISSING;
   SET OLD;
   IF A = 999 THEN A = .;
   IF B = 999 THEN B = .;
   IF C = 999 THEN C = .;
   IF D = 999 THEN D = .;
   IF E = 999 THEN E = .;
RUN;
```

Do you see a pattern here? Good. Here is the same program using arrays:

```
*------------------------------------------------------------*
| Example 1: Converting 999 to missing using an array         |
*------------------------------------------------------------*;
DATA MISSING;
   SET OLD;
   ARRAY X[5] A B C D E;
   DO I=1 TO 5;
      IF X[I]= 999 THEN X[I]= .;
   END;
   DROP I;
RUN;
```

OK, it's not that much shorter. But, if we had to recode hundreds of variables you would clearly see the advantage. Here's how the program works:

You first need to create an array to represent the five variables A, B, C, D, and E. You can choose an array name using the same rules you use for SAS variable names. However, be sure not to use the same name for a SAS array as for a variable in the same DATA step. In the previous example, the array name is X. Next, in square brackets [ ], curly brackets {}, or parentheses ( ), you enter the number of elements (variables) in the array. You may, if you are "counting challenged" (as is one of the authors), use an asterisk in place of the number. We usually use, and prefer, square brackets; much of the documentation from the SAS Institute uses burly brackets. Your choice.

Following the brackets is a list of SAS variable names. You may list them explicitly or use any of the SAS DATA step conventions that refer to a group of variable names, such as the single dash, double dash, or the reserved names _CHARACTER_ or _NUMERIC_. Be forewarned that an array cannot contain a mixture of character and numeric variables. A later example will demonstrate how to create an array of character variables.

Having created your array, you can refer to any of the array elements by using the array name and the appropriate subscript within brackets. In this first example, the element X[3] would represent the third element in the array, C. Just substituting an array element for a variable name in a DATA step would accomplish little. The most common use of an array is to place it in an iterative loop such as a DO loop, a DO WHILE, or a DO UNTIL structure. In the previous example, for each of the iterations of the DO loop, you are setting values of 999 to missing for each of the elements of the array and, therefore, each of the variables A through E.

If you are not familiar with DO loops, the syntax is:

```
DO COUNTER=START TO END BY INCREMENT;
    (lines of sas code)
END;
```

The SAS statements between the DO and END statement will be repeated according to the directions specified in the DO statement. In the first example, the index variable I was used as the counter, and the iteration went from 1 to 5. Since the INCREMENT value was omitted, it defaulted to 1. Also, don't forget to DROP DO loop counters in your DATA step.

C.   EXTENDING EXAMPLE 1 TO CONVERT ALL NUMERIC VALUES OF 999 TO MISSING

Here is a useful extension of Example 1 that converts values of 999 to missing for all numeric variables:

```
*----------------------------------------------------------------*
| Example 2: Converting 999 to missing for all numeric vars      |
*----------------------------------------------------------------*;
DATA ALLNUMS;
    SET ALL;
    ARRAY PRESTON[*] _NUMERIC_ ;
    DO I = 1 TO DIM(PRESTON);
        IF PRESTON[I] = 999 THEN PRESTON[I] = .;
    END;
    DROP I;
RUN;
```

First, you may wonder where the array name PRESTON came from. That's easy. It is the name of one of the authors' youngest boy, who loves to program (and has visited the SAS Institute). The reserved name _NUMERIC_ refers to all the numeric variables in the data set ALL. Since it might be a lot of trouble to count how many numeric variables there are, you use the asterisk (*) instead of the actual number. The only trick here is that you don't know the ending value for the DO loop. Luckily, the DIM function comes to the rescue. The argument of the DIM function is an array name, and it returns the number of elements in the array.

D.   CONVERTING THE VALUE OF N/A (NOT APPLICABLE) TO A CHARACTER MISSING VALUE

For this problem, you have a SAS data set called OLD where the character string 'N/A' (not applicable) was used in place of a character missing value (a blank). You want to convert the values of 'N/A' to missing for several character variables. As before, here is the program without arrays:

```
*----------------------------------------------------------------*
| Example 3: Converting 'N/A' to Missing for character vars      |
*----------------------------------------------------------------*;
DATA NOTAPPLY;
   SET OLD;
   IF S1 = 'N/A' THEN S1 = ' ';
   IF S2 = 'N/A' THEN S2 = ' ';
   IF S3 = 'N/A' THEN S3 = ' ';
   IF X = 'N/A' THEN X = ' ';
   IF Y = 'N/A' THEN Y = ' ';
   IF Z = 'N/A' THEN Z = ' ';
RUN;
```

And here is the same program using a character array:

```
*----------------------------------------------------------------*
| Example 3: Converting 'N/A' to Missing for character vars      |
*----------------------------------------------------------------*;
DATA NOTAPPLY;
   SET OLD;
   ARRAY RUSSELL[*] $ S1-S3 X Y Z;
   DO J = 1 TO DIM(RUSSELL);
      IF RUSSELL[J] = 'N/A' THEN RUSSELL[J] = ' ';
   END;
   DROP J;
RUN;
```

This time, I'm sure you guessed that the array name RUSSELL is the name of another son. (He is a PADI certified SCUBA diver and a musician.) Since you want to declare the array as a character array, you place a dollar sign ($) following the brackets. In this case, since you are reading the observations from an existing SAS data set, the variables S1, S2, S3, X, Y, and Z are already declared as character variables, and you could actually omit the $ in the array statement. However, a good programming practice is always to include a $ in the definition of an array of character variables. Now is a good time to mention that you can also indicate a LENGTH for the elements (if these variables don't already exist) of either a numeric or character array following immediately before the list of variables. In the examples so far, the variables have come from an already created SAS data set and had predefined lengths. To create an array of character variables Q1–Q50 with lengths of one byte, the syntax would be:

```
ARRAY Q[50] $ 1 Q1-Q50;
```

E.   CONVERTING HEIGHTS AND WEIGHTS FROM ENGLISH TO METRIC UNITS

For this example, you want to input three heights and five weights in English units and create new variables that represent these same values in metric units.

Here is the program without arrays:

```
*-------------------------------------------------------*
| Example 4: Metric conversion without using arrays     |
*-----------------------------------------------------*;
DATA CONVERT;
    INPUT HT1-HT3 WT1-WT5;
    HTCM1 = 2.54 * HT1;
    HTCM2 = 2.54 * HT2;
    HTCM3 = 2.54 * HT3;
    WTKG1 = WT1 / 2.2;
    WTKG2 = WT2 / 2.2;
    WTKG3 = WT3 / 2.2;
    WTKG4 = WT4 / 2.2;
    WTKG5 = WT5 / 2.2;
DATALINES;
(data goes here)
RUN;
```

And now, the array version:

```
*---------------------------------------------------*
| Example 4: Metric conversion using arrays         |
*-------------------------------------------------*;
DATA CONVERT;
    INPUT HT1-HT3 WT1-WT5;
    ARRAY HT[3];
```

[Continued]

```
    ARRAY HTCM[3];
    ARRAY WT[5];
    ARRAY WTKG[5];
    *** Yes, we know the variable names are missing, read on;
    DO I=1 TO 5;
        IF I LE 3 THEN HTCM[I] = 2.54 * HT[I];
        WTKG[I] = WT[I] / 2.2;
    END;
  DATALINES;
  (data goes here)
  RUN;
```

There are a few things to notice about this program. First, you may have observed that there are no variable names following the array names! (This is a shortcut that you may want to use to impress your boss.) Anytime you write an array statement where there are no variable names listed, the names default to the array name, followed by the range of numbers cited in the brackets. For example, the array statement:

```
    ARRAY QUES[3];
```

is equivalent to:

```
    ARRAY QUES[3] QUES1-QUES3;
```

Next, you have the problem that the number of heights and weights are not the same. There are several solutions to this problem. The previous example used a single DO loop with the index going from 1 to 5. An IF statement inside the loop ensured that the index for the height variables would not exceed 3. An alternative would be for two DO loops, one going from 1 to 3 and another going from 4 to 5. For example:

```
  DO I=1 TO 3;
      HTCM[I] = 2.54 * HT[I];
      WTKG[I] = WT[I] / 2.2;
  END;
  DO I=4 TO 5;
      WTKG[I] = WT[I] / 2.2;
  END;
```

Pay your money and take your choice!

F.   TEMPORARY ARRAYS

The arrays you have encountered thus far all represent a list of numeric or character variables. There is a special type of array, called a temporary array, that does not actually refer to a list of variables at all! Instead, you can declare an array to be temporary and use the array elements in their subscripted form in the DATA step. No real variables are created when you use a temporary array. You can also provide initial values for each of the array elements. Let's look at an example first, and then discuss the advantages and disadvantages of temporary arrays.

This first example uses a temporary array to hold the passing scores on five exams. Students' scores are then read and compared to these passing scores, and the number of failed courses is recorded. Now for the program:

```
*--------------------------------------------------------------*
| Example 5: Using a temporary array to determine the number   |
|     of tests passed                                          |
*--------------------------------------------------------------*;
DATA PASSING;
    ARRAY PASS[5] _TEMPORARY_ (65 70 65 80 75);
    ARRAY SCORE[5];
    INPUT ID $ SCORE[*];
    PASS_NUM = 0;
    DO I=1 TO 5;
        IF SCORE[I] GE PASS[I] THEN PASS_NUM + 1;
    END;
    DROP I;
DATALINES;
001 64 69 68 82 74
002 80 80 80 60 80
;
PROC PRINT DATA=PASSING;
    TITLE "Passing Data Set";
    ID ID;
    VAR PASS_NUM SCORE1-SCORE5;
RUN;
```

The main feature of this program is the array PASS, defined as a temporary array by the keyword _TEMPORARY_ following the brackets. Notice the five scores within parentheses following the _TEMPORARY_ keyword. These are initial values assigned to the five array elements PASS[1] through PASS[5]. Since the values of temporary array elements are automatically retained, the five passing scores are available for comparison to the student grades in every iteration of the DATA step. It is important to note that the variables PASS1 through PASS5 are not created by this array and do not exist in the data set.

Again, notice that we used the "shortcut" method of defining the SCORE array. As mentioned previously, the array statement:

```
ARRAY SCORE[5];
```

is equivalent to

```
ARRAY SCORE[5] SCORE1-SCORE5;
```

We just never seem to be able to pass up the chance to save a few keystrokes.

The most compelling reason to use temporary arrays is for efficiency. Also, you do not have to bother dropping useless variables.

G.   USING A TEMPORARY ARRAY TO SCORE A TEST

This example also uses temporary arrays. Instead of assigning initial values to the array elements when defining the array, the initial values are read as raw data, which causes some interesting problems and innovative solutions. The object of this sample program is to score a 10-question multiple-choice test. The first line of data contains the answer key, and the remaining lines contain student IDs and student answers to the 10 questions. Here is a test-scoring program that makes good use of temporary arrays:

```
*----------------------------------------------------------*
| Example 6: Using a temporary array to score a test       |
*----------------------------------------------------------*;
DATA SCORE;
    ARRAY KEY[10] $ 1 _TEMPORARY_;
    ARRAY ANS[10] $ 1;
    ARRAY SCORE[10] _TEMPORARY_;
    IF _N_ = 1 THEN
        DO I=1 TO 10;
            INPUT KEY[I] @;
        END;
    INPUT ID $ @5 (ANS1-ANS10) ($1.);
    RAWSCORE = 0;
    DO I=1 TO 10;
        SCORE[I] = (ANS[I] EQ KEY[I]);
        RAWSCORE + SCORE[I];
    END;
    PERCENT = 100*RAWSCORE/10;
    DROP I;
DATALINES;
A B C D E E D C B A
001 ABCDEABCDE
002 AAAAABBBBB
;
PROC PRINT DATA=SCORE;
    TITLE "SCORE Data Set";
    ID ID;
    VAR RAWSCORE PERCENT;
RUN;
```

Let's take it step by step. The two arrays KEY and SCORE are declared to be temporary. In addition, the elements of the KEY array are to be one-byte characters. The elements in the KEY array hold the answer key (which will be retained by the nature of temporary array elements), and the elements of the SCORE array will be either 1s (correct answer) or 0s (incorrect answer). The array ANS is not a temporary array, and the variables ANS1–ANS10 will contain the student answers to the 10 questions comprising the test.

Unlike the previous example, you do not initialize the values of the KEY array with the ARRAY statement. Instead, you read the values from the first line of data. That's what the group of code starting with IF N_ = 1 is all about. Since the first line of data is the answer key, this DO group will only execute once for the first line of data. Notice that instead of reading in the data with:

```
INPUT KEY[1] KEY[2] KEY[3] ... KEY[10];
```

The easier (and more generalizable form):

```
DO I=1 TO 10;
    INPUT KEY[I] @;
END;
```

is used instead. Remember that there are no variables with names KEY1, KEY2, etc., in this data set. Also note that it is not proper to write:

```
INPUT KEY[1]- KEY[10];
```

since the form BASEn–BASEm works only for real variables and not elements of arrays.

For the remaining lines of data, the statement IF N_ = 1 is false, and the program drops down to the INPUT ID $ ANS1-ANS10; statement. Scoring is performed by the somewhat unusual statement:

```
SCORE[I] = (ANS[I] EQ KEY[I]);
```

This statement causes the student answer ANS [I] to be compared to the value of KEY [I]. If they are equal, this part of the statement is true (equal to 1) and SCORE[I] is set equal to 1. Otherwise, SCORE[I] will be set to false (equal to 0). Finally, the statement:

```
RAWSCORE + SCORE[I];
```

accumulates the raw score for each student. You could have used a SUM function outside this loop like this:

```
RAWSCORE = SUM(OF SCORE[1] SCORE[2] ... SCORE[10]);
```

This, like the alternative INPUT statement discussed earlier is not as easy to generalize (for tests of different length) as the structure used here.

For those truly compulsive programmers (like one of the authors), you can omit the SCORE array entirely and simply code the following:

```
DO I=1 TO 10;
    RAWSCORE + (KEY][I] EQ ANS[I]);
END;
```

However, we wanted the excuse to show you a numeric temporary array.

H. SPECIFYING ARRAY BOUNDS

All of the arrays you have seen thus far had array elements starting from 1. So, for example, if you wanted an array to represent the five variables YR1993, YR1994, YR1995, YR1996, and YR1997, you could write your array statement like this:

```
ARRAY YR[5] YR1993-YR1997;
```

This is OK, but you would have to remember that YR[1] represents the variable YR1993, YR[2] represents the variable YR1994, and so on. You might like to have the array element YR[1993] associated with the variable YR1993 and YR[1994] associated with the variable YR1994 instead. You can specify starting and ending boundaries in your array statement by entering the starting value, a colon, and the ending value within the brackets following the array name. For the problem just discussed, the array statement:

```
ARRAY YR[1993:1997] YR1993-YR1997;
```

gives you just what you want. You separate the lower and upper bounds of the array index by a colon. Another application where specifying array bounds is useful is when you are counting from 0 instead of 1. For example, you may be measuring heart rate (HR) at times 0, 1, 2, and 3. A convenient array statement would be:

```
ARRAY HR[0:3] HR0-HR3;
```

Now that you see that array bounds do not have to run from 1 to n, you need to rethink the use of the DIM function, which returns the number of elements in an array. If the array starts from 1, the DIM function will also represent the upper bound of the array. However, in the previous array YR, the DIM function would return a 5! To extract the lower and upper bounds of an array, use the LBOUND and UBOUND functions instead. They return the lower and upper bounds of an array, respectively.

I. TEMPORARY ARRAYS AND ARRAY BOUNDS

Here is an interesting program that converts plain text to Morse code. The program uses a temporary array to store the Morse equivalents of the letters and subscripts the elements of the array, starting at 65 (since the RANK function, which returns the location of a letter in the ASCII collating sequence, returns a 65 for a capital "A", a 66 for a capital "B", etc.). Here then, is the program to convert plain text to Morse code:

```
*----------------------------------------------------------*
| Example 7: Converting plain text to morse code           |
*----------------------------------------------------------*;
DATA _NULL_;
    FILE PRINT;
    ARRAY M[65:90] $ 4 _TEMPORARY_
        ('.-' '-...' '-.-.' '-..' '.'
         '..-.' '--.' '...' '..' '.---'
         '-.-' '.-..' '--' '-.' '---'
         '.--.' '--.-' '.-.' '...' '-'
```

```
           '..-' '...-' '.--' '-..-' '-.--'
           '--..');
    INPUT LETTER $1. @@;
    LETTER = UPCASE(LETTER);
    IF LETTER EQ ' ' THEN DO;
        PUT ' ' @;
        RETURN;
    END;
    MORSE = M[RANK(LETTER)];
    PUT MORSE @;
DATALINES;
This is a test
;
```

J.   IMPLICITLY SUBSCRIPTED ARRAYS

Before leaving the topic of arrays, we should mention the alternate type of array that does not explicitly show the subscript when you refer to an array element in the DATA step. This was the original form of the array statement in version 5 that was superseded by the explicit subscript form that we have discussed until now. We strongly recommend that you use the explicit form when writing any new programs. However, since the implicit form is still supported and you may have to maintain older SAS code that contains this type of array, we will briefly show you how it works. The form the ARRAY statement is:

```
ARRAY ARRAYNAME(index_variable) list-of-sas-variables;
```

Length and $ attributes are also available and are placed before the list of SAS variables. When using the array name in a DATA step, the index variable is first set to a value (usually in a DO loop) and the array name is used without a subscript. For example:

```
*----------------------------------------------------------*
| Example 8: demonstrating the older implicit array        |
*----------------------------------------------------------*;
DATA OLDARRAY;
    ARRAY MOZART(I) A B C D E;
    INPUT A B C D E;
    DO I=1 TO 5;
        IF MOZART = 999 THEN MOZART = .;
    END;
    DROP I;
DATALINES;
(data lines)
;
```

Notice that inside the DO loop, the array name MOZART is used without a subscript. When I = 1, MOZART will represent the variable A; when I = 2, MOZART will represent the variable B; and so forth. A default subscript _I_ is used if no subscript is indicated in the ARRAY statement. The DO loop would then read:

```
DO _I_ = 1 to 5;
```

A very useful form of the DO loop "DO OVER" is available when the implicit subscript form of the array is used. DO OVER automatically computes the lower and upper bounds of the array and loops over all the elements in the array. The previous program can be written using a DO OVER structure like this:

```
*----------------------------------------------------------*
| Example 9: Demonstrating the Older Implicit ARRAY        |
*----------------------------------------------------------*;
DATA OLDARRAY;
    ARRAY MOZART A B C D E;
    INPUT A B C D E;
    DO OVER MOZART;
        IF MOZART = 999 THEN MOZART = .;
    END;
DATALINES;
(data lines)
;
```

As convenient as this may seem, we still recommend the explicit arrays.

Yes, you can live without arrays, but a thorough understanding of them can substantially reduce the size of a SAS program. We hope that the examples offered here will give you the courage to try arrays in your next program.

PROBLEMS

Remember, you can download all the data sets and programs for these problems from the web site: www.prenhall.com/cody

15.1 Rewrite this program using arrays:

```
DATA PROB15_1;
    INPUT (HT1-HT5)(2.) (WT1-WT5)(3.);
    DENS1 = WT1 / HT1**2;
    DENS2 = WT2 / HT2**2;
    DENS3 = WT3 / HT3**2;
    DENS4 = WT4 / HT4**2;
    DENS5 = WT5 / HT5**2;
DATALINES;
6862727074150090208230240
64  68  70140    150    170
;
```

15.2 Rewrite the following program using arrays:

```
DATA TEMPERATURE;
   INPUT TF1-TF10;
   TC1 = 5/9*(TF1 - 32);
   TC2 = 5/9*(TF2 - 32);
   TC3 = 5/9*(TF3 - 32);
   TC4 = 5/9*(TF4 - 32);
   TC5 = 5/9*(TF5 - 32);
   TC6 = 5/9*(TF6 - 32);
   TC7 = 5/9*(TF7 - 32);
   TC8 = 5/9*(TF8 - 32);
   TC9 = 5/9*(TF9 - 32);
   TC10 = 5/9*(TF10 - 32);
DATALINES;
32 212 -40 10 20 30 40 50 60 70
-10 0 10 20 30 40 50 60 70 80
;
PROC PRINT DATA=TEMPERATURE NOOBS;
   TITLE "Listing of Data Set TEMPERATURE";
RUN;
```

15.3 Rewrite the following program using arrays:

```
DATA OLDMISS;
   INPUT A B C X1-X3 Y1-Y3;
   IF A = 999 THEN A = .;
   IF B = 999 THEN B = .;
   IF C = 999 THEN C = .;
   IF X1 = 999 THEN X1 = .;
   IF X2 = 999 THEN X2 = .;
   IF X3 = 999 THEN X3 = .;
   IF Y1 = 777 THEN Y1 = .;
   IF Y2 = 777 THEN Y2 = .;
   IF Y3 = 777 THEN Y3 = .;
DATALINES;
1 2 3 4 5 6 7 8 9
999 4 999 999 5 999 777 7 7
;
```

15.4 Rewrite the following program using two arrays, one for the numeric variables, the other for the character variables.

```
DATA MIXED;
   INFORMAT A1-A3 B C $5.;
   INPUT X1-X3 Y Z A1-A3 B C;
   LX1 = LOG(X1);
```

[Continued]

```
    LX2 = LOG(X2);
    LX3 = LOG(X3);
    LY = LOG(Y);
    LZ = LOG(Z);
    IF A1 = '?' THEN A1 = ' ';
    IF A2 = '?' THEN A2 = ' ';
    IF A3 = '?' THEN A3 = ' ';
    IF B = '?' THEN B = ' ';
    IF C = '?' THEN C = ' ';
DATALINES;
10 20 30 40 50 ONE TWO THREE ? ?
11 22 33 44 55 ? LLL MMM ? VVV
;
PROC PRINT DATA=MIXED NOOBS;
    TITLE "Listing of Data Set MIXED";
RUN;
```

15.5 You are given the SAS data set SPEED, created by running the following program. Create a
new data set SPEED2 from SPEED, with some new variables. The new variables LX1–LX5
are the natural (base e) logs of the variables X1–X5, and the variables SY1–SY3 are the
square roots of the variables Y1–Y3. Use arrays to create the new variables. (NOTE : See
Chapter 17 for how to take a natural log of a number.) In case you don't want to turn to
Chapter 17 now, the statement to take the natural log of X and assign the value to a variable
LOG_X is: LOG_X=LOG(X);

```
***Program to create the data set SPEED;
DATA SPEED;
    INPUT X1-X5 Y1-Y3;
DATALINES;
1 2 3 4 5 6 7 8
11 22 33 44 55 66 77 88
;
```

*15.6 The passing grade on each of five tests was 65, 70, 60, 75, and 66. Run the following data set
to create a SAS data set called ANSWERS, which contains five test scores for each student
in the class (plus an ID number). Then, using a temporary array to hold the five passing
grades, count how many tests are passed by each student.

```
DATA ANSWERS;
    ***Passing grades: 65, 70, 60, 75, and 66;
    INPUT ID $ SCORE1-SCORE5;
DATALINES;
001 50 70 62 78 85
002 90 86 87 91 94
003 63 72 58 73 68
;
```

15.7 Rewrite the following program using arrays:

```
DATA PROB15_7;
   LENGTH C1-C5 $ 2;
   INPUT C1-C5 $ X1-X5 Y1-Y5;
   IF C1 = 'NA' THEN C1 = ' ';
   IF C2 = 'NA' THEN C2 = ' ';
   IF C3 = 'NA' THEN C3 = ' ';
   IF C4 = 'NA' THEN C4 = ' ';
   IF C5 = 'NA' THEN C5 = ' ';
   IF X1 = 999 OR Y1 = 999 THEN DO;
      X1 = .;
      Y1 = .;
   END;
   IF X2 = 999 OR Y2 = 999 THEN DO;
      X2 = .;
      Y2 = .;
   END;
   IF X3 = 999 OR Y3 = 999 THEN DO;
      X3 = .;
      Y3 = .;
   END;
   IF X4 = 999 OR Y4 = 999 THEN DO;
      X4 = .;
      Y4 = .;
   END;
   IF X5 = 999 OR Y5 = 999 THEN DO;
      X5 = .;
      Y5 = .;
   END;
DATALINES;
AA BB CC DD EE 1 2 3 4 5 6 7 8 9 10
NA XX NA YY NA 999 2 3 4 999 999 4 5 6 7
;
```

*15.8 An experiment was conducted where five times (TIME0 to TIME4) are measured in seconds. Using an array with subscripts of 0 to 4, convert the five times to minutes, rounded to the nearest tenth of a minute. Call the five new variables MINUTE0 to MINUTE4. Run the following DATA step to create a data set (EXPER) containing the five TIME variables:

```
DATA EXPER;
   INPUT TIME0-TIME4;
DATALINES;
100 200 300 400 500
55 110 130 150 170
;
```

Restructuring SAS Data Sets Using Arrays

A. Introduction
B. Creating a New Data Set with Several Observations per Subject from a Data Set with One Observation per Subject
C. Another Example of Creating Multiple Observations from a Single Observation
D. Going from One Observation per Subject to Many Observations per Subject Using Multidimensional Arrays
E. Creating a Data Set with One Observation per Subject from a Data Set with Multiple Observations per Subject
F. Creating a Data Set with One Observation per Subject from a Data Set with Multiple Observations per Subject Using a Multidimensional Array

A.   INTRODUCTION

This chapter describes how to restructure data sets by using arrays. First, what do we mean by the term restructuring? You may want to create multiple observations from a single observation (or vice versa) for several possible reasons: You may want to create multiple observations from a single observation to count frequencies or to allow for BY variable processing or to restructure SAS data sets for certain statistical analyses. Creating a single observation from multiple observations may make it easier for you to compute differences between values without resorting to LAG functions or perhaps to use the REPEATED statement in PROC GLM.

PROC TRANSPOSE may come to mind as a solution to these transforming problems, but using arrays in a DATA step can be more flexible and allow you to have full control over the transformation process.

B.   CREATING A NEW DATA SET WITH SEVERAL OBSERVATIONS PER SUBJECT FROM A DATA SET WITH ONE OBSERVATION PER SUBJECT

Suppose you have a data set called DIAGNOSE, with the variables ID, DX1, DX2, and DX3. The DX variables represent three diagnosis codes. The observations in data set DIAGNOSE are:

Data set DIAGNOSE			
ID	**DX1**	**DX2**	**DX3**
01	3	4	
02	1	2	3
03	4	5	
04	7		

As you can see, some subjects have only one diagnosis code, some two, and some all three. Suppose you want to count how many subjects have diagnosis 1, how many have diagnosis 2, and so on. You don't care if the diagnosis code is listed as DX1, DX2, or DX3. In the example here, you would have a frequency of one for diagnosis codes 1, 2, 5, and 7, and a frequency of two for diagnosis codes 3 and 4.

One way to accomplish this task is to restructure the data set DIAGNOSE, which has one observation per subject and three diagnosis variables, to a data set that has a single diagnosis variable and as many observations per subject as there are diagnoses for that subject. This new data set (call it NEW_DX) would look as follows:

Restructured Data Set (NEW_DX)	
ID	**DX**
01	3
01	4
02	1
02	2
02	3
03	4
03	5
04	7

It is now a simple job to count diagnosis codes using PROC FREQ on the single variable DX. Let us first write a SAS DATA step that accomplishes this task and does not use arrays. Here is the code:

```
*----------------------------------------------------------*
| Example 1A: Creating multiple observations from a single |
| observation without using an array                       |
*----------------------------------------------------------*;
DATA NEW_DX;
   SET DIAGNOSE;
   DX = DX1;
   IF DX NE . THEN OUTPUT;
   DX = DX2;
   IF DX NE . THEN OUTPUT;
   DX = DX3;
   IF DX NE . THEN OUTPUT;
   KEEP ID DX;
RUN;
```

As you read each observation from data set DIAGNOSE, you create from one to three observations in the new data set NEW_DX. The KEEP statement is needed since you only want the variables ID and DX in the new data set.

Notice the repetitive nature of the program, and your array light bulb should "turn on." Here is the program rewritten using arrays:

```
*----------------------------------------------------------*
| Example 1B: Creating multiple observations from a single |
| observation using an array                               |
*----------------------------------------------------------*;
DATA NEW_DX;
   SET DIAGNOSE;
   ARRAY DXARRAY[3] DX1 - DX3;
   DO I = 1 TO 3;
      DX = DXARRAY[I];
      IF DX NE . THEN OUTPUT;
   END;
   KEEP ID DX;
RUN;
```

In this program, you first create an array called DXARRAY, which contains the three numeric variables DX1, DX2, and DX3. The two lines of code inside the DO loop are similar to the repeated lines in the nonarray example with the variable names DX1, DX2, and DX3 replaced by the array elements. For a more detailed discussion of array processing, refer to the previous chapter. Note that PROC TRANSPOSE can also be used to restructure a SAS data set. *Longitudinal Data and SAS: A Programmer's Guide* (Cody, SAS Institute, Cary, NC, 2001. ISBN: 1-58025-924-30) provides detailed examples using PROC TRANSPOSE to restructure SAS data sets.

To count the number of subjects with each diagnosis code, you can now use PROC FREQ, like this:

```
PROC FREQ DATA=NEW_DX;
   TABLES DX / NOCUM;
RUN;
```

In this example, you saved only one line of SAS code. However, if there were more variables, DX1 to DX50 for example, the savings would be substantial.

C.   ANOTHER EXAMPLE OF CREATING MULTIPLE OBSERVATIONS FROM A SINGLE OBSERVATION

Here is an example that is similar to Example 1. You start with a data set that contains an ID variable and three variables S1, S2, and S3, which represent a score at times 1, 2, and 3, respectively. The original data set, called ONEPER, looks as follows:

Data Set ONEPER

ID	S1	S2	S3
01	3	4	5
02	7	8	9
03	6	5	4

You want to create a new data set called MANYPER, which looks like this:

Data Set MANYPER

ID	Time	Score
01	1	3
01	2	4
01	3	5
02	1	7
02	2	8
02	3	9
03	1	6
03	2	5
03	3	4

The program to restructure data set ONEPER to data set MANYPER is similar to the program in Example 1 except that you need to create the TIME variable in the restructured data set. This is easily accomplished by naming the DO loop counter TIME as follows:

```
*---------------------------------------------------------------*
| Example 2: Creating multiple observations from a single       |
| observation using an array                                    |
*---------------------------------------------------------------*;
DATA MANYPER;
   SET ONEPER;
   ARRAY S[3];
   DO TIME = 1 TO 3;
      SCORE = S[TIME];
      OUTPUT;
   END;
   KEEP ID TIME SCORE;
RUN;
```

As in several examples in Chapter 15, the ARRAY statement does not have a variable list. Remember that when this list is omitted, the variable names default to the array name, followed by the numbers from the lower bound to the upper bound. In this case, the statement

```
ARRAY S[3];
```

is equivalent to

```
ARRAY S[3] S1-S3;
```

Still going in the direction of creating multiple observations from a single observation, let us extend this program to include an additional dimension.

D. GOING FROM ONE OBSERVATION PER SUBJECT TO MANY OBSERVATIONS PER SUBJECT USING MULTIDIMENSIONAL ARRAYS

Suppose you have a SAS data set (call it WT_ONE) that contains an ID and six weights on each subject in an experiment. The first three values represent weights at times 1, 2, and 3 under condition 1; the next three values represent weights at times 1, 2, and 3 under condition 2. To clarify this, suppose that data set WT_ONE contained two observations:

Data Set WT_ONE

ID	WT1	WT2	WT3	WT4	WT5	WT6
01	155	158	162	149	148	147
02	110	112	114	107	108	109

You want a new data set called WT_MANY to look like this:

Data Set WT_MANY			
ID	**COND**	**Time**	**Weight**
01	1	1	155
01	1	2	158
01	1	3	162
01	2	1	149
01	2	2	148
01	2	3	147
02	1	1	110
02	1	2	112
02	1	3	114
02	2	1	107
02	2	2	108
02	2	3	109

A convenient way to make this conversion is to create a two-dimensional array, with the first dimension representing condition and the second representing time. So, instead of having a one-dimensional array like this:

```
ARRAY WEIGHT[6] WT1-WT6;
```

you could create a two-dimensional array like this:

```
ARRAY WEIGHT[2,3] WT1-WT6;
```

The comma between the 2 and 3 separates the dimensions of the array. This is a 2 by 3 array. Array element WEIGHT[2,3], for example, would represent a subject's weight under condition 2 at time 3.

Let us use this array structure to create the new data set that contains six observations for each ID. Each observation is to contain the ID and one of the six weights, along with two new variables, COND and TIME, which represent the condition and the time at which the weight was recorded. Here is the restructuring program:

```
*-----------------------------------------------------------------*
| Example 3: Using a multi-dimensional array to restructure       |
| a data set                                                      |
*-----------------------------------------------------------------*;
DATA WT_MANY;
   SET WT_ONE;
   ARRAY WTS[2,3] WT1-WT6;
```
                                                            [Continued]

```
      DO COND = 1 TO 2;
         DO TIME = 1 TO 3;
            WEIGHT = WTS[COND,TIME];
            OUTPUT;
         END;
      END;
      DROP WT1-WT6;
   RUN;
```

To cover all combinations of condition and time, you use "nested" DO loops (a DO loop within a DO loop). Here's how it works: COND is first set to 1 by the outer loop. Next, TIME is set to 1, 2, and 3 while COND remains at 1. Each time a COND and TIME combination is selected, a WEIGHT is set equal to the appropriate array element, and the observation is written out to the new data set.

E.   CREATING A DATA SET WITH ONE OBSERVATION PER SUBJECT FROM A DATA SET WITH MULTIPLE OBSERVATIONS PER SUBJECT

It's now time to reverse the restructuring process. We will do the reverse of Example 2 to demonstrate how to create a single observation from multiple observations. This time we start with data set MANYPER and create data set ONEPER. First the program, then the explanation:

```
   *--------------------------------------------------------------*
   | Example 4A: Creating a data set with one observation per     |
   | subject from a data set with multiple observations per       |
   | subject. (Caution: This program will not work if there       |
   | are any missing time values.)                                |
   *--------------------------------------------------------------*;
PROC SORT DATA=MANYPER;
   BY ID TIME;
RUN;
DATA ONEPER;
   ARRAY S[3] S1-S3;
   RETAIN S1-S3;
   SET MANYPER;
   BY ID;
   S[TIME] = SCORE;
   IF LAST.ID THEN OUTPUT;
   KEEP ID S1-S3;
RUN;
```

First, you sort data set MANYPER to be sure that the observations are in ID and TIME order. In this example, data set MANYPER is already in the correct order, but the SORT procedure makes the program more general. Next, you need to create an array containing the variables you want in the ONEPER data set, namely, S1, S2, and S3. You can "play computer" to see how this program works. The first observation in data set MANYPER is:

```
ID = 01     TIME = 1     SCORE = 3
```

Therefore, S[TIME] will be S[1], representing the variable S1 and set to the value of SCORE, which is 3. Since LAST.ID is false, the OUTPUT statement does not execute. However, the value of S1 is retained. In the next observation, time is 2 and SCORE is 4, so the variable S2 is assigned a value of 4. Finally, the third and last observation is read for ID 01. S3 is set to the value of SCORE, which is 5, and since LAST.ID is true, the first observation in data set ONEPER is written. Everything seems fine. Almost.

What if data set MANYPER did not have an observation at all three values of time for each ID? Use the data set MANYPER2 shown next to see what would happen:

Data Set MANYPER2		
ID	**Time**	**Score**
01	1	3
01	2	4
01	3	5
02	1	7
02	3	9
03	1	6
03	2	5
03	3	4

Notice that ID number 02 does not have an observation with TIME=2. What will happen if you run program Example 4A? Since you retained the values of S1, S2, and S3 and never replaced the value of S2 for ID number 02, ID number 02 will be given the value of S2 from the previous subject! Not what you want. You must always be careful when you retain variables. To be sure that this will not happen, you need to set the values of S1, S2, and S3 to missing each time you encounter a new subject. This is readily accomplished by checking the value of FIRST.ID. The corrected program is:

```
*------------------------------------------------------------*
| Example 4B: Creating a data set with one observation per   |
| subject from a data set with multiple observations per     |
| subject. (corrected version)                               |
*------------------------------------------------------------*;
                                             [Continued]
```

```
PROC SORT DATA=MANYPER2;
   BY ID TIME;
RUN;
DATA ONEPER;
   ARRAY S[3] S1-S3;
   RETAIN S1-S3;
   SET MANYPER2;
   BY ID;
   IF FIRST.ID THEN DO I = 1 TO 3;
      S[I] = .;
   END;
   S[TIME] = SCORE;
   IF LAST.ID THEN OUTPUT;
   KEEP ID S1-S3;
RUN;
```

This program will now work correctly whether or not there are missing TIME values.

F. CREATING A DATA SET WITH ONE OBSERVATION PER SUBJECT FROM A DATA SET WITH MULTIPLE OBSERVATIONS PER SUBJECT USING A MULTIDIMENSIONAL ARRAY

This example is the reverse of Example 3. That is, you want to start from data set WT_MANY and wind up with data set WT_ONE. The solution to this problem is similar to Example 4, except that we use a multidimensional array. Instead of writing the program in two steps, as we did in Examples 4A and 4B, we present the general solution that will work whether or not there are any missing observations in the data set. Here is the program:

```
*----------------------------------------------------------------*
| Example 5: Creating a data set with one observation per        |
|    subject from a data set with multiple observations per      |
|    subject using a Multi-dimensional array                     |
*----------------------------------------------------------------*;
PROC SORT DATA=WT_MANY;
   BY ID COND TIME;
RUN;
DATA WT_ONE;
   ARRAY WT[2,3] WT1-WT6;
   RETAIN WT1-WT6;
   SET WT_MANY;
   BY ID;
```

```
    IF FIRST.ID THEN
        DO I = 1 TO 2;
            DO J = 1 TO 3;
                WT[I,J] = .;
            END;
        END;
    WT[COND,TIME] = WEIGHT;
    IF LAST.ID THEN OUTPUT;
    KEEP ID WT1-WT6;
RUN;
PROC PRINT DATA=WT_ONE;
    TITLE 'WT_ONE Again';
RUN;
```

You have seen how to restructure SAS data sets, going from one to many or from many to one observation per subject, using arrays. You may want to keep these examples handy for the next time you have a restructuring job to be done.

PROBLEMS

Remember, you can download all the data sets and programs for these problems from the web site: www.prenhall.com/cody

16.1 We have a data set called FROG, which looks like this:

ID	X1	X2	X3	X4	X5	Y1	Y2	Y3	Y4	Y5
01	4	5	4	7	3	1	7	3	6	8
02	8	7	8	6	7	5	4	3	5	6

We want a data set that has an observation for each subject (ID) at each time interval (X1 represents X at time 1, etc.). Write a program, using arrays, to accomplish this objective. The new data set (TOAD) should look like:

ID	Time	X	Y
01	1	4	1
01	2	5	7
01	3	4	3
01	4	7	6
01	5	3	8
02	1	7	5
02	2	7	4
02	3	8	3
02	4	6	5
02	5	7	6

Run the following program to create data set FROG:

```
DATA FROG;
    INPUT ID X1-X5 Y1-Y5;
DATALINES;
01  4  5  4  7  3  1  7  3  6  8
02  8  7  8  6  7  5  4  3  5  6
;
```

16.2 A questionnaire asks each respondent to select up to four reasons why they like to stay at Motel XYZ. These reasons are selected from a list of 20 reasons. You want a frequency count for each of the 20 reasons. Data from the original questionnaire is stored in data set QUES, created by running the following program:

```
DATA QUES;
    INPUT ID $ REASON1-REASON4;
DATALINES;
001 3 6 13 17
002 8 3 4 .
003 20 2 . .
004 8 4 20 19
;
```

Write a DATA step to restructure this data set and then run PROC FREQ to determine the frequencies.

*16.3 We have a data set (called STATE) that contains an ID variable and up to five states (two-letter codes) where an individual may have visited last year. Three observations from this data set are shown:

ID	STATE1	STATE2	STATE3	STATE4	STATE5
1	NY	NJ	PA	TX	GA
2	NJ	NY	CA	XX	XX
3	PA	XX	XX	XX	XX

As you can see, "XX" was used as a missing value. Write a program to (a) read these records and replace the values of "XX" with blanks, and (b) compute frequency counts showing how many people visited each state. Present the frequency list in decreasing order of frequency (use the ORDER=FREQ option of PROC FREQ).

Run the following program to create data set STATE:

```
DATA STATE;
    INFORMAT STATE1-STATE5 $ 2;
    INPUT ID STATE1-STATE5;
DATALINES;
1  NY  NJ  PA  TX  GA
2  NJ  NY  CA  XX  XX
3  PA  XX  XX  XX  XX
;
```

16.4 You have a long thin data set (THIN) created by running the following DATA step. You want a wide data set as shown in the following listing. Using arrays, write a program to do this.

```
DATA THIN;
   INPUT ID $ TIME X @@;
DATALINES;
001 1 10   001 2 12   001 3 15
004 1 17
003 1 14   003 2 18   003 3 22   003 4 28
002 1 18   004 2 28
;
```

A listing of data set WIDE should look like this:

```
Listing of Data Set WIDE

ID      X1      X2      X3      X4

001     10      12      15       .
002     18       .       .       .
003     14      18      22      28
004     17      28       .       .
```

*16.5 You have inherited an old SAS program (shown here) and want to convert it to one using explicit array subscripts. Rewrite the program to do this.

```
DATA OLDFASH;
   SET BLAH;
   ARRAY JUNK(J) X1-X5 Y1-Y5 Z1-Z5;
   DO OVER JUNK;
      IF JUNK = 999 THEN JUNK=.;
   END;
   DROP J;
RUN;
```

A Review of SAS Functions:

PART I. FUNCTIONS OTHER THAN CHARACTER FUNCTIONS

A. Introduction
B. Arithmetic and Mathematical Functions
C. Random Number Functions
D. Time and Date Functions
E. The INPUT and PUT Functions: Converting Numerics to Character and Character to Numeric Variables
F. The LAG and DIF Functions

A.   INTRODUCTION

Throughout this book we have used functions to perform a variety of tasks. We use a LOG function to transform variables, and we use various DATE functions to convert a date into an internal SAS date. We will see in this chapter that the SAS programming language has a rich assortment of functions that can greatly simplify some complex programming problems. Take a moment to browse through this chapter to see what SAS functions can do for you. A complete reference to SAS functions can be found in *SAS Functions by Example* (Cody, SAS Institute, Cary, NC, 2004. ISBN: 1-59407-378-7).

B.   ARITHMETIC AND MATHEMATICAL FUNCTIONS

The functions you are probably most familiar with are ones that perform arithmetic or mathematical calculations. Remember that all SAS functions are recognized as such because the function name is always followed by a set of parentheses containing one or more arguments (or possibly no arguments, TODAY() for example). This way, the program can always differentiate between a variable name and a function. Here is a short program that computes a new variable, called LOGLOS, which is the natural log of LOS (length of stay). This is a common way to "pull in the tail" of a distribution skewed to the right. We have:

```
DATA FUNC_EG;
   INPUT ID SEX $ LOS HEIGHT WEIGHT;
   LOGLOS = LOG(LOS);
DATALINES;
1 M 5 68 155
2 F 100 62 98
3 M 20 72 220
;
```

The new variable (LOGLOS) will be in the data set FUNC_EG, and its values will be the natural (base e) log of LOS. Note that a zero value for LOS will result in a missing value for LOGLOS. When zeros are possible values, and you still want a log transformation, it is common practice to add a small number (usually .5 or 1) to the variable before taking the log.

We now list some of the more common arithmetic and mathematical functions and their purposes:

Function Name	Action
LOG	Base e log
LOG10	Base 10 log
SIN	Sine of the argument (in radians)
COS	Cosine (in radians)
TAN	Tangent (in radians)
ARSIN	Arcsine (inverse sine) of argument (in radians)
ARCOS	Arccosine (in radians)
ARTAN	Arctangent (in radians)
INT	Drops the fractional part of a number
SQRT	Square root

Some functions can accommodate more than one argument. For example, the ROUND function can take two arguments, separated by commas. The first argument is the number to be rounded, and the second argument indicates the round-off unit. Here are some examples:

```
ROUND (X,1)      Round X to the nearest whole number
ROUND (X,.1)     Round X to the nearest tenth
ROUND (X,100)    Round X to the nearest hundred
ROUND (X,20)     Round X to the nearest twenty
```

(NOTE: If you omit the second argument of the ROUND function, it rounds to the nearest integer.)

Other functions operate on a list of arguments. A good example of this is the MEAN function. If we have a series of variables (say A, B, and C) for each subject, and we want the mean of these three numbers, we write:

```
MEAN_X = MEAN(A, B, C);
```

If you want to use the notation VAR1-VARn with functions such as MEAN or SUM, you must precede the variable list by the word OF, like this:

```
MEAN_X = MEAN(OF X1-X5);
```

The reason for this is that without the word OF, SAS would subtract X5 from X1 and then take the mean (of the single value).

We may use any variable list following the word OF. An important difference between the MEAN function and the alternative expression:

```
MEAN_X = (X1 + X2 + X3 + X4 + X5)/5;
```

is that the MEAN function returns the mean of the nonmissing values. Thus, if we had a missing value for X5, the function would return the mean of X1, X2, X3, and X4. Only if all the variables listed as arguments of the MEAN function are missing will the MEAN function return a missing value. Our equation for the previous mean would return a missing value if any of the X values were missing.

The MIN, MAX, SUM, STD, and STDERR functions work the same way as the MEAN function, except that a minimum, maximum, sum, standard deviation, or a standard error, respectively, is computed instead of a mean.

Two useful functions are N and NMISS. They return, as you would expect, the number of nonmissing (N) or the number of missing (NMISS) values in a list of variables. Suppose we have recorded 100 scores for each subject and want to compute the mean of those 100 scores. However, we want to compute the mean score only if there are 75 or more nonmissing responses. Without the N function, we would have to do a bit of programming. Using the N function, the computation is much simpler:

```
DATA EASYWAY;
    INPUT (X1-X100)(2.);
    IF N(OF X1-X100) GE 75 THEN
    AVE = MEAN(OF X1-X100);
DATALINES;
(lines of data)
;
```

The NMISS function is used in a similar fashion.

C. RANDOM NUMBER FUNCTIONS

In Chapter 6, we saw how we could use random numbers to assign subjects to groups. The function RANUNI will generate uniform random numbers in the range from 0 to 1. Random-number generators (more properly called pseudorandom number generators) require an initial number, called a seed, used to calculate the first random number. From then on, each random number is used in some way to generate the next. A zero seed will cause the function to use a seed derived from the time clock, thus generating a different random series each time it is used. RANUNI can also be seeded with any number of your choosing. If you supply the seed, the function will generate the same series of random numbers each time. A simple example follows where we use a uniform random number to put a group of subjects in random order:

```
DATA SHUFFLE;
    INPUT NAME : $20.;
    X = RANUNI(0);
DATALINES;
CODY
SMITH
MARTIN
LAVERY
THAYER
;
PROC SORT DATA=SHUFFLE;
    BY X;
RUN;
PROC PRINT DATA=SHUFFLE;
    TITLE "Names in Random Order";
    VAR NAME;
RUN;
```

To generate a series of random numbers from n to m, we need to scale our 0 to 1 random numbers accordingly. To generate a series of random numbers from 1 to 100, we can write:

```
X = 1 + 99*RANUNI(0);
```

To generate integers from 1 to 10, you can use the expression:

```
X = INT(RANUNI(0)*10 + 1);
```

You might expect this expression to give you integers from 1 to 11, but remember that the largest value of RANUNI is 1, and getting this value exactly would be very rare. So, even if the value of RANUNI is .99999, multiplying by 10 and adding 1 and throwing away any fractional part yields integers from 1 to 10. (Go ahead and try generating, say, 10,000 numbers using the previous expression and running PROC FREQ to see the frequency of the generated values. They should be about equal. Take care in using the ROUND function instead of the INT function for this purpose. If you round to the nearest integer, values from 0 to .5 give you a zero, values from .5 to 1.5 give you a 1, and so forth. Thus, there will be fewer zeros and fewer maximum values if you use ROUND. Take care.)

For purposes of statistical modeling, we might also want a series of random numbers chosen from a normal distribution (mean = 0, variance = 1). The RANNOR function will generate such variables. The allowable seeds for RANNOR follow the same rules as for RANUNI. You can scale these values to any mean and standard deviation. To generate normally distributed values with a mean of M and a standard deviation of S, you would use:

```
NORMAL = RANNOR(seed)*S + M;
```

D. TIME AND DATE FUNCTIONS

We saw some examples of date functions in Chapter 4. We summarize the time and date functions here.

There are several extremely useful date functions. One, MDY (month, day, year), will convert a month, day, and year value into a SAS date variable. Suppose, for example, that you have a data set with variables MONTH, DAY, and YEAR and you want to create a SAS date value. We could use the MDY function to compute the date, thus:

```
SAS_DATE = MDY(MONTH,DAY,YEAR);
```

While we are on the topic of dates, remember that a SAS date constant is always represented by a two-digit day, a three-letter month, and a two- or four-digit year, all placed within single or double quotes, followed by an upper- or lowercase 'D'. For example, to compute the age of a person as of May 15, 2004, who had a date of birth stored in the variable DOB, you would write:

```
AGE = YRDIF(DOB,'15MAY2004'D,'ACTUAL');
```

The YRDIF function computes the number of years from the first argument to the second. The value of 'ACTUAL' for the third argument tells SAS to use the actual number of days in each month and to count leap years.

The MDY function can also be used to create a SAS date where the dates are not in a form that can be read using any of the SAS date informats. For example, the following program reads in values for month, day, and year from noncontiguous columns and computes a SAS date:

```
DATA DATES;
   INPUT ID 1-3 MONTH 4-5 DAY 10-11 YEAR 79-80;
   DATE = MDY (MONTH,DAY,YEAR);
   DROP MONTH DAY YEAR;
   FORMAT DATE MMDDYY8.;
DATALINES;
(data lines)
;
```

There are several date functions that extract information from a SAS date. For example, the YEAR function returns a four-digit year from a SAS date. The MONTH function returns a number from 1 to 12, which represents the month for a given date. There are two functions that return day information. The DAY function returns the day of the month (i.e., a number from 1 to 31) and the WEEKDAY function returns the day of the week (a number from 1 to 7, 1 being Sunday). As an example, suppose we want to see distributions by month and day of the week for hospital admissions. The variable ADMIT is a SAS date variable:

```
PROC FORMAT;
    VALUE DAYWK 1='SUN' 2='MON' 3='TUE' 4='WED' 5='THU'
                6='FRI' 7='SAT';
    VALUE MONTH 1='JAN' 2='FEB' 3='MAR' 4='APR' 5='MAY' 6='JUN'
                7='JUL' 8='AUG' 9='SEP' 10='OCT' 11='NOV' 12='DEC';
RUN;
DATA HOSP;
    INPUT @1 ADMIT MMDDYY8. etc. ;
    DAY = WEEKDAY(ADMIT);
    MONTH = MONTH(ADMIT);
    FORMAT ADMIT MMDDYY8. DAY DAYWK. MONTH MONTH.;
DATALINES;
(data lines)
;
PROC GCHART DATA=HOSP;
    VBAR DAY / DISCRETE;
    VBAR MONTH / DISCRETE;
RUN;
```

Later in this chapter, look for a shortcut method for producing the day of the week or month name in the discussion of the PUT function.

Besides working with date values, SAS has a corresponding set of functions to work with "time." For example, we can read a time in "military" form (from 00:00 to 24:00), in hh:mm:ss (hours, minutes, seconds), or in hh:mm (hours and minutes) format using the time8. informat. We can then extract hour, minute, or second information from the time variable using the HOUR, MINUTE, or SECOND functions, just the way we used the YEAR, MONTH, and WEEKDAY functions previously.

Before we leave the date and time functions, let's discuss two very useful functions, INTCK and INTNX. They may save you pages of SAS coding. INTCK returns the number of intervals between any two dates. Valid, interval choices are: DAY, WEEK, MONTH, QTR, YEAR, HOUR, MINUTE, and SECOND. The syntax of the INTCK function is:

```
INTCK(interval,start,end);
```

Interval is one of the choices here, placed within single or double quotes; start is the starting date; and end is the ending date. As an example, suppose we want to know now many quarters our employees have worked. If START is the SAS variable that holds the starting date, the number of quarters worked would be:

```
NUM_QTR = INTCK('QTR',START,TODAY());
```

(NOTE: the TODAY function, which has no argument, returns today's date.)

Since the algorithms used to compute the number of intervals can be confusing, we recommend reading the chapter on SAS date functions in *SAS Functions by Example* (see reference in section A).

The INTNX function can be thought of as the inverse of the INTCK function; it returns a date, given an interval, starting date, and the number of intervals elapsed. Suppose we know the date of hire and want to compute the date representing the start of the third quarter. You would use:

```
DATE3RD = INTNX('QTR',HIRE,3);
FORMAT DATE3RD MMDDYY8.;
```

If HIRE were 01/01/90, 01/05/90, or 03/30/90, the value of DATE3RD would be 10/01/90. If a person was hired on April 1, 1990, his/her third-quarter date would be 01/01/91.

E.   THE INPUT AND PUT FUNCTIONS: CONVERTING NUMERIC TO CHARACTER AND CHARACTER TO NUMERIC VARIABLES

While the INPUT and PUT functions (not to be mistaken for INPUT and PUT statements) have many uses, one common application is to convert numeric and character values.

The PUT function uses the formatted value of a variable to create a new variable. For example, suppose we have recorded the AGE of each subject. We also have a format that places the ages into four groups:

```
PROC FORMAT;
   VALUE AGEGRP LOW-20='1' 21-40='2' 41-60='3' 61-HIGH='4';
RUN;
DATA PUTEG;
   INPUT AGE @@;
   AGE4 = PUT (AGE,AGEGRP.);
DATALINES;
5 10 15 20 25 30 66 68 99
;
```

In this example, the variable AGE4 is a character variable with values of '1', '2', '3', or '4'. As another example, suppose we want a variable to contain the three-letter day of the week abbreviations (MON, TUE, etc.). One of the SAS built-in formats is

WEEKDATEw., which returns values such as WEDNESDAY, SEPTEMBER 12, 1990 (if we use WEEKDATE29.). The format WEEKDATE3. is the first three letters of the day name (SUN, MON, etc.). To create our character day variable, we can use the PUT function:

```
DAYNAME = PUT(DATE,WEEKDATE3.);
```

There are some useful tricks that can be accomplished using the PUT function. Consider one file that has Social Security numbers as nine-digit numerics. Another file has the Social Security numbers coded as 11-digit character strings (123-45-6789). Our job is to merge the two files, based on the Social Security numbers. There are many ways to solve this problem, either removing the dashes in the character representation of the Social Security number and converting the result to a numeric value (see the next paragraph) or converting the nine-digit number to a character string and placing the dashes in the proper places. One method is to use the fact that SAS has a built-in format, SSN11., that formats nine-digit numerics to 123-45-6789-style Social Security numbers. Therefore, using the PUT function, we can create a character variable in the form 123-45-6789, like this:

```
SS = PUT(ID,SSN11.);
```

In general, to convert a numeric variable to a character variable, we can use the PUT function with the appropriate format. If we have a file where group is a numeric variable and we want a character variable instead, we can write:

```
GROUPCHR = PUT(GROUP,1.);
```

We use the INPUT function in a similar manner, except that we can "reread" a variable according to a new informat. The most common use of the function is to convert character values into numeric values. There are several examples of such conversions in the last section of Chapter 12, "Reading 'Unstructured' Data." We will show you two examples here.

In the first example, we convert a character representation of a Social Security number (in the form 123-45-6789) into a nine-digit numeric, the reverse of the previous PUT example. First, we have to remove the dashes from the character variable. We use the COMPRESS function to do this. COMPRESS takes the form:

```
CHAR_VAR = COMPRESS(CHAR_VAR,'list_of_characters');
```

If the second argument, list_of_characters, is omitted, the COMPRESS function will, by default, remove blanks from a character value. To remove the dashes from a Social Security (SS) number, we write:

```
CHAR_VAR = COMPRESS(SS,'-');
```

To create a numeric variable, we can use the INPUT function to perform the character to numeric conversion:

```
ID = INPUT(COMPRESS(SS,'-'),9.);
```

where ID is a SAS numeric variable.

For the second example, we want to read data values that may either represent groups ('A' or 'B') or numeric scores. Since we don't know if we will be reading a character or a number, we read every value as a character and test if it is an 'A' or a 'B'. If not, we assume it is a score and use the INPUT function to convert the character variable to numeric. Here is the code:

```
DATA FREEFORM;
   INPUT TEST $ @@;
   RETAIN GROUP;
   IF TEST='A' OR TEST='B' THEN DO;
      GROUP = TEST;
      DELETE;
   END;
   ELSE SCORE = INPUT(TEST, 5.);
   DROP TEST;
DATALINES;
A 45 55 B 87 A 44 23 B 88 99
;
PROC PRINT DATA=FREEFORM NOOBS;
   TITLE "Listing of Data Set FREEFORM";
RUN;
```

To help make this example clearer, the data set formed by running this program is shown here:

```
Listing of Data Set FREEFORM

GROUP       SCORE

  A           45
  A           55
  B           87
  A           44
  A           23
  B           88
  B           99
```

As you can see, the INPUT function provides a flexible way of reading data.

F. THE LAG AND DIF FUNCTIONS

A "lagged" value is one from an earlier time. In SAS, we may want to compare a data value from a current observation with a value from a previous observation. We may also want to look back several observations. Without the LAG function, this is a difficult task—with it, it's simple. The value returned by the LAG function is the value of the argument the last time the function was executed (see a more complete explanation in Chapter 19, Section H). In more complicated DATA steps, this can be very tricky. If we execute the LAG function for each observation, then the value of the LAG function will be the value of the argument in the previous observation. The value LAGn (X) where n is a number, is the value from the nth previous execution of the LAG function. A common application of the LAG function is to take differences between observations. Suppose each subject was measured twice, and each measurement was entered as a separate observation. We want to take the value of X at time 1 and subtract it from the value of X at time 2. We proceed as follows:

Data set ORIG looks like this:

SUBJ	TIME	X
1	1	4
1	2	6
2	1	7
2	2	2
etc.		

To subtract the X at time 1 from the X at time 2, we write:

```
DATA LAGEG;
   SET ORIG;
   ***Note: Data Set ORIG is Sorted by SUBJ and TIME;
   DIFF = X - LAG(X);
   IF TIME = 2 THEN OUTPUT;
RUN;
```

You could shorten this program even further by using the DIFn function, which returns the difference between a value from the current observation and the nth previous observation. The previous calculation would be:

```
DIFF = DIF(X);
```

Chapter 19, Section H, shows how to use the LAG function to compute moving averages.

PROBLEMS

Remember, you can download all the data sets and programs for these problems from the web site: www.prenhall.com/cody

17.1 You have a SAS data set called HOSP, that contains a patient ID, gender, date of birth (DOB), date of service (DOS), length of stay (LOS), systolic blood pressure (SBP), diastolic blood pressure (DBP), and heart rate (HR). Run the following program to create this data set:

```
DATA HOSP;
    INFORMAT ID $3. GENDER $1. DOB DOS MMDDYY8.;
    INPUT ID GENDER DOB DOS LOS SBP DBP HP;
    FORMAT DOB DOS MMDDYY10.;
DATALINES;
1 M 10/21/46 3/17/97 3 130 90 68
2 F 11/1/55 3/1/97 5 120 70 72
3 M 6/6/90 1/1/97 100 102 64 88
4 F 12/21/20 2/12/97 10 180 110 86
;
```

Create a new data set (NEW_HOSP) that contains all of the variables in HOSP, plus the following:

(a) The base 10 log of LOS (call it LOG_LOS).

(b) The patient's age, as of his/her last birthday, on the date of service (call it AGE_LAST).

(c) A new variable (X) computed as the square root of the mean of the systolic and diastolic blood pressures, rounded to the nearest tenth.

17.2 First run the following program to create a temporary SAS data set called FUNCTIONS.

```
DATA FUNCTIONS;
    INPUT @1  SUBJECT      $3.
          @4  DOB      MMDDYY10.
          @14 VISIT    MMDDYY10.
          @23 (SCORE1-SCORE6)(2.);
    FORMAT DOB VISIT MMDDYY10.;
DATALINES;
00110/21/195011/11/2004908070757688
00205/05/200312/20/200499  98  9790
00307/15/194107/06/2004        9896
00406/24/193709/25/2004777879808182
;
```

Using this data set, create a new data set (QUES2) with the following variables: SUBJ, AGE_AT_VISIT (age at the time of the visit, rounded to the nearest tenth of a year), AGE_JAN (age as of January 1, 2004), AVE_SCORE (the mean of SCORE1 to SCORE6. This variable should be computed only if there are four or more nonmissing scores.), MIN_SCORE (the lowest, nonmissing score), MAX_SCORE (the largest score), and DAY (day of the week corresponding to the VISIT date). Format this variable so that it prints as Mon, Tue, Wed, etc.

17.3 A data set (MANY) contains the variables X1–X5, Y1–Y5. First, run the following program to create this data set:

```
DATA MANY;
    INPUT X1-X5 Y1-Y5;
DATALINES;
1 2 3 4 5 6 7 8 9 10
3 . 5 . 7 5 . . . 15
9 8 . . . 4 4 4 4 1
;
```

Write a program to include the following in data set MANY:

(a) The mean (average) of the Xs (call it MEAN_X) and the mean of the Ys (call it MEAN_Y).

(b) The minimum value of the Xs (call it MIN_X) and the minimum value of the Ys (call it MIN_Y).

(c) A new variable (CRAZY) that is the maximum of the Xs times the minimum of the Ys times the sum of (the number of nonmissing Xs plus the number of missing Ys). In other words:

CRAZY=(Maximum of X'S) × (Minimum of Y's) ×
(Number of nonmissing X's + Number of missing Y's).

(d) Compute a variable MEAN_X_Y that is the mean of all the Xs and Ys (the mean of all 10 numbers) with the provision that there be three or more nonmissing X values and four or more nonmissing Y values.

17.4 Run the following program to create the SAS data set BIG. Using this data set, create a SAS data set TEN_PERCENT that is an approximate 10% random sample of BIG.

```
DATA BIG;
   DO SUBJ = 1 TO 100;
      X = INT(RANUNI(123)*100 + 1);
      OUTPUT;
   END;
RUN;
```

*17.5 Create a SAS data set (UNI) that contains 1000 random numbers in the range of 1 to 5. Use the INT function to give you only integers, and be sure that there is an equal likelihood of choosing each of the five integers. Be careful that the values of 1 and 5 have the same probability of being chosen as 2, 3, and 4. Run PROC FREQ to count the number of 1s, 2s, and so forth, and compute chi-square using the tables option TESTP=(list-of-proportions). This TABLES option allows you to enter a list of test proportions for a one-way test. For this problem, the tables statement would read:

```
TABLES N / NOCUM TESTP=(.2 .2 .2 .2 .2);
```

By the way, there is also a TESTF option where you enter frequencies (counts) rather than proportions.

17.6 Using data set BIG from problem 17.4, create a data set EXACT that contains exactly 20 observations from BIG, at random.

17.7 Use the data set HOSP from problem 17.1 and create two vertical bar charts: one for the day of the week (formatted please) and one for the month of the year (no need to format) of the date of service.

17.8 Write a DATA step that creates a SAS data set (ASSIGN) containing 100 observations with variable SUBJ (from 1 to 100) and GROUP, which is either 'A' or 'B', where there is a 50-50 chance of being in either group. Run PROC FREQ to determine how many subjects are in group 'A' and how many are in group 'B'. They may not be equal.

17.9 Run the following program to create a SAS data set called MIXED:

```
DATA MIXED;
   INPUT X Y A $ B $;
DATALINES;
1 2 3 4
5 6 7 8
;
```

Create a new data set (NUMS) containing all four variables (you can use a new name for A and B) with only numeric variables.

*17.10 Run the following program to create a SAS data set NUM_CHAR. Using this data set, create a new SAS data set (CORRECT) as follows: X, Y, and Z are numeric variables (yes, you have to use the same variable names. HINT: Swap and DROP); DATE is a real SAS date (i.e., the number of days from January 1, 1960, format it with DATE9.); NUMERAL is a character variable (make the length 8); CHAR_DATE is a character variable with the value of DOB (but as a character string in the MMDDYY10. format).

```
DATA NUM_CHAR;
    INPUT X $ Y $ Z $ DATE : $10. NUMERAL DOB : DATE9.;
    FORMAT DOB MMDDYY10.;
DATALINES;
10 20 30 10/21/1946 123 09SEP2004
1 2 3 11/11/2004 999 01JAN1960
;
```

*17.11 Using the data set NEW_HOSP created in problem 17.1, create a new character variable called AGEGROUP, with the following groupings:

```
1 = Ages less than 20 (but not missing)
2 = 20 to 40
3 = 41 to highest
```

Do this with a user-defined format and a PUT function.

C H A P T E R   1 8

A Review of SAS Functions:

PART II. CHARACTER FUNCTIONS

A.   INTRODUCTION

SAS software is rich in its assortment of functions that deal with character data. This class of functions is sometimes called STRING functions. In this chapter, we demonstrate some of the more useful string functions.

Some of the functions we discuss are: VERIFY, TRANSLATE, TRANWRD, COMPRESS, COMPBL, LENGTH, SUBSTR, INPUT, PUT, SCAN, TRIM, UPCASE, LOWCASE, REPEAT, ‖ (concatenation), INDEX, INDEXC, AND SOUNDEX. Did you realize there were so many string functions? For more information on SAS character functions, we refer you to *SAS Functions by Example,* (Cody, SAS Institute, Cary, NC, 2004). Let's get started.

B. HOW LENGTHS OF CHARACTER VARIABLES ARE SET IN A SAS DATA STEP

Before we actually discuss these functions, we need to understand how SAS assigns storage lengths to character variables and what the LENGTH function does for us. Look at the following program:

```
DATA EXAMPLE1;
    INPUT GROUP $ @10 STRING $3.;
    LEFT = 'X '; *X AND 4 BLANKS;
    RIGHT = ' X'; *4 BLANKS AND X;
    C1 = SUBSTR(GROUP,1,2) ;
    C2 = REPEAT(GROUP,1);
    LGROUP = LENGTH(GROUP) ;
    LSTRING = LENGTH(STRING) ;
    LLEFT = LENGTH(LEFT);
    LRIGHT = LENGTH(RIGHT);
    LC1 = LENGTH(C1);
    LC2 = LENGTH(C2);
DATALINES;
ABCDEFGH 123
XXX 4
Y 5
;
PROC CONTENTS DATA=EXAMPLE1 POSITION;
    TITLE "Output from PROC CONTENTS";
RUN;
PROC PRINT DATA=EXAMPLE1 NOOBS;
    TITLE "Listing of Example 1";
RUN;
```

One purpose of this example is to clarify the term LENGTH. If you look at the output from PROC CONTENTS, each of the variables is listed, along with a TYPE and LENGTH. Take a moment and look at a partial listing from PROC CONTENTS:

```
The CONTENTS Procedure

Variables in Creation Order

#    Variable    Type    Len

1    GROUP       Char      8
2    STRING      Char      3
```

```
 3    LEFT      Char      2
 4    RIGHT     Char      2
 5    C1        Char      8
 6    C2        Char    200
 7    LGROUP    Num       8
 8    LSTRING   Num       8
 9    LLEFT     Num       8
10    LRIGHT    Num       8
11    LC1       Num       8
12    LC2       Num       8
```

The column labeled LEN (length) is the number of bytes needed to store the values for each of the variables listed. By default, all the numeric variables are stored in eight bytes.

But what about the storage lengths of the character variables? Look first at the two variables listed in the INPUT statement: GROUP and STRING. Since this is the first mention of these variables in this DATA step, their lengths are assigned by the rules governing INPUT statements. Since no columns or informats were associated with the variable GROUP, its length is set to eight bytes, the default length for character variables in this situation. The variable STRING uses a $3. INFORMAT so its length will be set to three bytes. The length of LEFT and RIGHT are determined by the assignment statement. The storage lengths of C1 and C2 are more difficult to understand.

The variable C1 is defined to be a substring of the variable GROUP. The SUBSTR function takes the form:

```
SUBSTR(char_var,start,length);
```

This function says to take a substring from char_var, starting at the position indicated by the start argument, for a length indicated by the length argument. Why then, is the length of C1 equal to 8 and not 2? The SAS compiler determines lengths at compile time. Since the starting position and length arguments of the SUBSTR function can be variable expressions, the compiler must set the length of C1 equal to the largest possible value it can ever attain, the length of GROUP.

The same kind of logic controls the length of C2, defined by the REPEAT function. Since the number of addition replications is defined by the second argument of the REPEAT function, and this argument can be a variable expression, the compiler sets the length of C2 to the largest possible value, 200. Why 200? Because that was the maximum length of a character variable in version 6 of the SAS system, and it seemed unreasonable to make the default length 32,767, the maximum length of a character variable from version 8 on.

There is a lesson here: Always use a LENGTH statement for any character variables that do not derive their length elsewhere. For example, to set the length of C2 to 16, you would write:

```
LENGTH C2 $ 16;
```

The LENGTH function does not, as you might guess, return the storage length of a character variable. Instead, it returns the position of the right-most nonblank character. Thus, trailing blanks are excluded in its computation.

The value of LLEFT and LRIGHT are 1 and 5, respectively, for every observation. This demonstrates that the trailing blanks in LEFT are not counted by the LENGTH function, while the leading blanks in RIGHT are. The following table summarizes the lengths for the remaining variables:

GROUP	LGROUP	STRING	LSTRING	C1	LC1	C2		LC2
ABCDEFGH	8	123	3	AB	2	ABCDEFGHABCDEFGH		16
XXX	3		1	XX	2	XXX	XXX	11
Y	1		1	Y	1	Y	Y	9

The values of LGROUP and LSTRING are straightforward. The value of LC1 is 1 for the third observation since C1 is only 1 byte in length in the third observation. The values for LC2 are more complicated. The REPEAT function says to take the original value and repeat it n times. So, for the first observation, LC2 is 16 (2 times 8). For observations 2 and 3, the trailing blanks come into play. In observation 2, the value of GROUP is 'XXXbbbbb' (where the b's stand for blanks). When we repeat this string one additional time, we get: 'XXXbbbbbXXXbbbbb'. Not counting the trailing blanks, we have a length of 8 + 3 = 11. Using the same logic for the third observation, we have a 'Y' followed by seven blanks, repeated once. Not counting the last seven trailing blanks, we have a length of 8 + 1 = 9.

With these preliminaries out of the way, we can now begin our tour of some of the very useful string functions available in SAS software.

C. WORKING WITH BLANKS

This example demonstrates how to convert multiple blanks to a single blank. Suppose you have some names and addresses in a file. Some of the data-entry clerks placed extra spaces between the first and last names and in the address fields. You prefer to store all names and addresses with single blanks. Here is an example of how this conversion is accomplished:

```
DATA EXAMPLE2;
    INPUT #1 @1 NAME $20.
          #2 @1 ADDRESS $30.
          #3 @1  CITY $15.
             @20 STATE $2.
             @25 ZIP $5.;
    NAME = COMPBL(NAME);
    ADDRESS = COMPBL(ADDRESS);
    CITY = COMPBL(CITY);
DATALINES;
```

```
RON CODY
89 LAZY BROOK ROAD
FLEMINGTON NJ 08822
BILL BROWN
28 CATHY STREET
NORTH CITY NY 11518
;
PROC PRINT DATA=EXAMPLE2;
    TITLE "Example 2";
    ID NAME;
    VAR ADDRESS CITY STATE ZIP;
RUN;
```

Thus, a seemingly difficult task is accomplished in a single line by using the COMPBL (COMPress BLank) function, compressing multiple blanks to a single blank. How useful!

D.   HOW TO REMOVE CHARACTERS FROM A STRING

A more general problem is the removal (deletion) of selected characters from a string. For example, suppose you want to remove blanks, parentheses, and dashes from a phone number that has been stored as a character value. Here comes the COMPRESS function to the rescue! The COMPRESS function can remove any number of specified characters from a character variable. The following program uses the COMPRESS function twice. The first time to remove blanks from the string; the second time to remove blanks plus the other characters previously mentioned. Here is the code:

```
DATA EXAMPLE3;
    INPUT PHONE $ 1-15;
    PHONE1 = COMPRESS(PHONE);
    PHONE2 = COMPRESS(PHONE,'(-) ');
DATALINES;
(908)235-4490
(201) 555-77 99
;
PROC PRINT DATA=EXAMPLE3;
TITLE "Listing of Example 3";
RUN;
```

The variable PHONE1 has just blanks removed. Notice that the COMPRESS function does not have a second argument here. When it is omitted, the COMPRESS function removes only blanks. For the variable PHONE2, the second argument of the COMPRESS function contains a list of the characters to remove: left parenthesis, dash, right parenthesis, and blank. This string is placed within single or double quotes.

E. **CHARACTER DATA VERIFICATION**

A common task in data processing is to validate data. For example, you may want to be sure that only certain values are present in a character variable. In the following example, only the values 'A', 'B', 'C', 'D', and 'E' are valid data values. A very easy way to test if there are any invalid characters present is as follows:

```
DATA EXAMPLE4;
   INPUT ID      $ 1-4
         ANSWER $ 5-9;
     P = VERIFY(ANSWER,'ABCDE');
     OK = (P EQ 0);
DATALINES;
001 ACBED
002 ABXDE
003 12CCE
004 ABC E
;
PROC PRINT DATA=EXAMPLE4 NOOBS;
TITLE "Listing of Example 4";
RUN;
```

The "workhorse" of this example is the VERIFY function, which is a bit complicated. It inspects every character in the first argument, and if it finds any value not in the verify string (the second argument), it will return the position of the first offending value. If all the values of the string are located in the verify string, a value of 0 is returned. In the first observation, P will be 0 and OK will be 1; in the second observation, P will be 3 (the position of the 'X') and OK will be 0; in the third observation, P will be 1 and OK will be 0; finally, in the fourth observation, P will be 4 and OK will be 0.

Another way to solve the same problem is the following: Suppose someone gave you the data set EXAMPLE5, created by running the following short DATA step:

```
DATA EXAMPLE5;
    INPUT STRING $3.;
DATALINES;
ABC
EBX
aBC
VBD
;
```

To list any observation with values for STRING that are not the letters A–E or blank, the following DATA step could be used:

```
DATA _NULL_;
    SET EXAMPLE5;
    FILE PRINT;
    CHECK = 'ABCDE';
    IF VERIFY(STRING,CHECK) NE 0 THEN
        PUT 'Error in Record ' _N_ STRING=;
RUN;
```

If some values of STRING are less than three characters in length, the previous program will flag them as errors (VERIFY will return the position of the first trailing blank). If you want to prevent this from happening, you can use the TRIM function to remove the trailing blanks before looking for invalid values, like this:

```
IF VERIFY(TRIM(STRING),CHECK) NE 0 THEN
    PUT 'Error in Record ' _N_ STRING=;
```

F.   SUBSTRING EXAMPLE

We mentioned in the Introduction that a substring is a part of a longer string (although it can actually be the same length, but this would not be too useful). In this example, you have ID codes that contain a state abbreviation in the first two positions. Furthermore, in positions 7–9 is a numeric code. You want to create two new variables: one containing the two-digit state codes, and the other containing a numeric variable constructed from the three numerals in positions 7, 8, and 9. Here goes:

```
DATA EXAMPLE6;
    INPUT ID $ 1-9;
    LENGTH STATE $ 2;
    STATE = SUBSTR(ID,1,2);
    NUM = INPUT(SUBSTR(ID,7,3),3.);
DATALINES;
NYXXXX123
NJ1234567
;
PROC PRINT DATA=EXAMPLE6 NOOBS;
    TITLE "LISTING OF EXAMPLE 6";
RUN;
```

Creating the state code is easy. We use the SUBSTR function. The first argument is the variable from which we want to extract the substring; the second argument is the starting position of the substring; and the last argument is the length of the substring (not the ending position as you might guess). Also note the use of the LENGTH statement to set the length of STATE to 2 bytes.

Extracting the three-digit number code is more complicated. First, we use the SUBSTR function to extract the three numerals (numerals are character representations of numbers). However, the result of a SUBSTR function is always a character value. To convert the character value to a number, we use the INPUT function. The INPUT function takes the first argument and "reads" it as if it were coming from a file, according to the INFORMAT listed as the second argument. So, for the first observation, the SUBSTR function would return the string '123', and the INPUT function would convert this to the number 123. As a point of interest, you may use a longer INFORMAT as the second argument without any problems. For example, the INPUT function could have been written as:

```
INPUT(SUBSTR(ID,7,3),8.);
```

and everything would have worked out fine. This fact is useful in situations where you do not know the length of the string ahead of time.

G.   USING THE SUBSTR FUNCTION ON THE LEFT-HAND SIDE OF THE EQUAL SIGN

There is a particularly useful and somewhat obscure use of the SUBSTR function that we would like to discuss next. You can use this function to place characters in specific locations within a string by placing the SUBSTR function on the left-hand side of the equal sign (in the older SAS manuals we believe this was called a SUBSTR pseudo function).

Suppose you have some systolic blood pressures (SBP) and diastolic blood pressures (DBP) in a SAS data set. You want to print out these values and star high values with an asterisk. Here is a program that uses the SUBSTR function on the left of the equal sign to do that:

```
DATA EXAMPLE7;
   INPUT SBP DBP @@;
   LENGTH SBP_CHK DBP_CHK $ 4;
   SBP_CHK = PUT(SBP,3.);
   DBP_CHK = PUT(DBP,3.);
   IF SBP GT 160 THEN SUBSTR(SBP_CHK,4,1) = '*';
   IF DBP GT 90 THEN SUBSTR(DBP_CHK,4,1) = '*';
DATALINES;
120 80 180 92 200 110
;
PROC PRINT DATA=EXAMPLE7 NOOBS;
   TITLE "Listing of Example 7";
RUN;
```

We first need to set the lengths of SBP_CHK and DBP_CHK to four (three spaces for the value plus one for the possible asterisk). Next, we use a PUT function to perform a numeric to character conversion. The PUT function is, in some ways, similar to the INPUT function. It "writes out" the value of the first argument, according to the FORMAT specified in the second argument. By "write out" we actually mean assign the value to the variable on the left of the equal sign. The SUBSTR function then places an asterisk in the fourth position when a value of SBP is greater than 160 or a value of DBP is greater than 90.

H. DOING THE PREVIOUS EXAMPLE ANOTHER WAY

It is both interesting and instructive to obtain the previous results without using the SUB-STR function on the left-hand side of the equal sign. We are not doing this just to show you a hard way to accomplish something we already did. Rather, this alternative solution uses a number of character functions that can be demonstrated. Here is the program:

```
DATA EXAMPLE8;
   INPUT SBP DBP @@;
   LENGTH SBP_CHK DBP_CHK $ 4;
   SBP_CHK = PUT(SBP,3.);
   DBP_CHK = PUT(DBP,3.);
   IF SBP GT 160 THEN SBP_CHK = SUBSTR(SBP_CHK,1,3) || '*';
   IF DBP GT 90 THEN DBP_CHK = TRIM(DBP_CHK) || '*';
DATALINES;
120 80 180 92 200 110
;
PROC PRINT DATA=EXAMPLE8 NOOBS;
   TITLE "Listing of Example 8";
RUN;
```

This program is not really more complicated but maybe just not as elegant as the first program. This program uses the concatenation operator (||) to join the three-character blood-pressure value with an asterisk. Since SBP_CHK and DBP_CHK were both assigned a length of 4, we wanted to be sure to concatenate, at most, the first 3 bytes with the asterisk. Just for didactic purposes, we did this two ways. For the SBP_CHK variable, we used a SUBSTR function to extract the first 3 bytes. For the DBP_CHK variable, the TRIM function was used. The TRIM function, as previously mentioned, removes trailing blanks from a character string.

I. UNPACKING A STRING

To save disk storage, you may wish to store several single-digit numbers in a longer character string. For example, storing five numbers as numeric variables with the default 8 bytes each would take up 40 bytes of disk storage per observation. Even reducing this to 3 bytes each would result in 15 bytes of storage. If, instead, you store the five digits as a single character value, you need only 5 bytes. (NOTE: With today's multigigabyte hard drives, this may seem unnecessary. However, it is still worthwhile to think about and to minimize storage when possible.)

That is fine, but at some point, you may need to get the numbers back out for computation purposes. Here is a nice way to do this:

```
DATA EXAMPLE9;
   INPUT STRING $ 1-5;
DATALINES;
12345
8 642
;
DATA UNPACK;
   SET EXAMPLE9;
   ARRAY X[5];
   DO J = 1 TO 5;
      X[J] = INPUT(SUBSTR(STRING,J,1),1.);
   END;
   DROP J;
RUN;
PROC PRINT DATA=UNPACK NOOBS;
   TITLE "Listing of UNPACK";
RUN;
```

We first created an array to hold the five numbers X1 to X5. Don't be alarmed if you don't see any variables listed on the ARRAY statement. ARRAY X[5]; is equivalent to ARRAY X[5] X1-X5; We use a DO loop to cycle through each of the five starting positions corresponding to the five numbers we want. As mentioned before, since the result of the SUBSTR function is a character value, we need to use the INPUT function to perform the character to numeric conversion.

J.   PARSING A STRING

"Parsing" a string means to take it apart based on some rules. In the example to follow, five separate character values are placed together on a line, with either a space, a comma, a semicolon, a period, or an explanation mark between them. You would like to extract the five values and assign them to five character variables. With the SCAN function this difficult task is simplified:

```
DATA EX_10;
   INPUT LONG_STR $ 1-80;
   ARRAY PIECES[5] $ 10 PIECE1-PIECE5;
   DO I = 1 TO 5;
      PIECES[I] = SCAN(LONG_STR,I,',;.! ');
   END;
```

```
    DROP LONG_STR I;
DATALINES4;
THIS LINE,CONTAINS!FIVE.WORDS
ABCDEFGHIJKL XXX;YYY
;;;;
PROC PRINT DATA=EX_10 NOOBS;
    TITLE "Listing of Example 10";
RUN;
```

Before we discuss the SCAN function, a brief word about DATALINES4 and the four semicolons ending our data. If you have data values that include semicolons, you cannot use a simple DATALINES (or CARDS) statement since the semicolon would signal the end of your data. Instead, the statement DATALINES4 (or CARDS4) is used, causing the program to continue reading data values until four semicolons are read.

The function:

```
SCAN(char_var,n,'list-of-delimiters');
```

returns the nth "word" from the char_var, where a "word" is defined as anything between two delimiters. If there are fewer than n words in the character variable, the SCAN function will return a blank. If n is negative, SCAN will process the string from right to left.

By placing the SCAN function in a DO loop, we can pick out the nth word in the string.

K.   LOCATING THE POSITION OF ONE STRING WITHIN ANOTHER STRING

Two somewhat similar functions, INDEX and INDEXC, can be used to locate a string or one of several strings within a larger string. For example, if you have a string 'ABCDEFG' and want the location of the letters DEF (starting position 4), the following INDEX function could be used:

```
INDEX('ABCDEFG','DEF');
```

This would return a value of 4. If you want to know the starting position of any one of several characters, the INDEXC function can be used. As an example, if you wanted the starting position of either 'B', 'C', 'F', or 'G' in the string 'ABCDEFG', you would code:

```
INDEXC('ABCDEFG','BCFG');
```

or equivalently:

```
INDEXC('ABCDEFG','B','C','F','G');
```

The function would return a value of 2, the starting position of 'B'. Here is a short program that demonstrates these two functions:

```
DATA EX_11;
   INPUT STRING $ 1-10;
   FIRST = INDEX(STRING,'XYZ');
   FIRST_C = INDEXC(STRING,'X','Y','Z');
DATALINES;
ABCXYZ1234
1234567890
ABCX1Y2Z39
ABCZZZXYZ3
;
PROC PRINT DATA=EX_11 NOOBS;
   TITLE "Listing of Example 11";
RUN;
```

FIRST and FIRST_C for each of the 4 observations are:

```
Listing of Example 11

   STRING          FIRST          FIRST_C

ABCXYZ1234          4              4
1234567890          0              0
ABCX1Y2Z39          0              4
ABCZZZXYZ3          7              4
```

When the search fails, both functions return a zero.

L.   CHANGING LOWERCASE TO UPPERCASE AND VICE VERSA

The two companion functions UPCASE and LOWCASE do just what you would expect—they convert lowercase to uppercase and vice versa. These two functions are especially useful when data-entry clerks are careless and a mixture of upper- and lowercase values are entered for the same variable. You may want to place all of your character variables in an array and UPCASE (or LOWCASE) them all. Here is an example of such a program:

```
DATA EX_12;
   LENGTH A B C D E $ 1;
   INPUT A B C D E X Y;
DATALINES;
```

```
M f P p D 1 2
m f m F M 3 4
;
DATA UPPER;
   SET EX_12;
   ARRAY ALL_C[*] _CHARACTER_;
   DO I = 1 TO DIM(ALL_C);
      ALL_C[I] = UPCASE(ALL_C[I]);
   END;
   DROP I;
RUN;
PROC PRINT DATA=UPPER NOOBS;
   TITLE "Listing of UPPER";
RUN;
```

This program uses the _CHARACTER_ keyword to select all the character variables. The result of running this program is to convert all values for the variables A, B, C, D, and E to uppercase. The LOWCASE function could be used in place of the UPCASE function if you wanted all your character values in lowercase.

M.   SUBSTITUTING ONE CHARACTER FOR ANOTHER

A very handy character function is TRANSLATE. It can be used to convert one character to another in a string. For example, suppose you recorded multiple choices on a test as 1, 2, 3, 4, or 5, which represented the letters 'A' through 'E', respectively. When you print out the character values, you want to see the letters rather than the numerals. While formats along with PUT statements would accomplish this very nicely, it also serves as an example for the TRANSLATE function. Here is the code:

```
DATA EX_13;
   INPUT QUES : $1. @@;
   QUES = TRANSLATE(QUES,'ABCDE','12345');
DATALINES;
1 4 3 2 5
5 3 4 2 1
;
PROC PRINT DATA=EX_13 NOOBS;
   TITLE "LISTING OF EXAMPLE 13";
RUN;
```

The syntax for the TRANSLATE function is:

```
TRANSLATE(char_var,to_string,from_string);
```

Each value in from_string is translated to the corresponding value in the to_string.

Another interesting application of the TRANSLATE function is the creation of dichotomous numeric variables from character variables. For example, you may wish to set values of 'N' to 0 and values of 'Y' to 1. Although this is easily done with IF-THEN/ELSE statements, let's see if we can do it using the TRANSLATE function. Here goes:

```
DATA EX_14;
   LENGTH CHAR $ 1;
   INPUT CHAR @@;
   X = INPUT(TRANSLATE(UPCASE(CHAR),'01','NY'),1.);
DATALINES;
N Y n y A B 0 1
;
PROC PRINT DATA=EX_14 NOOBS;
   TITLE "Listing of Example 14";
RUN;
```

The UPCASE function sets all values to uppercase. Next, the TRANSLATE function converts values of 'N' to '0' and 'Y' to '1'. Finally, the INPUT function converts the numerals '0' and '1' to the numbers 0 and 1, respectively.

N.   SUBSTITUTING ONE WORD FOR ANOTHER IN A STRING

The TRANWRD (translate word) function can perform a search and replace operation on a string variable. For example, you may want to standardize addresses by converting the words 'Street', 'Avenue', and 'Road' to the abbreviations 'St.', 'Ave.', and 'Rd.', respectively. Look at the following program:

```
DATA EX_15
DATA CONVERT;
   INPUT @1 ADDRESS $20. ;
   ***Convert Street, Avenue, and Boulevard to
      their abbreviations;
   ADDRESS = TRANWRD (ADDRESS,'Street','St.');
   ADDRESS = TRANWRD (ADDRESS,'Avenue','Ave.');
   ADDRESS = TRANWRD (ADDRESS,'Road','Rd.');
DATALINES;
```

```
89 Lazy Brook Road
123 River Rd.
12 Main Street
;
PROC PRINT DATA=CONVERT;
   TITLE "Listing of Data Set CONVERT";
RUN;
```

The syntax of the TRANWRD function is:

```
TRANWRD(char_var,'find_string','replace_string');
```

That is, the function will replace every occurrence of find_string with replace_string. Notice that the order of the find and replace strings are reversed compared to the TRANSLATE function where the to_string comes before the from_string as arguments to the function. In this example, 'Street' will be converted to 'St.', 'Avenue' to 'Ave.', and 'Road' to 'Rd.'. The following listing confirms this fact:

```
Listing of Data Set CONVERT

Obs          ADDRESS

 1           89 Lazy Brook Rd.
 2           123 River Rd.
 3           12 Main St.
```

O. CONCATENATING (JOINING) STRINGS

Since we are discussing strings, we should mention the concatenation operation. Although this is not a function, it is a useful string operation, and this seems as good as anywhere to tell you about it! In computer jargon, "concatenate" means to join. So, if we concatenate the string 'ABC' with the string 'XYZ', the result is 'ABCXYZ'. Pretty clever, eh? Things can get "a bit sticky" if we forget that SAS character variables may be padded on the right with blanks to fill out the predefined length of the variable. If this is the case, we can use the TRIM function to remove trailing blanks before we concatenate the strings. The concatenation operator is || (two vertical bars). Suppose we have a SAS data set with a separate variable for first name (FIRST) and last name (LAST), and we want a single variable NAME consisting of the last name, a comma, and a blank, followed by the first name. Take a look at the following program that first creates the SAS data set with variables FIRST and LAST and then the following

DATA step creates the NAME variable:

```
DATA ONE;
   LENGTH FIRST LAST $ 15;
   INPUT FIRST LAST;
DATALINES;
Ron Cody
Elizabeth Cantor
Ralph Fitzpatrick
;
DATA CONVERT;
   SET ONE;
   LENGTH NAME $ 32;
   NAME = TRIM(LAST) || ', ' || FIRST;
RUN;
PROC PRINT DATA=CONVERT NOOBS;
   TITLE "Listing of Data Set CONVERT";
RUN;
```

The output listing is:

```
Listing of Data Set CONVERT

FIRST                LAST                 NAME

Ron                  Cody                 Cody, Ron
Elizabeth            Cantor               Cantor, Elizabeth
Ralph                Fitzpatrick          Fitzpatrick, Ralph
```

A simpler solution available in SAS version 9 and above is to use the CATX as follows:

```
NAME = CATX(', ',LAST,FIRST);
```

CATX strips leading and trailing blanks from the strings to be concatenated, and it inserts the separator value indicated in the first argument to the function between the two strings.

P.   SOUNDEX CONVERSION

SAS software provides a soundex function that returns the soundex equivalent of a name. Soundex equivalents of names allow you to match two names from two different sources even though they might be spelled differently. Great care is needed when using this function since many very dissimilar-looking names may translate to the same soundex equivalent.

The soundex equivalent of most names will result in strange-looking codes such as C3 or A352. Here is a sample program and the results of the soundex translations:

```
DATA EX_16;
   LENGTH NAME1-NAME3 $ 10;
   INPUT NAME1-NAME3;
   S1 = SOUNDEX(NAME1);
   S2 = SOUNDEX(NAME2);
   S3 = SOUNDEX(NAME3);
DATALINES;
cody Kody cadi
cline klein clana
smith smythE adams
;
PROC PRINT DATA=EX_16 NOOBS;
   TITLE "Listing of Example 16";
RUN;
```

This program will result in the following listing:

```
Listing of Example 16

NAME1           NAME2           NAME3       S1      S2      S3

cody            Kody            cadi        C3      K3      C3
cline           klein           clana       C45     K45     C45
smith           smythE          adams       S53     S53     A352
```

Q.   SPELLING DISTANCE: THE SPEDIS FUNCTION

Another useful function for performing "fuzzy" matches between text strings is the SPEDIS (spelling distance) function. This is a fairly complicated function, and we refer you to *SAS Functions by Example* (Cody, SAS Institute, Cary, NC, 2004) for a more detailed discussion of the text-matching functions. The SPEDIS function returns the spelling distance between two text strings. If the strings match exactly, the function returns a 0. For each type of spelling mistake (such as interchanging two letters or adding an extra letter, etc.) the function assigns penalty points. The farther apart two strings are, the larger the value returned by the SPEDIS function. The following program demonstrates this function:

```
DATA SPELLING_DISTANCE;
   INFORMAT WORD_ONE WORD_TWO $15.;
   INPUT WORD_ONE WORD_TWO;
   DISTANCE = SPEDIS(WORD_ONE,WORD_TWO);
DATALINES;
```

[Continued]

```
Exact Exact
Mistake Mistaken
abcde acbde
abcde uvwxyz
123-45-6789 123-54-6789
;
PROC PRINT DATA=SPELLING_DISTANCE NOOBS;
   TITLE "Listing of Data Set SPELLING_DISTANCE";
RUN;
```

Here is the listing:

```
Listing of Data Set SPELLING_DISTANCE

WORD_ONE          WORD_TWO          DISTANCE

Exact             Exact                0
Mistake           Mistaken             7
abcde             acbde               10
abcde             uvwxyz             130
123-45-6789       123-54-6789          4
```

PROBLEMS

Remember, you can download all the data sets and programs for these problems from the web site: www.prenhall.com/cody

18.1 First, create a data set called CHAR1 by running the following DATA step:

```
DATA CHAR1;
   INPUT @1  STRING1 $1.
         @2  STRING2 $5.
         @3  STRING3 $8.
         @11 (C1-C5) ($1.);
DATALINES;
XABCDE12345678YNYNY
YBBBBB12V56876yn YY
ZCCKCC123-/. ,WYNYN
;
```

Create a new data set ERROR containing any observations in CHAR1 that meet any one of the following conditions:

1. A value other than an 'X', 'Y', or 'Z' for the variable STRING1.
2. A value other than an 'A', 'B', 'C', 'D', or 'E' in STRING2.
3. A value other than an upper- or lowercase 'N' or 'Y' for the variables C1–C5. (NOTE: Blank values of C1–C5 will place the observation in the ERROR data set.)

18.2 Run the following DATA step to create the data set ADDRESS. Using this as input, do the following:

    1. Replace all multiple blanks by a single blank.
    2. Change all Street to St., Road to Rd., Place to Pl.
    3. Remove all Mr. Dr. Mrs. and Ms.
    4. Left adjust all character values.

```
DATA ADDRESS;
   INPUT #1 @1 LINE1 $50.
         #2 @1 LINE2 $50.
         #3 @1 LINE3 $50.;
DATALINES;
Mr. Jason     Simmons
123   Sesame Street
Madison, WI
Dr.   Justin  Case
78       River Road
Flemington, NJ
Ms.   Marilyn Crow
777 Jewell    Place
;
```

18.3 Using the data set CHAR1 from problem 18.1, create a new variable (NEW3) from the variable STRING3 based on the following rules: First, remove embedded blanks and the characters '–', '/', '.', and ','. Next, substitute the letters A–H for the numerals 1–8. Finally, set NEW3 equal to a missing value if there are any characters other than A–H (trailing blanks are OK) in the string. The value of NEW3 for the three observations should be: 'ABCDEFGH', missing, and 'ABC'.

18.4 Run the following program to create data set STRING. Create variable X (numeric) from the first two positions in the string and variable Y (numeric) from the third position (for example, in line 1, X would be equal to 12 and Y would be equal to 3). Extract characters 4 and 5 and call it STATE (character). Make sure the STATE value is uppercase. Positions 6–10 are used to create five numeric variables N1–N5. If you are bold, you can use an array to create N1 to N5.

```
DATA STRING;
   INPUT STRING $10.;
DATALINES;
123nj76543
892NY10203
876pA83745
;
```

18.5 For the variables C1–C5 in data set CHAR1 (problem 18.1), change all lowercase values to uppercase, and set any remaining values other than 'Y', 'N', or blank to missing. You may wish to use an ARRAY to solve this problem.

*18.6   Run the following DATA step to create a SAS data set called GOOD_BAD. Valid characters in the string ANSWER are ABC, abc, and 0123456789. Note that ANSWER is 40 bytes in length, and you want to ignore trailing blanks (embedded blanks are to be flagged as errors). Write a SAS DATA step to read in observations from GOOD_BAD, and create two data sets, VALID and INVALID. Place any invalid values of ANSWER in INVALID and place valid values in VALID. Lines 1 and 3 will wind up in VALID; the remaining lines will be in INVALID. (HINT: The TRIM and VERIFY functions wll be useful.)

```
DATA GOOD_BAD;
    INPUT ANSWER $40.;
DATALINES;
1324AcB876acccCCC
123 456
aabbccAABBCC123123
abcde12345
invalid
;
```

18.7   You are given a data set (PHONE) with phone numbers stored in a variety of formats. Run the following program to create the PHONE data set and create a new variable, NUMBER, that is a numerical variable derived from the character variable CHAR_NUM with all extraneous symbols and blanks removed. Assume that the longest phone number contains 10 digits.

```
DATA PHONE;
    INPUT CHAR_NUM $20.;
DATALINES;
(908)235-4490
(800) 555 - 1 2 1 2
203/222-4444
;
```

18.8   You have two data sets, ONE and TWO. ONE has a variable called ID, which is a character variable in the form 123-45-6789. TWO has a variable called ID, which is numeric (nine digits). You want to merge data sets ONE and TWO by the ID variables to create a new data set (BOTH) with ID (numeric and formatted with the SSN11. format), NAME, and SCORE. You can run the following short program to create the two data sets:

```
DATA ONE;
    INPUT @1  ID   $11.
          @12 NAME $15.;
DATALINES;
```

```
123-45-6789Jeff Smith
111-22-3333Stephen King
999-88-7777Jan Chambliss
;
DATA TWO;
    INPUT @1  ID 9.
          @11 SCORE 3.;
DATALINES;
999887777 100
123456789  65
111223333  59
;
```

18.9 The data set (EXPER) created by running the following program contains the variables ID, GROUP, and DOSE. Create two new variables as follows: One is a 2-byte character variable created from the second and third bytes of ID. The other is a variable created by concatenating GROUP and DOSE. This variable should be 6 bytes long with a '-' between the GROUP and DOSE values, and it should not contain any embedded blanks. Call the two variables SUB_ID and GRP_DOSE, respectively:

```
DATA EXPER;
    INPUT ID      $ 1-5
          GROUP   $ 7
          DOSE    $ 9-12;
DATALINES;
1NY23 A HIGH
3NJ99 B HIGH
2NY89 A LOW
5NJ23 B LOW
;
```

*18.10 You have lines of data containing subject numbers (SUBJ) and weights (WEIGHT). The weight values are either in kilograms, in which case they are followed by kg (in upper- or lowercase, with or without a period). Weights in pounds either have nothing following the number or lbs. (again, in upper- or lowercase, with or without a period, and with or without the 'S'). Read these lines of data and create a SAS data set (WT) with all weights in pounds (numeric). Note that one kilogram is equal to 2.2 pounds. Here are some sample lines of data (note there are several subject numbers and weights on each line). Some hints: Read in all of the weights as a character variable. Change all of these values to uppercase. To get the number part of each weight, you can compress all of the characters (KGLBS and period) in one statement. You only have to test to see if there is a 'K' in the string to see if you need to multiply by 2.2. Remember to convert from character to numeric (don't let SAS do an automatic conversion).

```
1 50kg  2 120  3 121Lbs.  4 88KG.  5 200
6 80kG  7 250lb
```

*18.11 Using the data set EXPER from problem 18.9, create a new variable (ID2) from ID. Make this new variable 6 bytes in length and place an asterisk in the sixth byte if the fourth byte of ID has a value of 5 or more. To solve this problem, create a numeric variable based on the fourth byte of ID, and check if it is greater than or equal to 5. Do not check this byte as a character value against the numerals 5–9.

*18.12 **(a)** Write a SAS program to read the following lines of data. You want this data set to contain a variable called NUMBER, which is a two-digit number following the first occurrence of the string 'XYZ' in each of the lines. For example, in line 1, the value of NUMBER is 78; in line 2, it has a missing value; in line 3 NUMBER is 12 and in line 4 it is 98.

```
ABC123XYZ7823
NONE HERE
XYZ12345
12345XYZ9876
```

(b) This one is a bit harder. Write a SAS program to read the following four lines of data. This time, you are looking for up to two, one-digit numbers following the letters 'X', 'Y', or 'Z'. Call the first number N1 and the second number N2. Note that these variables should be SAS numeric values. HINT: In line 1, N1 is 7 and N2 is 5; in line 2, N1 and N2 are both missing. In line 3, N1 is 8 and N2 is missing; in line 4, N1 is 8 and N2 is 9.

```
B4Y7999Z5V8
NONE HERE
ONLY ONE X8
Z8Z9
```

*18.13 Merge the two data sets created by running the following program using the GENDER, date of birth (DOB), and the SOUNDEX equivalent of the NAME to determine matches. Keep only those observations where a match is successful. Call the new data set COMBINED. There should be three observations in data set COMBINED.

```
DATA ONE;
    INPUT @1  GENDER$1.
          @2 DOBMMDDYY8.
          @10 NAME$11.
          @21 STATUS$1.;
    FORMAT DOB MMDDYY8.;
DATALINES;
```

```
M10/21/46CADYA
F11/11/50CLINEB
M11/11/52SMITHA
F10/10/80OPPENHEIMERB
M04/04/60JOSEA
;
DATA TWO;
    INPUT @1  GENDER$1.
          @2  DOBMMDDYY8.
          @10 NAME$11.
          @21 WEIGHT3.;
    FORMAT DOB MMDDYY8.;
DATALINES;
M10/21/46CODY160
F11/11/50CLEIN102
F11/11/52SMITH101
F10/10/80OPPENHAIMER120
M02/07/60JOSA220
;
```

18.14 Run the following program to create the SAS data set VERSE. The variable LONG_STRING contains several words separated by blanks or blanks and commas. Count how many times the word 'the' appears in each line. Ignore case. First do this by breaking LONG_STRING into words and testing each word (do this in a loop—exit the loop when there are no more words in the string). Assume the longest word is 10 bytes in length. If you have version 9 or above, use the COUNT function to do the same task.

```
DATA VERSE;
    INPUT LONG_STRING $50.;
DATALINES;
The time was early and the sky was still dark
This line does not contain any
Four the last time, the man walked the plank
;
```

Selected Programming Examples

A. INTRODUCTION

This chapter contains a number of common applications and serves two functions: One is to allow you to use any of the programs here, with modification, if you have a similar application; the other is to demonstrate SAS programming techniques.

B. EXPRESSING DATA VALUES AS A PERCENTAGE OF THE GRAND MEAN

A common problem is to express data values as percentages of the grand mean, rather than in the original raw units. In this example, we record the heart rate (HR), systolic blood pressure (SBP), and diastolic blood pressure (DBP) for each subject. We want to express these values as percentages of the mean HR, SBP, and DBP for all subjects. For example, in the following data set, the mean heart rate for all subjects is 70 (mean of 80, 70, and 60). The first subject's score of 80, expressed as a percentage, would be $100\% \times 80/70 = 114.28\%$.

The approach here will be to use PROC MEANS to compute the means and to output them in a data set, which we then combine with the original data so that we can perform the required computation.

```
DATA TEST;
   INPUT HR SBP DBP;
DATALINES;
80 160 100
70 150 90
60 140 80
;
PROC MEANS NOPRINT DATA=TEST;
   VAR HR SBP DBP;
   OUTPUT OUT=MOUT(DROP=_TYPE_ _FREQ_)
        MEAN=M_HR M_SBP M_DBP;
RUN;
DATA NEW;
   SET TEST;
   IF _N_ = 1 THEN SET MOUT;
   HRPER = 100*HR/M_HR;
   SBPPER = 100*SBP/M_SBP;
   DBPPER = 100*DBP/M_DBP;
   DROP M_HR M_SBP M_DBP;
RUN;
PROC PRINT DATA=NEW NOOBS;
   TITLE "Listing of Data Set NEW";
RUN;
```

Description

We use the NOPRINT option with PROC MEANS because we do not want the procedure to print anything but, rather, to create a data set of means. In this case, the output data set from PROC MEANS will consist of one observation. The variables _TYPE_ and _FREQ_, which are normally added to the output data set, are dropped (using a DROP= data set option) since they are not needed. _TYPE_ is useful when a CLASS statement is used with PROC MEANS. We show examples of this later. The single observation in the data set MOUT is shown here:

```
Listing of Data Set MOUT

M_HR     M_SBP     M_DBP

 70       150        90
```

We want to add the three variables M_HR, M_SBP, and M_DBP to every observation in the original data set so that we can divide each value by the mean and multiple by 100%. Since our original data set has three observations and the mean data set contains only one observation, we use a trick: We use the SAS internal variable _N_ to conditionally execute the SET statement. The program data vector (PDV) now contains the variables HR, SBP, DBP, M_HR, M_SBP, and M_DBP. Any variables coming in from a SET statement are automatically retained so the values of M_HR, M_SBP, and M_DBP will not change and will be available for every iteration of the DATA step. We can now divide HR by M_HR, SBP by M_SBP, and DBP by M_DBP. Each time we bring in another observation from the original (TEST) data set, the values of M_HR, M_SBP, and M_DBP will remain in the program data vector because they are retained. The final data set created by this program (NEW) is shown next:

```
Listing of Data Set NEW

HR    SBP    DBP    HRPER     SBPPER    DBPPER

80    160    100    114.286   106.667   111.111
70    150    90     100.000   100.000   100.000
60    140    80     85.714    93.333    88.889
```

C.   EXPRESSING A VALUE AS A PERCENTAGE OF A GROUP MEAN

This example is an extension of the previous problem. Here we have two groups (A and B), and we want to express each measurement as a percentage of the GROUP mean.

```
DATA TEST;
    INPUT GROUP $ HR SBP DBP @@;
DATALINES;
A 80 160 100 A 70 150 90 A 60 140 80
B 90 200 180 B 80 180 140 B 70 140 80
;
PROC SORT DATA=TEST;
    BY GROUP;
RUN;
PROC MEANS DATA=TEST NOPRINT NWAY;
    CLASS GROUP;
    VAR HR SBP DBP;
    OUTPUT OUT=MOUT(DROP=_TYPE_ _FREQ_)
           MEAN=M_HR M_SBP M_DBP;
RUN;
DATA NEW;
    MERGE TEST MOUT;
    BY GROUP;
    HRPER = 100*HR/M_HR;
    SBPPER = 100*SBP/M_SBP;
    DBPPER = 100*DBP/M_DBP;
    DROP M_HR M_SBP M_DBP;
RUN;
```

```
PROC PRINT DATA=NEW NOOBS;
   TITLE "Listing of Data Set NEW";
RUN;
```

Description

Since the output data set from PROC MEANS contains GROUP, we can MERGE it with the original data set, using GROUP as the BY variable. Data set MOUT will contain two observations, one for each value of GROUP. The contents of MOUT are shown here:

```
Listing of Data Set MOUT

GROUP    M_HR     M_SBP      M_DBP

  A       70     150.000     90.000
  B       80     173.333    133.333
```

Data set NEW, which contains both the original values and the percentage values is shown next:

```
Listing of Data Set NEW

GROUP    HR    SBP    DBP     HRPER      SBPPER     DBPPER

  A      80    160    100    114.286    106.667    111.111
  A      70    150     90    100.000    100.000    100.000
  A      60    140     80     85.714     93.333     88.889
  B      90    200    180    112.500    115.385    135.000
  B      80    180    140    100.000    103.846    105.000
  B      70    140     80     87.500     80.769     60.000
```

D. PLOTTING MEANS WITH ERROR BARS

When we plot a set of means, we may want to include error bars, which represent one standard error above and below the mean. The following program shows how we can use PROC GPLOT to produce a graph of means with error bars. NOTE: Please refer to the appropriate SAS Graph™ manual (or the Online Doc™) for more details on PROC GPLOT and the SYMBOL statement options.

```
DATA ORIG;
   INPUT SUBJ TIME DBP SBP;
DATALINES;
```

[Continued]

```
1 1 70 120
1 2 80 130
1 3 84 136
2 1 82 132
2 2 84 138
2 3 92 144
;
SYMBOL1 VALUE=NONE I=STD1MT COLOR=BLACK LINE=1 WIDTH=2;
PROC GPLOT DATA=ORIG;
    TITLE "Plot of Means with Error Bars";
    FOOTNOTE JUSTIFY=LEFT HEIGHT=1
        "Bars represent plus and minus one standard error";
    PLOT (SBP DBP)*(TIME);
RUN;
QUIT;
```

Here is the first of the two plots (SBP by TIME) produced by this program:

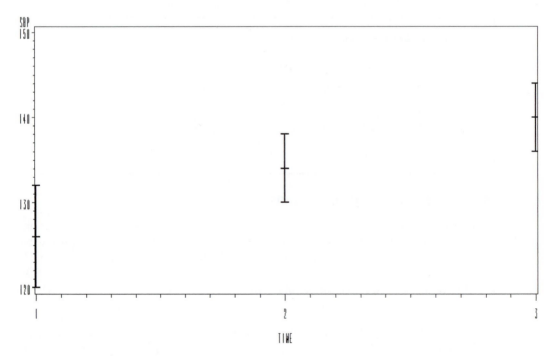

Plot of Means with Error Bars

Bars represent plus and minus one standard error

Description

The options on the SYMBOL statement specify the following:

VALUE = NONE	Do not use any plotting symbols for the raw data values
I = STD1MT	The interpolation method is standard error; the 1 specifies one standard error (alternatives are 2 and 3), M computes the standard error of the mean, and T adds a top and bottom to each line.
COLOR = BLACK	Sets the color of the plot to black
LINE = 1	Specifies a line style
WIDTH = 2	Controls the line thickness (higher numbers are darker)

E.   USING A MACRO VARIABLE TO SAVE CODING TIME

Programmers are always looking for a way to make their programs more compact (and to avoid typing). While there is an extensive macro language as part of the SAS system, we will use only a macro variable in this example. A macro variable can be assigned a value with a %LET function. The expression to the right of the equals sign will be assigned to the macro variable. We precede the macro variable name with an ampersand (&) in the program so that the system knows we are referring to a macro variable. In this example, we use a macro variable to take the place of a variable list. Any time we want to refer to the variables ONE, TWO, and THREE, we can use the macro variable LIST instead. (See Chapter 11, Section C for another example of a macro variable.)

```
%LET LIST = ONE TWO THREE;

DATA TEST;
    INPUT &LIST FOUR;
DATALINES;
1  2  3  4
4  5  6  6
;
PROC FREQ DATA=TEST;
   TABLES &LIST;
RUN;
```

F.   COMPUTING RELATIVE FREQUENCIES

In this example, we have an ICD (International Classification of Diseases) code for each subject as well as the year that diagnosis was made. We want to know what percentage of the observations contain a particular code for each year. That is, we want to express the frequency of each ICD code as a percentage of all codes for that year. For example, in the following data, there were three recorded codes in 1950 (note they are not in YEAR order). Code 450 occurred twice, giving us a relative frequency of 2/3, or 66.6%. We will use an output data set from PROC FREQ to compute these relative frequencies, as follows:

```
DATA ICD;
   INPUT ID YEAR ICD;
DATALINES;
001 1950 450
002 1950 440
003 1951 460
004 1950 450
005 1951 300
;
PROC FREQ DATA=ICD;
   TABLES YEAR*ICD / OUT=ICDFREQ(DROP=PERCENT) NOPRINT;
   ***Data set ICDFREQ contains the counts
   for each CODE in each YEAR;
   TABLES YEAR / OUT=TOTAL(DROP=PERCENT) NOPRINT;
   ***Data set TOTAL contains the total number
   of obs for each YEAR;
RUN;
DATA RELATIVE;
   MERGE ICDFREQ TOTAL(RENAME=(COUNT=TOT_CNT));
   ***We need to rename COUNT in one of the two data sets
   so that we can have both values in data set RELATIVE;
   BY YEAR;
   RELATIVE = 100*COUNT/TOT_CNT;
RUN;
PROC PRINT DATA=RELATIVE;
   TITLE "Relative Frequencies of ICD Codes by Year";
RUN;
```

Description

The first PROC FREQ TABLES statement creates an output data set (ICDFREQ) that looks like the following:

```
Listing of Data Set ICDFREQ

YEAR    ICD     COUNT

1950    440      1
1950    450      2
1951    300      1
1951    460      1
```

Notice that the data set created by PROC FREQ contains all the TABLES variables as well as the variable COUNT (named by PROC FREQ). The COUNT variable in this data set tells us how many times a given ICD code appeared in each year.

Next, we need the total number of ICD codes for each year to be used in the denominator to create a relative incidence of each ICD code for each year. A TABLES statement with only YEAR as a TABLE variable will give us the number of ICD codes there were for each year. Data set TOTAL is shown next:

```
Listing of Data Set TOTAL

YEAR     COUNT

1950       3
1951       2
```

All we have to do now is to merge the two data sets so that we can divide the COUNT variable in the ICDFREQ data set by the COUNT variable in the TOTAL data set. When we do the merging, we will rename COUNT in the TOTAL data set to TOT_CNT since we can't have two values for a single variable in one observation. Finally, we can divide COUNT by TOT_CNT to obtain our desired result.

```
Relative Frequencies of ICD Codes by Year

Obs    YEAR    ICD     COUNT    TOT_CNT    RELATIVE

 1     1950    440       1         3        33.3333
 2     1950    450       2         3        66.6667
 3     1951    300       1         2        50.0000
 4     1951    460       1         2        50.0000
```

G.   COMPUTING COMBINED FREQUENCIES ON DIFFERENT VARIABLES

In this example, questionnaires are issued to people to determine to which chemicals they are sensitive. Each subject replies yes or no (1 or 0) to each of the 10 chemicals in the list. We want to list the chemicals in decreasing order of sensitivity. If we compute frequencies for each of the 10 variables, we will be unable to display a list showing the chemicals in decreasing order of frequency. Our first step is to restructure the data set into one with up to 10 observations per subject. Each observation will include a chemical number (from 1 to 10) indicating which chemical was selected. Here is the program:

```
PROC FORMAT;
    VALUE SYMPTOM 1='ALCOHOL'
                  2='INK'
                  3='SULPHUR'
                  4='IRON'
                  5='TIN'
                                                    [Continued]
```

```
                    6='COPPER'
                    7='DDT'
                    8='CARBON'
                    9='SO2'
                   10='NO2';
RUN;
DATA SENSI;
   INPUT ID 1-4 (CHEM1-CHEM10)(1.);
   ARRAY CHEM[*]   CHEM1-CHEM10;
   DO I=1 TO 10;
      IF CHEM[I] = 1 THEN DO;
         SYMPTOM = I;
         OUTPUT;
      END;
   END;
   KEEP ID SYMPTOM;
   FORMAT SYMPTOM SYMPTOM.;
DATALINES;
00011010101010
00021000010000
00031100000000
00041001001111
00051000010010
;
PROC FREQ DATA=SENSI ORDER=FREQ;
   TABLES SYMPTOM / NOCUM;
RUN;
```

Description

For a more detailed description of how to restructure data sets using arrays, see the relevant section of Chapter 16. Data set SENSI will have as many observations per person as the number of 1s on the list of chemicals (CHEM1 to CHEM10). The variable SYMPTOM is set equal to the DO loop counter, I, which tells us which of the 10 chemicals was selected. The observations from data set SENSI are shown here to help clarify this point:

```
Listing of Data Set SENSI

ID         SYMPTOM

 1         ALCOHOL
 1         SULPHUR
 1         TIN
 1         DDT
 1         SO2
```

```
Listing of Data Set SENSI

ID          SYMPTOM

  2         ALCOHOL
  2         COPPER
  3         ALCOHOL
  3         INK
  4         ALCOHOL
  4         IRON
  4         DDT
  4         CARBON
  4         SO2
  4         NO2
  5         ALCOHOL
  5         COPPER
  5         SO2
```

Notice that the formatted values for SYMPTOM are displayed since we assigned a format to the variable. A simple PROC FREQ will now tell us the frequencies for each of the 10 chemicals. The ORDER=FREQ option of PROC FREQ will produce a frequency list in decreasing order of frequencies. The output from PROC FREQ is as follows:

```
Listing of Data Set SENSI

The FREQ Procedure

SYMPTOM     Frequency      Percent

ALCOHOL            5        27.78
SO2                3        16.67
COPPER             2        11.11
DDT                2        11.11
CARBON             1         5.56
INK                1         5.56
IRON               1         5.56
NO2                1         5.56
SULPHUR            1         5.56
TIN                1         5.56
```

H.   COMPUTING A MOVING AVERAGE

Suppose we record the COST of an item each day. For this example, we want to compute a moving average of the variable COST for a three-day interval. On day 3, we will average the COST for days 1, 2, and 3; on day 4, we will average the COST for days 2, 3, and 4, etc.

```
***Program to compute a moving average;
DATA MOVING;
    INPUT COST @@;
    DAY + 1;
    COST1 = LAG(COST);
    COST2 = LAG2(COST);
    IF _N_ GE 3 THEN DO;
        MOV_AVE = MEAN(COST, COST1, COST2);
        OUTPUT;
    END;
    DROP COST1 COST2;
DATALINES;
1 2 3 4 . 6 8 12 8
;
PROC PRINT DATA=MOVING NOOBS;
    TITLE 'Listing of Data Set MOVING';
RUN;
```

The data set MOVING is:

```
Listing of Data Set MOVING

COST    DAY    MOV_AVE

   3      3    2.00000
   4      4    3.00000
   .      5    3.50000
   6      6    5.00000
   8      7    7.00000
  12      8    8.66667
   8      9    9.33333
```

Description

The LAG function returns the value of the argument the last time the function was executed. If you place the LAG function in a DATA step where it will be executed for every iteration of the DATA step, it will give you the value of the argument from the previous observation. There are also a family of LAG functions, LAG1, LAG2, LAG3, etc., that will return the value of the argument from the nth previous execution of the function. Thus, LAG (which is equivalent to LAG1) will return the value from a previous observation; LAG2 will return the value from the next earlier observation, and so forth. Notice how the use of a moving average "smooths out" the abrupt change on day 8. A note of caution here: It is usually inadvisable to execute the LAG function conditionally. Consider the following example:

```
DATA NEVER;
    INPUT X @@;
    IF X GT 3 THEN X_LAG = LAG(X);
DATALINES;
5 7 2 1 4
;
```

What are the values of X_LAG in the five observations? Answer: Missing, 5, missing, missing, and 7! Read the definition of the LAG function carefully, and see if you can understand what is happening here.

I.   SORTING WITHIN AN OBSERVATION

We use PROC SORT to sort observations in a SAS data set. However, we may have occasion to sort within an observation. (NOTE: In version 9.1 and above, look for the two call routines SORTN and SORTC, which will sort values in an observation.) In the example that follows, five values (L1–L5) are recorded for each subject. We want to arrange the five values from lowest to highest.

```
DATA SORT;
    INPUT L1-L5;
    ARRAY S[5];
    DO I = 1 TO 5;
        S[I] = ORDINAL(I,OF L1-L5);
    END;
    DROP I;
DATALINES;
5 2 9 1 3
6 . 22 7 0
;
PROC PRINT DATA=SORT NOOBS;
    TITLE "Listing of Data Set SORT";
RUN;
```

Description

The key to this program is the ORDINAL function. The first argument of this function selects the nth largest value in the list of variables listed in the remaining arguments.

This function, unlike other descriptive statistics functions, does not ignore missing values (they are considered smaller than any nonmissing value). Also remember that if you include a list of variables in the form VARn–VARm, the list must be preceded by the word OF. Since the DO loop counter is the first argument of the ORDINAL function in this program, the elements of the S array are the values L1 to L5 in increasing order.

Since the ORDINAL function works only with numeric values, if you want to sort character values, you would have to use the RANK function to determine the order of a character value in the collating sequence and then use the previous program on the numerical values.

J. COMPUTING COEFFICIENT ALPHA (OR KR-20) IN A DATA STEP

This program computes a test statistic called coefficient alpha, which, for a test item that is dichotomous, is equivalent to the Kuder-Richardson formula 20. This statistic is available in PROC CORR with option ALPHA (Cronbach's Alpha—see Chapter 11, Section F.). You may still want to use the following code to compute your KR-20. At the very least, it serves as a good programming example. The formula for Cronbach's alpha is:

$$\text{Alpha (or KR-20 if dichotomous)} = \frac{k}{k-1}\left(1 - \frac{\Sigma \text{Item variances}}{\text{Test variance}}\right)$$

where k is the number of items in the test.

The key here is to output a data set that contains the item and test variances. Here is the program:

```
***Copy of the SCORE DATA set from Chapter 11, except the
   variables S1 to S5 are added to the KEEP list. Variables
   S1-S5 which are the scored responses for each of
   the 5 items on a test. S=1 for a correct answer, S=0 for an
   incorrect response;
DATA SCORE;
   ARRAY ANS[5] $ 1 ANS1-ANS5; ***Student answers;
   ARRAY KEY[5] $ 1 KEY1-KEY5; ***Answer key;
   ARRAY S[5] 3 S1-S5; ***Score array 1=right,0=wrong;
   RETAIN KEY1-KEY5;
   ***Read the answer key;
   IF _N_ = 1 THEN INPUT (KEY1-KEY5)($1.);
   ***Read student responses;
   INPUT @1  ID 1-9
         @11 (ANS1-ANS5)($1.);
   ***Score the test;
   DO I=1 TO 5;
      S[I] = (KEY[I] EQ ANS[I]);
   END;
```

```
***Compute Raw and Percentile scores;
RAW=SUM (OF S1-S5);
PERCENT=100*RAW / 5;
KEEP ID RAW PERCENT S1-S5;
LABEL ID       = 'Social Security Number'
      RAW       = 'Raw Score'
      PERCENT = 'Percent Score';
DATALINES;
ABCDE
123456789 ABCDE
035469871 BBBBB
111222333 ABCBE
212121212 CCCDE
867564733 ABCDA
876543211 DADDE
987876765 ABEEE
;
PROC MEANS NOPRINT DATA=SCORE;
   VAR S1-S5 RAW;
   OUTPUT OUT=VAROUT
          VAR=VS1-VS5 VRAW;
RUN;
DATA _NULL_;   ①
   FILE PRINT;   ②
   SET VAROUT;
   SUMVAR = SUM(OF VS1-VS5);   ③
   KR20 = (5/4)*(1-SUMVAR/VRAW);   ④
   PUT KR20=;   ⑤
RUN;
```

We use PROC MEANS to output a data set containing the item variances. The keyword VAR = computes variances for the variables listed in the VAR statement. This data set contains only one observation. In order to sum the item variances, we need to use another DATA step. You may not be familiar with the special SAS data set name _NULL_ ① This reserved data set name instructs the SAS system to process the observations as they are encountered but not to write them to a temporary or permanent SAS data set. This saves time and, perhaps, money. Line ③ computes the sum of the item variances, and line ④ is the formula for coefficient alpha.

We get the program to print the results by using a PUT statement ⑤. The results of this PUT are sent to the same place that normal SAS output goes, because of the FILE PRINT statement ②.

Finally, we present a macro to compute KR-20 from any data set containing scored items (0s and 1s) and a raw score. If you are not comfortable with macros, by all means, skip this section. Here it is:

```
%MACRO KR20(DSN,    /* Data set holding the scored test */
            N,      /* Number of items on the test      */
            RAW,    /* Name of the raw score variable   */
            ARRAY   /* Name of the array holding scored
                       responses                        */
            );
   PROC MEANS NOPRINT DATA=&DSN;
      VAR &ARRAY.1-&ARRAY&N &RAW;
      OUTPUT OUT=VAROUT
             VAR=VS1-VS&N VRAW;
   RUN;
   DATA _NULL_;
      FILE PRINT;
      SET VAROUT;
      SUMVAR = SUM(OF VS1-VS5);
      KR20 = (&N/%EVAL(&N - 1))*(1-SUMVAR/VRAW);
      PUT KR20=;
   RUN;
%MEND KR20;
```

For example, calling this macro for the previous problem, you would write:

`%KR20(SCORE,5,RAW,S)`

PROBLEMS

As a special treat to you, this chapter does not contain any SAS programming problems (since they are not really appropriate). So your assignment, should you decide to accept it, is to go out and have a good time (on us of course)!

CHAPTER 20

Syntax Examples

A.   INTRODUCTION

This chapter presents examples of each of the procedures previously listed. Instead of a generalized, hard-to-understand syntax reference, we present several concrete examples that cover the majority of options and statements that you may want for each procedure.

Simply replace our variable names and data set names with your own, and you are ready to go. (SAS procedure names, procedure options, statements and statement options are shown in bold.)

B.   PROC ANOVA

One-way Design:

```
PROC ANOVA DATA=MYDATA;
  CLASS RACE;
  MODEL SCORE FINAL = RACE;
  MEANS RACE / SNK; ***SNK is Student Newman Keuls;
RUN;
```

Two-way Factorial Design (Balanced Only):

```
PROC ANOVA DATA=MYDATA;
  CLASS GROUP DOSE;
  MODEL HR SBP DBP = GROUP | DOSE;
  MEANS GROUP | DOSE / DUNCAN; ***DUNCAN multiple range test;
RUN;
```

One-way Repeated Measures Design (Using the REPEATED Statement):

```
PROC ANOVA DATA=MYDATA;
  MODEL SCORE1-SCORE4=/ NOUNI;
  REPEATED TIME;
RUN;
```

One-way Repeated Measures Design (without Using the REPEATED Statement):

```
PROC ANOVA DATA=MYDATA;
  CLASS SUBJ TIME;
  MODEL SCORE = SUBJ TIME;
  MEANS TIME / SNK;
RUN;
```

For more advanced factorial and repeated measures designs, refer to Chapters 7 and 8.

C.   PROC APPEND

```
PROC APPEND BASE=BIG DATA=NEWDATA;
RUN;
```

D.   PROC CHART

Vertical Bar Chart (Frequencies):

```
PROC CHART DATA=MYDATA;
  VBAR GENDER;
RUN;
PROC CHART DATA=MYDATA;
  VBAR DAY / DISCRETE;
RUN;
```

Horizontal Bar Chart:

```
PROC CHART DATA=MYDATA;
  HBAR GROUP;
RUN;
```

Bar Chart Where Bars Represent Sums or Means of a Variable:

```
PROC CHART DATA=MYDATA;
  VBAR REGION / SUMVAR=SALES TYPE=SUM;
RUN;
```

For more details on PROC CHART see Chapter 2.

E.   PROC CONTENTS

```
PROC CONTENTS DATA=MYDATA POSITION;
RUN;
```

```
LIBNAME XXX 'C:\SASDATA';
PROC CONTENTS DATA=XXX._ALL_ POSITION;
RUN;
```

F.   PROC CORR

Correlation Matrix:

```
PROC CORR DATA=MYDATA NOSIMPLE;
  VAR A B C X Y Z;
RUN;
```

Correlate One Variable with Several Others:

```
PROC CORR DATA=MYDATA NOSIMPLE;
  WITH QUES1-QUES50;
  VAR GRADE;
RUN;
```

For additional examples, see Chapter 5.

G.   PROC DATASETS

```
LIBNAME XXX 'C:\SASDATA';
PROC DATASETS LIBRARY=XXX;
  MODIFY MYDATA;
    LABEL CANDID = 'CANDIDATE';
    RENAME OLDWT = WEIGHT;
    FORMAT COST DOLLAR7.;
RUN;
```

H.   PROC FACTOR

Principal Components Analysis (with rotation):

```
PROC FACTOR DATA=FACTOR PREPLOT PLOT ROTATE=VARIMAX
          NFACTORS=2 OUT=FACT SCREE;
  VAR QUES1-QUES6;
RUN;
```

Factor Analysis with Squared Multiple Correlations and Oblique Rotation:

```
PROC FACTOR DATA=FACTOR PREPLOT PLOT ROTATE=PROMAX
          NFACTORS=2 OUT=FACT SCREE;
  VAR QUES1-QUES6;
  PRIORS SMC;
RUN;
```

See Chapter 10 for more examples and explanations.

I.   PROC FORMAT

Temporary Character and Numeric Formats:

```
PROC FORMAT;
  VALUE $GENDER 'M' = 'Male'
                'F' = 'Female';
  VALUE LIKERT    1 = 'Strongly Disagree'
                  2 = 'Disagree'
                  3 = 'No Opinion'
                  4 = 'Agree'
                  5 = 'Strongly Agree';
  VALUE WTGRP     LOW-<20 = 'Zero to 20'
                  20-<40 = '20 TO 40'
                  40-HIGH = '40 and Above';
  VALUE $CODES    'A','C','E' = 'Group 1'
                  'X','Y','Z' = 'Group 2';
  VALUE NUMS      1,4-8       = 'Range One'
                  2,3,9-11    = 'Range Two';
RUN;
```

Permanent Formats:

```
LIBNAME XXX 'C:\SASDATA';
OPTIONS FMTSEARCH=(XXX);
PROC FORMAT LIBRARY=XXX;
  VALUE $YESNO '1' = 'Yes'
               '0' = 'No';
RUN;
```

J.   PROC FREQ

One-way Frequencies:

```
PROC FREQ DATA = MYDATA ORDER=FREQ;
  TABLES GENDER RACE GROUP / NOCUM;
RUN;
```

Two-way Frequencies (with Request for Chi-square):

```
PROC FREQ DATA=MYDATA;
  TABLES TREAT*OUTCOME / CHISQ;
RUN;
```

Three-way Frequencies (with a Request for All Statistics):

```
PROC FREQ DATA=MYDATA;
  TABLES STRATA*GROUP*OUTCOME / ALL;
RUN;
```

See Chapter 3 for more examples.

K.   PROC GCHART

```
PATTERN VALUE=EMPTY COLOR=BLACK;
```

For a Vertical Bar Chart:

```
PROC GCHART DATA=MYDATA;
  VBAR RACE;
RUN;
```

For a Horizontal Bar Chart:

```
PROC GCHART DATA=MYDATA;
  HBAR RACE;
RUN;
```

To Plot a Distribution of One Variable Within Another Variable:

```
PATTERN VALUE=L2 COLOR=BLACK;
PROC GCHART DATA=E_MART;
  VBAR SALES / GROUP=DEPT MIDPOINTS=4500 TO 6000 BY 1000;
  FORMAT SALES DOLLAR8.0;
RUN;
```

See Chapter 2 for more examples.

L.   PROC GLM

One-way Design:

```
PROC GLM DATA=MYDATA;
  CLASS TREAT;
  MODEL Y = TREAT;
```

```
   CONTRAST 'A VS. B AND C' TREAT 2 -1 -1;
   CONTRAST 'B VS. C' TREAT 0 1 -1;
MEANS TREAT / SNK;
RUN;
```

Two-way Factorial Design (Balanced or Unbalanced):

```
PROC GLM DATA=MYDATA;
  CLASS TREAT GENDER;
  MODEL Z = TREAT | GENDER;
  LSMEANS TREAT | GENDER / SCHEFFE ALPHA=.1;
RUN;
```

For more examples of factorial designs, contrast statements, and repeated measures designs, see Chapters 7 and 8.

M.   PROC GPLOT

For a Simple Scatter Plot of y by x:

```
PROC GPLOT DATA=MYDATA;
  PLOT Y * X;
RUN;
```

For a Simple Scatter Plot of y by x with Dots as Plotting Symbols:

```
SYMBOL VALUE=DOT COLOR=BLACK;
PROC GPLOT DATA=MYDATA;
  PLOT Y * X;
RUN;
```

N.   PROC LOGISTIC

```
PROC LOGISTIC DATA=LOGISTIC DESCENDING;
  MODEL ACCIDENT = AGE VISION DRIVE_ED /
                   SELECTION = FORWARD
                   CTABLE PPROB=(0 to 1 by .1)
                   LACKFIT
                   RISKLIMITS;
RUN;
```

For other examples of PROC LOGISTIC, see Chapter 9, Section F.

O.   PROC MEANS

Descriptive Statistics on All Subjects Together:

```
PROC MEANS DATA=MYDATA N MEAN STD STDERR MAXDEC=2;
  VAR HR SBP DBP;
RUN;
```

Descriptive Statistics Broken Down by One Variable:

```
PROC MEANS DATA=MYDATA N MEAN STD STDERR MIN MAX MAXDEC=2;
  CLASS GROUP;
  VAR X Y Z;
RUN;
```

Creating an Output Data Set Containing Means and Variances:

```
PROC MEANS DATA=MYDATA NOPRINT NWAY;
  CLASS GENDER GROUP;
  ID SUBJ;
  VAR X Y Z;
  OUTPUT OUT=SUMMARY
       MEAN=M_X M_Y M_Z
       VAR =V_X V_Y V_Z;
RUN;
```

For more examples of PROC MEANS with and without creating an OUTPUT data set, see Chapter 2.

P. PROC NPAR1WAY

```
PROC NPAR1WAY DATA=MYDATA WILCOXON;
  CLASS GROUP;
  VAR WEIGHT;
  EXACT WILCOXON;
RUN;
```

See Chapter 6, Section D, for more details.

Q. PROC PLOT

Simple X–Y Plot:

```
PROC PLOT DATA=MYDATA;
  PLOT Y*X;
RUN;
```

Choosing o's as Plotting Symbols:

```
PROC PLOT DATA=MYDATA;
  PLOT Y*X = 'o';
RUN;
```

Using the Value of Gender ('M' or 'F') as the Plotting Symbol:

```
PROC PLOT DATA=MYDATA;
  PLOT Y*X = GENDER;
RUN;
```

For more details, see Chapter 2, Section G.

R.   **PROC PRINT**

Simple Listing with Variable Names as Column Headings:

```
PROC PRINT DATA=MYDATA;
  TITLE 'This is the Title of My Report';
  ID SUBJ_ID;
  VAR DATE HR SBP DBP;
RUN;
```

Simple Listing with Variable Labels as Column Headings:
(NOTE: In this example, the LABEL statement is included in the procedure. In other situations, the labels may be assigned in the DATA step and you would not need an additional LABEL statement in the PROC. The formats may have also been assigned previously.)

```
PROC PRINT DATA=MYDATA LABEL;
  TITLE1 'Fancier Report';
  TITLE2 'Compiled by J. Smith and R. Cody';
  TITLE3 '----------------------';
  FOOTNOTE 'Printed on recycled paper';
  ID SUBJ_ID;
  VAR DATE COST SBP DBP;
  FORMAT DATE MMDDYY8. COST DOLLAR8. SBP DBP 4.;
  LABEL SUBJ_ID = 'Subject ID'
        DATE    = 'Date of Visit'
        COST    = 'Cost of Treatment';
RUN;
```

S.   **PROC RANK**

Create a New Data Set with R_X Representing the Rank of X:

```
PROC RANK DATA=MYDATA OUT=RANKDATA;
  VAR X;
  RANKS R_X;
RUN;
```

Using PROC RANK to Split the Group in Two (Median Split):

```
PROC RANK DATA=MYDATA OUT=NEWDATA GROUPS=2;
  VAR X;
RUN;
```

T.   **PROC REG**

Simple Linear Regression:

```
PROC REG DATA=MYDATA;
  MODEL Y = X;
  RANKS R_X;
RUN;
```

Two-variable Regression:

```
PROC REG DATA=MYDATA;
  MODEL LOSS = DOSAGE EXERCISE / P R;
RUN;
```

Stepwise Multiple Regression:

```
PROC REG DATA=MYDATA;
  MODEL OUTCOME = INCOME SES AGE IQ /
              SELECTION = STEPWISE;
RUN;
```

See Chapter 9 for more examples of PROC REG.

U.   PROC SORT

Sorting a Data Set "In Place":

```
PROC SORT DATA=MYDATA;
  BY ID DATE;
RUN;
```

Example Creating an Output Data Set (Using KEEP and WHERE= Data Set Options):

```
PROC SORT DATA=MYDATA(KEEP=ID HR SBP DBP GENDER
                      WHERE=(SBP GT 140)) OUT=OUTDATA;
  BY ID DATE;
RUN;
```

V.   PROC TTEST

```
PROC TTEST DATA=MYDATA;
  CLASS GENDER;
  VAR HR SBP DBP;
RUN;
```

Using PROC TTEST to perform a paired T-Test:

```
PROC TTEST DATA=PAIRED;
  PAIRED BEFORE * AFTER;
RUN;
```

W.   PROC UNIVARIATE

```
PROC UNIVARIATE DATA=MYDATA NORMAL PLOT;
  VAR X Y Z;
RUN;
```

Solutions to Problems

CHAPTER 1

1.1 (a)

```
DATA COLLEGE;
    INPUT ID AGE GENDER $ GPA CSCORE;
DATALINES;
1 18 M 3.7 650
2 18 F 2.0 490
3 19 F 3.3 580
4 23 M 2.8 530
5 21 M 3.5 640
;
```

(b)

```
PROC MEANS DATA=COLLEGE;
    VAR GPA CSCORE;
RUN;
```

(c) Between the INPUT and DATALINES lines insert:

$$INDEX = GPA + 3*CSCORE/500;$$

Add to the end of the program:

```
PROC SORT DATA=COLLEGE;
    BY INDEX;
RUN;

PROC PRINT DATA=COLLEGE;
    TITLE "Students in Index Order"; /* optional */
    ID ID;
    VAR GPA CSCORE INDEX;
RUN;
```

1.3 (a)

```
DATA TAXPROB;
   INPUT SS SALARY AGE RACE $;
   FORMAT SS SSN11.; /* See Chapter 3 about FORMATS */
DATALINES;
123874414 28000 35 W
646239182 29500 37 B
012437652 35100 40 W
018451357 26500 31 W
;
PROC MEANS DATA=TAXPROB N MEAN MAXDEC=0;
   TITLE "Descriptive Statistics for Salary and Age";
   VAR SALARY AGE;
RUN;
```

(b) Add a line after the INPUT statement:

$$TAX = .30 * SALARY;$$

Add to the end of the program:

```
PROC SORT DATA=TAXPROB;
   BY SS;
RUN;
PROC PRINT DATA=TAXPROB;
   TITLE "Listing of Salary and Taxes";
   ID SS;
   VAR SALARY TAX;
RUN;
```

1.5

```
1    DATA MISTAKE;
2    INPUT ID 1-3 TOWN 4-6 REGION 7-9 YEAR 11-12 BUDGET 13-14
3    VOTER-TURNOUT 16-20
4    DATALINES;
     00104050422 12345
     (more data lines)
5    ;
6    PROC MEANS DATA=MISTAKE;
7    VAR ID REGION VOTER-TURNOUT;
8    N,STD,MEAN;
9    RUN;
```

Line 2: Not a mistake, but you might want ID to be a character variable.
Line 3: Variable name cannot contain a dash. Semicolon missing at the end of the line.
Line 7: We probably don't want the mean ID. Also, would be more meaningful to use PROC FREQ for a categorical variable such as REGION.
Line 8: Options for PROC MEANS go on the PROC line between the word MEANS and the semicolon. The options must have a space between them, not a comma.
Corrected Program:

```
DATA MISTAKE;
    INPUT ID $ 1-3 TOWN 4-6 REGION 7-9 YEAR 11-12 BUDGET 13-14
          VOTER_TURNOUT 16-20;
DATALINES;
00104005  0422 12345
(more data lines)
;
PROC MEANS DATA=MISTAKE N MEAN STD;
    VAR BUDGET VOTER_TURNOUT;
RUN;
```

1.7 We have a SAS data set with the variables AGE, GENDER, RACE, INCOME, MARITAL, and HOME (homeowner versus renter).

Code Book Variable Name	Col(s)	Description and Formats
AGE	1	Age group of subject
		1 = 10–19 2 = 20–29 3 = 30–39
		4 = 40–49 5 = 50–59 6 = 60+
GENDER	2	Gender, 1 = male 2 = female
RACE	3	Race, 1 = white 2 = African Am.
		3 = Hispanic 4 = other
INCOME	4	Income group, 1 = 0 to $9,999
		2 = 10,000 to 19,999
		3 = 20,000 to 39,000
		4 = 40,000 to 59,000
		5 = 60,000 to 79,000
		6 = 80,000 and over
MARITAL	5	Marital status, 1 = single 2 = married
		3 = separated 4 = divorced 5 = widowed
HOME	6	Homeowner or renter, 1 = homeowner
		0 = renter

```
DATA CEO;
    INPUT AGE        $ 1
          GENDER     $ 2
          RACE       $ 3
          INCOME     $ 4
          MARITAL    $ 5
          HOME       $ 6;
DATALINES;
311411
122310
411221
(more data lines)
;
PROC FREQ DATA=CEO ORDER=FREQ;
    *Note, the ORDER = FREQ option will list the frequencies in
    decreasing frequency order, i.e. the most frequent first;
    TITLE 'Frequencies and Contingency Tables for CEO Report';
    TABLES AGE GENDER RACE INCOME MARITAL HOME
    AGE*GENDER*RACE INCOME*AGE*GENDER MARITAL*HOME;
    *or whatever other combinations you are interested in;
RUN;
PROC GCHART DATA=CEO;
    TITLE "Histograms";
    VBAR AGE GENDER RACE INCOME MARITAL HOME;
RUN;
```

1.9

```
DATA PROB1_9;
    INPUT ID RACE $ SBP DBP HR;
DATALINES;
001    W    130    80    60
002    B    140    90    70
003    W    120    70    64
004    W    150    90    76
005    B    124    86    72
;
PROC SORT DATA=PROB1_9;
    BY SBP;
RUN;
PROC PRINT DATA=PROB1_9;
    TITLE "Race and Hemodynamic Variables";
    ID ID;
    VAR RACE SBP DBP;
RUN;
```

CHAPTER 2

2.1

```
PROC FREQ DATA=COLLEGE;
   TABLES GENDER;
RUN;
```

2.3

```
DATA TAXPROB;
   INPUT SS SALARY AGE RACE $;
   IF AGE GE 0 AND AGE LE 35 THEN AGE_GROUP = 1;
   ELSE IF AGE GT 35 THEN AGE_GROUP = 2;
   FORMAT SS SSN11.; /* See Chapter 3 about FORMATS */
DATALINES;
123874414 28000 35 W
646239182 29500 37 B
012437652 35100 40 W
018451357 26500 31 W
;
PROC FREQ DATA=TAXPROB;
   TITLE "Problem 2-3";
   TABLES RACE AGE_GROUP / NOCUM;
RUN;
```

2.5

```
DATA PROB2_5;
   LENGTH GROUP $ 1;
   INPUT X Y Z GROUP $;
DATALINES;
2    4    6    A
3    3    3    B
1    3    7    A
7    5    3    B
1    1    5    B
2    2    4    A
5    5    6    A
;
```

(a)

```
PROC GCHART DATA=PROB2_5;
   VBAR GROUP;
RUN;
```

(b)

```
PROC GPLOT DATA=PROB2_5;
   PLOT Y*X;
RUN;
```

(c)

```
PROC SORT DATA=PROB2_5;
   BY GROUP;
RUN;
PROC GPLOT DATA=PROB2_5;
   BY GROUP;
   PLOT Y*X;
RUN;
```

Don't forget that you must have your data set sorted by the BY variables before you can use a BY statement in a PROC.

2.7

```
DATA LIVER;
   INPUT SUBJ DOSE REACT LIVER_WT SPLEEN;
DATALINES;
1    1    5.4   10.2   8.9
2    1    5.9    9.8   7.3
3    1    4.8   12.2   9.1
4    1    6.9   11.8   8.8
5    1   15.8   10.9   9.0
6    2    4.9   13.8   6.6
7    2    5.0   12.0   7.9
8    2    6.7   10.5   8.0
9    2   18.2   11.9   6.9
10   2    5.5    9.9   9.1
;
```
                                                    [Continued]

```
PROC SORT DATA=LIVER;
   BY DOSE;    *Note, optional since already in dose order;
RUN;
PROC UNIVARIATE DATA=LIVER NORMAL PLOT;
   TITLE "Distributions for Liver Data";
   VAR REACT -- SPLEEN;
   *** The notation first -- last is a short
       way of referring to all the variables from
       first to last IN THE ORDER THEY ARE in the
       SAS data set;
RUN;
PROC UNIVARIATE DATA=LIVER NORMAL PLOT;
   BY DOSE;
   TITLE "Distributions for Liver Data by Dose";
   VAR REACT -- SPLEEN;
RUN;
```

2.9

```
1    DATA 123;
2        INPUT AGE STATUS PROGNOSIS DOCTOR GENDER STATUS2
3            STATUS3;
4    (data lines)
         ;
5    PROC CHART DATA=123 BY GENDER;
6        VBAR STATUS
7        VBAR PROGNOSIS;
8    RUN;
9    PROC PLOT DATA=123;
10       DOCTOR BY PROGNOSIS;
11   RUN;
```

Line 1: Invalid data set name, cannot start with a number.

Line 2: Not really an error, but it would be better to list GENDER with the other demographic variables.

Line 2: Again, not an error, but an ID variable is desirable.

Lines 2 and 3: Boy, we're picky. If you have STATUS2 and STATUS3, STATUS should be STATUS1.

DATALINES; or CARDS; statement missing between lines 3 and 4.

Line 5: Two things wrong here: One, If you use a BY variable, the data set must be sorted in order of the BY variable; two, a semicolon is missing between PROC CHART and BY GENDER.

Line 6: Missing a semicolon at the end of the line.

Line 7: In case you thought this was an error, it isn't. You can have two (or more) VBAR statements with one PROC CHART.

Line 8: Missing the keyword PLOT before the plot request. Also, the plot request is of the form Y*X not Y BY X.

2.11 (a)

```
DATA SALES;
    INPUT PERSON $ TARGET $ VISITS CALLS UNITS;
DATALINES;
Brown       American   3    12    28000
Johnson     VRW        6    14    33000
Rivera      Texam      2    6     8000
Brown       Standard   0    22    0
Brown       Knowles    2    19    12000
Rivera      Metro      4    8     13000
Rivera      Uniman     8    7     27000
Johnson     Oldham     3    16    8000
Johnson     Rondo      2    14    2000
;
PROC MEANS DATA=SALES N SUM MEAN STD MAXDEC=0;
    CLASS PERSON;
    TITLE "Sales Figures for Each Salesperson";
    VAR VISITS CALLS UNITS;
RUN;
```

(b)

```
PROC GPLOT DATA=SALES;
    TITLE "Sales Plots";
    PLOT VISITS*CALLS = PERSON;
RUN;
```

(c)

```
PROC CHART DATA=SALES;
    TITLE "Distribution of Units Sold by Salesperson";
    VBAR PERSON / SUMVAR=UNITS TYPE=SUM;
RUN;
        or
PROC CHART DATA=SALES;
    TITLE "Distribution of Units Sold by Salesperson";
    VBAR UNITS / GROUP=PERSON;
RUN;
```

The first **PROC CHART** in part (c) will produce a single bar for each salesperson, the height representing the total (sum) of the units sold. The alternate statements will produce an actual frequency distribution of the number of units sold, for each salesperson, in a side-by-side fashion.

2.13

```
DATA PROB2_13;
   INPUT ID TYPE $ SCORE;
DATALINES;
1   A   44
1   B   9
1   C   203
2   A   50
2   B   7
2   C   188
3   A   39
3   B   9
3   C   234
;
PROC MEANS DATA=PROB2_13;
   CLASS TYPE;
   VAR SCORE;
RUN;
```

If you use a BY statement instead of a CLASS statement, remember to sort your data set first.

CHAPTER 3

3.1

```
PROC FORMAT;
   VALUE FGROUP 1 = 'CONTROL'
                2 = 'DRUG A'
                3 = 'DRUG B';
RUN;
```

3.3

```
PROC FORMAT;
   VALUE $GENDER 'M' = 'Male'
                 'F' = 'Female';
```

```
    VALUE $PARTY '1' = 'Republican'
                 '2' = 'Democrat'
                 '3' = 'Not Registered';
    VALUE YESNO 0 = 'No' 1 = 'Yes';
RUN;
DATA SURVEY;
    INPUT ID      1-3
          GENDER $ 4
          PARTY  $ 5
          VOTE     6
          FOREIGN  7
          SPEND    8;
    LABEL PARTY   = 'Political Party'
          VOTE    = 'Vote in Last Election?'
          FOREIGN = 'Agree with Government Policy?'
          SPEND   = 'Should we Increase Domestic Spending?';
    FORMAT GENDER $GENDER.
           PARTY $PARTY.
           VOTE FOREIGN SPEND YESNO.;
DATALINES;
007M1110
013F2101
137F1001
117 1111
428M3110
017F3101
037M2101
;
PROC FREQ DATA=SURVEY;
    TITLE "Political Survey Results";
    TABLES GENDER PARTY VOTE FOREIGN SPEND;
    TABLES VOTE*(SPEND FOREIGN) / CHISQ;
RUN;
```

3.5

```
***Method 1 ***;
DATA DEMOG;
    INPUT WEIGHT HEIGHT GENDER $;
    *Create weight groups;
    IF 0 LE WEIGHT LT 101 THEN WTGRP = 1;
    ELSE IF 101 LE WEIGHT LT 151 THEN WTGRP = 2;
    ELSE IF 151 LE WEIGHT LE 200 THEN WTGRP = 3;
    ELSE IF WEIGHT GT 200 THEN WTGRP = 4;
    *Create height groups;
                                        [Continued]
```

```
    IF 0 LE HEIGHT LE 70 THEN HTGRP = 1;
    ELSE IF HEIGHT GT 70 THEN HTGRP = 2;
DATALINES;
155 68 M
98 60 F
202 72 M
280 75 M
130 63 F
.   57 F
166 . M
;
PROC FREQ DATA=DEMOG;
    TABLES WTGRP*HTGRP;
RUN;
```

(NOTE: You may use $<=$ instead of LE, $<$ instead of LT, and $>$ instead of GT.)

```
***Method 2 ***;
PROC FORMAT;
    VALUE WTFMT 0-100   = '1'
                101-150 = '2'
                151-200 = '3'
                201-HIGH = '4';
    VALUE HTFMT 0-70    = '1'
                71-HIGH = '2';
RUN;
DATA DEMOG;
    INPUT WEIGHT HEIGHT GENDER $;
DATALINES;
155 68 M
98 60 F
202 72 M
280 75 M
130 63 F
.   57 F
166 . M
;
PROC FREQ DATA=DEMOG;
    TITLE "Problem 3-5";
    TABLES WEIGHT*HEIGHT;
    FORMAT WEIGHT WTFMT. HEIGHT HTFMT.;
RUN;
```

3.7

```
DATA ASTHMA;
    INPUT ASTHMA $ SES $ COUNT;
DATALINES;
YES LOW 40
NO LOW 100
YES HIGH 30
NO HIGH 130
;
PROC FREQ DATA=ASTHMA;
    TITLE "Relationship between Asthma and SES";
    TABLES SES*ASTHMA / CHISQ;
    WEIGHT COUNT;
RUN;
```

Chi-square = 4.026, p = .045.

3.9

```
DATA VITAMIN;
    INPUT V_CASE $ V_CONT $ COUNT;
    LABEL V_CASE = 'Case Use Vitamins'
          V_CONT = 'Control Use Vitamins';
    ***Note: Values of V_CASE and V_CONT chosen so that 1-YES will
            come before 2-NO in the table;
DATALINES;
1-YES  1-YES  100
1-YES  2-NO    50
2-NO   1-YES   90
2-NO   2-NO   200
;
PROC FREQ DATA=VITAMIN;
    TITLE "Matched Case-control Study";
    TABLES V_CASE * V_CONT / AGREE;
    WEIGHT COUNT;
RUN;
```

Chi-square (McNemar) = 11.429, p = .001. More likely to develop disease X if you do not use vitamins. (Remember, it is only the discordant pairs (yes/no or no/yes) that contribute to the McNemar chi-square.)

3.11

```
DATA VDT_USE;
   INPUT GROUP $ VDT $ COUNT;
DATALINES;
CASE 1-YES 30
CASE 2-NO 50
CONTROL 1-YES 90
CONTROL 2-NO 200
;
PROC FREQ DATA=VDT_USE;
   TITLE "Case-control study of VDT Use";
   TABLES VDT * GROUP / CHISQ CMH;
   WEIGHT COUNT;
RUN;
```

Chi-square $= 1.1961$, $p = .2741$ OR $= 1.333$, 95% CI $(.7956, 2.2349)$.

3.13

```
DATA CLASS;
   INPUT TYPE : $10. PROBLEM $ COUNT;
DATALINES;
1-STANDARD 1-YES 30
1-STANDARD 2-NO 220
2-PROOFED  1-YES 20
2-PROOFED 2-NO 280
;
PROC FREQ DATA=CLASS;
   TITLE "Sound Proofing Study";
   TABLES TYPE * PROBLEM / CHISQ CMH;
   WEIGHT COUNT;
RUN;
```

RR $= 1.800$ (room noise increases the incidence of problems), 95% CI $(1.057, 3.065)$.

```
PROC FORMAT;
   VALUE SIZE 1 = 'Small'
              2 = 'Medium'
              3 = 'Large'
              4 = 'Gigantic';
RUN;
```

```
DATA CLASS;
    INPUT SIZE PROBLEM $ COUNT @@;
    FORMAT SIZE SIZE.;
DATALINES;
11-YES 31 2-NO 122 1-YES 62 2-NO 22
31-YES 173 2-NO 384 1-YES 804 2-NO 120
;
PROC FREQ DATA=CLASS;
    TITLE "Relationship Between Class Size and Behavior";
    TABLES PROBLEM * SIZE / CHISQ;
    WEIGHT COUNT;
RUN;
```

Chi-square test for trend = 6.038, p = .014.
NOTE: The chi-square for the 2 by 4 table is 6.264, with p = .094.

3.17

```
DATA TEMP;
    INPUT T_CONTRL $ GROUP : $10. COLD $ COUNT;
DATALINES;
1-POOR SMOKERS 1-YES 30
1-POOR SMOKERS 2-NO 50
1-POOR NONSMOKERS 1-YES 40
1-POOR NONSMOKERS 2-NO 100
2-GOOD SMOKERS 1-YES 20
2-GOOD SMOKERS 2-NO 55
2-GOOD NONSMOKERS 1-YES 35
2-GOOD NONSMOKERS 2-NO 150
;
PROC FREQ DATA=TEMP;
    TITLE "Relationship Between Temperature Control and Colds";
    TABLES GROUP * T_CONTRL * COLD / ALL;
    WEIGHT COUNT;
RUN;
```

The overall RR for the combined tables = 1.468.
The 95% CI is (1.086, 1.985).
The p-value is .013.

3.19

```
PROC FORMAT;
    VALUE PROB 1 = 'Cold'
               2 = 'Flu'
               3 = 'Trouble Sleep'
```

[Continued]

```
                        4 = 'Chest Pain'
                        5 = 'Muscle Pain'
                        6 = 'Headache'
                        7 = 'Overweight'
                        8 = 'High BP'
                        9 = 'Hearing Loss';
RUN;
DATA PATIENT;
   INPUT SUBJ    1-2
         PROB1    3
         PROB2    4
         PROB3    5
         HR      6-8
         SBP     9-11
         DBP    12-14;
DATALINES;
1112778130 80
178782180110
03162120 78
426168130 80
8958120 76
994882178100
;
PROC MEANS DATA=PATIENT N MEAN STD MAXDEC=1;
   TITLE "Statistics from Patient Data Base";
   VAR HR SBP DBP;
RUN;
```

For part (b) add:

```
***Solution without arrays;
DATA PROBLEM;
   SET PATIENT;
   PROB = PROB1;
   IF PROB NE . THEN OUTPUT;
   PROB = PROB2;
   IF PROB NE . THEN OUTPUT;
   PROB = PROB3;
   IF PROB NE . THEN OUTPUT;
   FORMAT PROB PROB.;
   KEEP PROB;
RUN;
```

```
*Solution with arrays;
DATA PROBLEM;
   SET PATIENT;
   ARRAY XPROB[3] PROB1-PROB3;
   DO I = 1 TO 3;
      PROB = XPROB[I];
      IF PROB NE . THEN OUTPUT;
   END;
   FORMAT PROB PROB.;
   KEEP PROB;
RUN;
PROC FREQ DATA=PROBLEM;
   TABLES PROB;
RUN;
```

3.21 Line 3: The formats cannot be assigned to variables before they have been defined. Therefore, move lines 5 through 8 to the beginning of the program (before line 1).
Line 11: PROC FREQ uses the keyword TABLES, not VAR, to specify a list of variables.
Line 11: You cannot use the CHISQ option unless a two-way table (or higher order) is specified. That is, we could have written:

```
PROC FREQ DATA=IGOOFED;
   TABLES GENDER*RACE / CHISQ;
RUN;
```

Line 14: You cannot use a BY statement unless the data set has been sorted first by the same variable.

CHAPTER 4

4.1

```
DATA PROB4_1;
   INPUT @1 ID 3.
         @5 (DOB ST_DATE END_DATE)(MMDDYY8.)
         /****************************************************
            Note: The line above is using a variable list
            and an informat list.  An alternative is to
            list each variable with a column pointer and
            an informat
         ****************************************************/
                                                  [Continued]
```

```
            @29 SALES 4.;
    AGE = YRDIF(DOB,ST_DATE,'ACTUAL');
    *For section E, substitute the line below for AGE;
    AGE = INT(YRDIF(DOB,ST_DATE,'ACTUAL'));
    LENGTH = YRDIF(ST_DATE,END_DATE,'ACTUAL');
    SALES_YR = SALES / LENGTH;
    *For section, E substitute the line below for SALES_YR;
    SALES_YR = ROUND ((SALES/LENGTH),10);
    FORMAT DOB ST_DATE END_DATE MMDDYY10.
            SALES_YR DOLLAR8.;
DATALINES;
001 10211946111219801228198887343
002 09131955020219800204198880123
005 06061940031219810312200040000
003 07051944111519801113200009544
;
PROC PRINT DATA=PROB4_1;
    TITLE "Report for Homework Problem 4-1";
    ID ID;
    VAR DOB AGE LENGTH SALES_YR;
RUN;
```

4.3

```
DATA RATS;
    INPUT @1 RAT_NO 1.
        @3 DOB DATE9.
        @13 DISEASE DATE9.
        @23 DEATH DATE9.
        @33 GROUP $1.;
    BIR_TO_D = DISEASE - DOB;
    DIS_TO_D = DEATH - DISEASE;
    AGE = DEATH - DOB;
    FORMAT DOB DISEASE DEATH MMDDYY10.;
DATALINES;
1 23MAY1990 23JUN1990 28JUN1990 A
2 21MAY1990 27JUN1990 05JUL1990 A
3 23MAY1990 25JUN1990 01JUL1990 A
4 27MAY1990 07JUL1990 15JUL1990 A
5 22MAY1990 29JUN1990 22JUL1990 B
6 26MAY1990 03JUL1990 03AUG1990 B
7 24MAY1990 01JUL1990 29JUL1990 B
8 29MAY1990 15JUL1990 18AUG1990 B
;
```

```
PROC MEANS DATA=RATS MAXDEC=1 N MEAN STD STDERR;
   CLASS GROUP;
   VAR BIR_TO_D DIS_TO_D AGE;
RUN;
```

4.5

```
DATA PATIENTS;
   INPUT  @1  ID        3.
          @4  DATE MMDDYY6.
          @10 HR        3.
          @13 SBP       3.
          @16 DBP       3.
          @19 DX        3.
          @22 DOCFEE    4.
          @26 LABFEE    4.;
   FORMAT DATE MMDDYY8.;
DATALINES;
00710218307012008001400400150
00712018307213009002000500200
00909038306611007013700300000
00507058307414008201300900000
00501158208018009601402001500
00506188207017008401400800400
00507038306414008401400800200
;
PROC SORT DATA=PATIENTS;
   BY ID DATE;
RUN;
DATA PROB4_5;
   SET PATIENTS;
   BY ID;
  *Omit the first VISIT for each patient;
  IF NOT FIRST.ID;
RUN;
PROC MEANS DATA=PROB4_5 MEAN MAXDEC=2;
   CLASS ID;
   VAR HR SBP DBP;
RUN;
```

4.7

```
PROC SORT DATA=PATIENTS; ***From problem 4-5;
   BY ID;
RUN;
```

[Continued]

```
DATA PROB4_7;
   SET PATIENTS;
   BY ID;
   *Omit patients with only one visit;
IF FIRST.ID AND LAST.ID THEN DELETE;
RUN;
PROC MEANS DATA=PROB4_7 NOPRINT NWAY;
   CLASS ID;
   VAR HR SBP DBP;
   OUTPUT OUT = PAT_MEAN MEAN= / AUTONAME;
RUN;
```

4.9

```
***Program to create data set BLOOD;
DATA BLOOD;
   LENGTH GROUP $ 1;
   INPUT ID GROUP $ TIME WBC RBC @@;
DATALINES;
1 A 1 8000 4.5    1 A 2 8200 4.8    1 A 3 8400 5.2
1 A 4 8300 5.3    1 A 5 8400 5.5
2 A 1 7800 4.9    2 A 2 7900 5.0
3 B 1 8200 5.4    3 B 2 8300 5.4    3 B 3 8300 5.2
3 B 4 8200 4.9    3 B 5 8300 5.0
4 B 1 8600 5.5
5 A 1 7900 5.2    5 A 2 8000 5.2    5 A 3 8200 5.4
5 A 4 8400 5.5
;
PROC MEANS DATA=BLOOD NWAY NOPRINT;
   CLASS ID;
   ID GROUP;
   VAR WBC RBC;
   OUTPUT OUT=TEMP(WHERE=(_FREQ_ GT 2)
                  DROP=_TYPE_)
                  MEAN = M_WBC M_RBC;
RUN;
PROC PRINT DATA=TEMP NOOBS;
   TITLE "Listing of data set TEMP";
RUN;
```

4.11

```
Replace the OUTPUT statement of PROC MEANS with:
OUTPUT OUT=TEMP(WHERE=(_FREQ_ GT 2)
               DROP=_TYPE_)
```

```
                    MEAN=
                    STD=  /AUTONAME;
RUN;
```

CHAPTER 5

5.1 (a)

```
DATA PROB5_1;
   INPUT X Y Z;
DATALINES;
1 3 15
7 13 7
8 12 5
3 4 14
4 7 10
;
PROC CORR DATA=PROB5_1;    /* x vs. y r =   .965 p = .0078 */
   VAR X;                  /* x vs. z r = −.975 p = .0047 */
   WITH Y Z;
RUN;
```

(b)

```
PROC CORR DATA=PROB5_1;    /* y vs. z r = −.963 p = .0084 */
   VAR X Y Z;
RUN;
```

5.3

```
DATA PRESSURE;
   INPUT AGE SBP;
DATALINES;
15 116
20 120
25 130
30 132
40 150
50 148
;
PROC CORR DATA=PRESSURE;
   VAR AGE SBP;
RUN;
```

5.5

```
***a;
PROC REG DATA=PROB5_1;
    MODEL Y = X;
RUN;
/**************************************
 Intercept = .781, prob>|t| = .5753
 Slope = 1.524, prob>|t| = .0078
**************************************/
```

5.7

```
DATA PROB5_7;
    INPUT X Y Z;
    LX = LOG(X);
    LY = LOG(Y);
    LZ = LOG(Z);
DATALINES;
1 3 15
7 13 7
8 12 5
3 4 14
4 7 10
;
PROC CORR DATA=PROB5_7;
    VAR LX LY LZ;
RUN;
```

5.9 (a)

```
PROC PLOT DATA=PROB5_1;
    PLOT Y*X;
RUN;
```

(b)

```
SYMBOL1 VALUE=DOT COLOR=BLACK;
PROC REG DATA=PROB5_1;
   MODEL Y = X;
   PLOT Y*X;
RUN;
***Alternative using PROC GPLOT;
SYMBOL VALUE=DOT COLOR=BLACK I=RL;
PROC GPLOT DATA=PROB5_1;
   TITLE "Plot of Problem 5-1 Data and Regression Line";
   PLOT Y*X;
RUN;
```

5.11 (a–c)

```
DATA PROB5_11;
   INPUT COUNTY POP HOSPITAL FIRE_CO RURAL $;
DATALINES;
1    35   1    2    YES
2    88   5    8    NO
3     5   0    1    YES
4    55   3    3    YES
5    75   4    5    NO
6   125   5    8    NO
7   225   7    9    YES
8   500  10   11    NO
;
PROC UNIVARIATE DATA=PROB5_11 NORMAL PLOT;
   TITLE "Checking the Distributions";
   VAR POP HOSPITAL FIRE_CO;
RUN;
PROC CORR DATA=PROB5_11 NOSIMPLE PEARSON SPEARMAN;
   TITLE "Correlation Matrix";
   VAR POP HOSPITAL FIRE_CO;
RUN;
```

Because of the outliers in the population variable, we prefer the Spearman correlation for this problem.

(d) We can use the output from UNIVARIATE to find the medians and do the recoding. In Chapter 6 we will see that PROC RANK can be used to produce a median cut automatically by using the GROUPS=2 option. For now, we will recode the variables using formats. You can also create new variables in the DATA step with IF statements.

```
PROC FORMAT;
   VALUE POP LOW-81  = 'Below median'
             82-HIGH = 'Above Median';
   VALUE HOSPITAL LOW-4  = 'Below Median'
                  5-HIGH = 'Above Median';
   VALUE FIRE_CO LOW-6  = 'Below Median'
                 7-HIGH = 'Above Median';
RUN;
PROC FREQ DATA=PROB5_11; ***Data set from above;
   TITLE "Cross Tabulations";
   FORMAT POP POP. HOSPITAL HOSPITAL. FIRE_CO FIRE_CO.;
   TABLES RURAL*(POP HOSPITAL FIRE_CO) / CHISQ;
RUN;
```

5.13 Line 1: Incorrect data set name, cannot contain a dash.

Lines 3-5: These lines will recode missing values to 1, which we probably do not want to do. the correct form of these statements is:

IF X LE 0 AND X NE . THEN X = 1;

Line 10: The options PEARSON and SPEARMAN do not follow a slash. The line should read:

PROC CORR DATA=MANY_ERR PEARSON SPEARMAN;

Line 11: The correct form for a list of variables where the "root" is not the same is:

VAR X -- LOGZ;

Remember, the single dash is used for a list of variables such as

ABC1 - ABC25.

CHAPTER 6

6.1

```
DATA HEADACHE;
   INPUT TREAT $ TIME @@;
DATALINES;
```

```
A 40 A 42 A 48 A 35 A 62 A 35
T 35 T 37 T 42 T 22 T 38 T 29
;
PROC TTEST DATA=HEADACHE;
   TITLE "Headache Study";
   CLASS TREAT;
   VAR TIME;
RUN;
```

Not significant at the .05 level (t = 1.93, p = .083).

6.3

```
PROC NPAR1WAY DATA=HEADACHE WILCOXON;
   TITLE "Nonparametric Comparison";
   CLASS TREAT;
   VAR TIME;
   EXACT WILCOXON;
RUN;
```

Sum of ranks for A = 48.5; for B, 29.5.
Exact two-sided p = .1385
Approximation using a normal approximation with a continuity correction z = 1.45, p = .146.

6.5

```
***Use a paired t-test;
   DATA PAIR;
      INPUT SUBJ A_TIME B_TIME;
   DATALINES;
   1 20 18
   2 40 36
   3 30 30
   4 45 46
   5 19 15
   6 27 22
   7 32 29
   8 26 25
   ;
   PROC TTEST DATA=PAIR;
      TITLE "Paired T-Test";
      PAIRED A_TIME * B_TIME;
   RUN;
   t = 3.00, p = .0199; drug B works faster.
```

6.7

```
PROC FORMAT;
   VALUE GROUP 0 = 'A' 1 = 'B' 2 = 'C';
RUN;
DATA RANDOM;
   INPUT SUBJ @@;
   GROUP = RANUNI(0); *NOTE: CAN ALSO USE UNIFORM FUNCTION;
DATALINES;
001 137 454 343 257 876 233 165 002
;
PROC RANK DATA=RANDOM OUT=RANKED GROUP=3;
   VAR GROUP;
RUN;
PROC SORT DATA=RANKED;
   BY SUBJ;
RUN;
PROC PRINT DATA=RANKED;
   TITLE "Listing of Subject Numbers and Group Assignments";
   FORMAT GROUP GROUP.;
   ID SUBJ;
   VAR GROUP;
RUN;
```

6.9 Line 11: Correct procedure name is TTEST.

CHAPTER 7

7.1

```
DATA BRANDTST;
   DO BRAND = 'A','N','T';
      DO SUBJ = 1 TO 8;
         INPUT TIME @;
         OUTPUT;
      END;
   END;
DATALINES;
8 10 9 11 10 10 8 12
4 7 5 5 6 7 6 4
12 8 10 10 11 9 9 12
;
```

```
PROC ANOVA DATA=BRANDTST;
    TITLE "Brand of Tennis Shoe Comparison";
    CLASS BRAND;
    MODEL TIME = BRAND;
    MEANS BRAND / SNK;
RUN;
```

$F = 28.89$, $p = .0001$; N is significantly lower than either T or A ($p < .05$). T and A are not significantly different ($p > .05$).

7.3

```
DATA BOUNCE;
    DO AGE = 'NEW', 'OLD';
        DO BRAND = 'W','P';
            DO I = 1 TO 5;
                INPUT BOUNCES @;
                OUTPUT;
            END;
        END;
    END;
    DROP I;
DATALINES;
67 72 74 82 81 75 76 80 72 73
46 44 45 51 43 63 62 66 62 60
;
PROC ANOVA DATA=BOUNCE;
    TITLE "Two-way ANOVA (AGE by BRAND) for Tennis Balls";
    CLASS AGE BRAND;
    MODEL BOUNCES = AGE | BRAND;
    MEANS AGE | BRAND;
RUN;
```

NOTE: A simpler INPUT statement could have been used:

INPUT BRAND $ AGE $ BOUNCES;

With the data listed one number per line such as:

W NEW 67

P NEW 75

etc.

Both main effects (AGE and BRAND) are significant ($p = .0001$ and $.0002$, respectively). The interaction is also significant, $p = .0002$.

7.5 (a)

```
DATA SODA;
   INPUT BRAND $ AGEGRP RATING  @@;
DATALINES;
C 1 7  C 1 6  C 1 6  C 1 5  C 1 6  P 1 9  P 1 8
P 1 9  P 1 9  P 1 9  P 1 8  C 2 9  C 2 8  C 2 8
C 2 9  C 2 7  C 2 8  C 2 8  P 2 6  P 2 7  P 2 6
P 2 6  P 2 5
;
PROC GLM DATA=SODA;
   TITLE "Two-way Unbalanced ANOVA";
   CLASS BRAND AGEGRP;
   MODEL RATING = BRAND | AGEGRP;
   MEANS BRAND | AGEGRP;
RUN;
```

(b)

```
PROC MEANS DATA=SODA NWAY NOPRINT;
   CLASS BRAND AGEGRP;
   VAR RATING;
   OUTPUT OUT=MEANS MEAN=;
RUN;
SYMBOL1 VALUE=CIRCLE I=JOIN COLOR=BLACK;
SYMBOL2 VALUE=SQUARE I=JOIN COLOR=BLACK;
PROC GPLOT DATA=MEANS;
    PLOT M_RATING*AGEGRP = BRAND;
RUN;
```

(c)

```
PROC SORT DATA=SODA;
   BY AGEGRP;
RUN;
PROC TTEST DATA=SODA;
   BY AGEGRP;
   CLASS BRAND;
   VAR RATING;
RUN;
```

7.7

```
PROC TTEST DATA=BRANDTST;
    WHERE BRAND = 'A' OR BRAND = 'T';
    /* Alternative: WHERE BRAND IN ('A','T'); */
    /* WHERE BRAND NE 'N'; is not as desirable, since
    in a more general data set, there may be missing
    or miscoded values */
    CLASS BRAND;
    VAR TIME;
RUN;
```

7.9 Line 4: Since this is a two-way unbalanced design, PROC GLM should be used instead of PROC ANOVA

7.11 (a)

```
DATA PROB7_6;
    DO GROUP = 'A','B','C';
        INPUT M_SCORE AGE @;
        OUTPUT;
    END;
DATALINES;
90 16 92 18 97 18
88 15 88 13 92 17
72 12 76 12 88 16
82 14 78 14 92 17
65 12 90 17 99 17
74 13 68 12 82 14
;
PROC ANOVA DATA=PROB7_6;
    CLASS GROUP;
    MODEL M_SCORE AGE = GROUP;
    MEANS GROUP / SNK;
RUN;
```

(b)

```
PROC GLM DATA=PROB7_6;
    TITLE "Testing Assumption of Homogeneity of Slope";
    CLASS GROUP;
    MODEL M_SCORE = AGE GROUP AGE*GROUP;
RUN;
```

Interaction term not significant. OK to do analysis of covariance

(c)

```
PROC GLM DATA=PROB7_6;
    TITLE "Analysis of Covariance";
    CLASS GROUP;
    MODEL M_SCORE = AGE GROUP;
    LSMEANS GROUP / PDIFF;
RUN;
```

In the unadjusted analysis, the groups are significantly different (p = .0479) and the ages are nearly significant (p = .0559). The null hypothesis that the slopes are equal among the three groups is not rejected (AGE*GROUP interaction p = .1790). Adjusting for age, the group differences on math scores disappear completely (p = .7606).

CHAPTER 8

8.1

```
DATA SHIRT;
    INPUT (JUDGE BRAND COLOR WORK OVERALL)(1.);
    INDEX = (3*OVERALL + 2*WORK + COLOR) / 6.0;
DATALINES;
11836
21747
31767
41846
12635
22534
32546
42436
13988
23877
33978
43887
;
PROC ANOVA DATA=SHIRT;
    TITLE "Problem 8-1";
    CLASS JUDGE BRAND;
    MODEL COLOR WORK OVERALL INDEX = JUDGE BRAND;
    MEANS BRAND / SNK;
RUN;
```

8.3

```
DATA WATER;
   INPUT ID 1-3 CITY $ 4 RATING 5;
DATALINES;
00118
00126
00138
00145
00215
00226
00235
00244
00317
00324
00336
00344
00417
00425
00437
00443
;
PROC ANOVA DATA=WATER;
   TITLE "Problem 8-3";
   CLASS ID CITY;
   MODEL RATING = ID CITY;
   MEANS CITY / SNK;
RUN;
```

8.5

```
PROC FORMAT;
   VALUE CITY 1 = 'New York'
              2 = 'New Orleans'
              3 = 'Chicago'
              4 = 'Denver';
RUN;
DATA PROB8_5;
   INPUT JUDGE 1-3 @;
   DO CITY = 1 TO 4;
      INPUT TASTE 1. @;
      OUTPUT;
   END;
   FORMAT CITY CITY.;
DATALINES;
                                              [Continued]
```

```
0018685
0025654
0037464
0047573
;
***Same PROC ANOVA statements as problem 8-3 except for the
Data Set Name;
***Solution using the REPEATED statement of PROC ANOVA;
DATA REPEAT;
    INPUT ID 1-3 @4 (RATING1-RATING4)(1.);
DATALINES;
0018685
0025654
0037464
0047573
;
PROC ANOVA DATA=REPEAT;
   MODEL RATING1-RATING4 = /NOUNI;
   REPEATED CITY;
RUN;
```

The unadjusted comparison shows that the cities are not all equal ($p = .0067$). Using the Greenhouse-Geisser correction, the p-value is .0375 and, using the Huynh-Feldt correction, the p-value is .0108. Therefore, you should feel comfortable in rejecting the null hypothesis at the .05 level, regardless of which correction (if any) you use.

8.7

```
DATA RATS;
    INPUT GROUP $ RATNO DISTAL PROXIMAL;
DATALINES;
N 1 34 38
N 2 28 38
N 3 38 48
N 4 32 38
D 5 44 42
D 6 52 48
D 7 46 46
D 8 54 50
;
PROC ANOVA DATA=RATS;
   CLASS GROUP;
   MODEL DISTAL PROXIMAL = GROUP / NOUNI;
   REPEATED LOCATION 2;
RUN;
```

Although the main effects are significant (GROUP p = .01, LOCATION p = .0308), the interaction term is highly significant (GROUP*LOCATION interaction F = 31.58, p = .0014). We should look carefully at the interaction graph to see what exactly is going on.

8.9 The DO loops are in the wrong order, and the OUTPUT statement is missing. Lines 2 through 8 should read:

```
DO SUBJ = 1 TO 3;
    DO GROUP = 'CONTROL','DRUG';
        DO TIME = 'BEFORE','AFTER';
            INPUT SCORE @;
            OUTPUT;
        END;
    END;
END;
```

There are no other errors.

CHAPTER 9

9.1

```
DATA TOMATO;
    DO LIGHT = 5,10,15;
        DO WATER = 1,2;
            DO I = 1 TO 3;
                INPUT YIELD @;
                OUTPUT;
            END;
        END;
    END;
    DROP I;
DATALINES;
12 9 8 13 15 14 16 14 12 20 16 16 18 25 20 25 27 29
;
PROC REG DATA=TOMATO;
    TITLE "Question 9-1";
    MODEL YIELD = LIGHT WATER;
RUN;
```

9.3

```
DATA LIBRARY;
    INPUT BOOKS ENROLL DEGREE AREA;
DATALINES;
 4    5   3    20
 5    8   3    40
10   40   3   100
 1    4   2    50
 5    2   1   300
 2    8   1   400
 7   30   3    40
 4   20   2   200
 1   10   2     5
 1   12   1   100
;
PROC REG DATA=LIBRARY;
    MODEL BOOKS = ENROLL DEGREE AREA / SELECTION=FORWARD;
RUN;
```

9.5

```
DATA PROB9_5;
    INPUT GPA HS_GPA BOARD IQ;
DATALINES;
3.9    3.8    680    130
3.9    3.9    720    110
3.8    3.8    650    120
3.1    3.5    620    125
2.9    2.7    480    110
2.7    2.5    440    100
2.2    2.5    500    115
2.1    1.9    380    105
1.9    2.2    380    110
1.4    2.4    400    110
;
PROC REG DATA=PROB9_5;
    MODEL GPA = HS_GPA BOARD IQ / SELECTION = MAXR;
RUN;
```

9.7

```
DATA PEOPLE;
   INPUT HEIGHT WAIST LEG ARM WEIGHT;
DATALINES;
(data lines)
;
PROC CORR DATA=PEOPLE NOSIMPLE;
   VAR HEIGHT -- WEIGHT;
RUN;
PROC REG DATA=PEOPLE;
   MODEL WEIGHT = HEIGHT WAIST LEG ARM /
   SELECTION = STEPWISE;
RUN;
```

You may also use FORWARD, BACKWARD, or MAXR instead of STEPWISE;

9.9 Ha! No errors here. As a matter of fact, you can use this program for problem 9.7.

9.11

```
PROC LOGISTIC DATA=SMOKING ORDER=FORMATTED;
   TITLE "SMOKING, ASBESTOS, AND SES IN THE MODEL";
   CLASS SES (PARAM=REF REF='2-Medium');
   MODEL OUTCOME = SMOKING ASBESTOS SES / CTABLE PPROB=0 TO 1 BY .1
                                          OUTROC=ROCDATA;
RUN;
SYMBOL1 V=DOT I=SM60 COLOR=BLACK WIDTH=2;
PROC GPLOT DATA=ROCDATA;
   TITLE "ROC Curve";
   PLOT _SENSIT_ * _1MSPEC_ ;
   LABEL _SENSIT_ = 'Sensitivity'
         _1MSPEC_ = '1 - Specificity';
RUN;
```

9.13

```
*Program Name: PROB9_13.SAS in C:\APPLIED
Purpose: Solution to homework problem 9-13
Date: April, 2004;
                                            [Continued]
```

```
PROC FORMAT;
   VALUE YES_NO   0 = 'No'
                  1 = 'Yes';
RUN;
DATA LOGISTIC;
   INPUT ACCIDENT DRINK PREVIOUS  @@;
   LABEL ACCIDENT = 'Accident in Last Year?'
      DRINK = 'Drinking Problem?'
      PREVIOUS = 'Accident in Previous Year?';
   FORMAT ACCIDENT DRINK PREVIOUS YES_NO.;
DATALINES;
1 0 1  1 1 1  1 1 1  1 0 0  1 1 1  0 0 1  0 1 0  0 1 0
0 0 1  0 0 0  0 1 0  0 0 1  0 1 0  0 0 0  0 0 0  1 1 0
1 0 1  1 1 1  1 1 1  1 0 1  1 1 1  1 1 0  1 0 1  1 1 0
1 1 1  0 1 1  0 1 1  0 0 1  0 0 1  0 1 0  0 0 0  0 0 1
0 1 0  0 0 0  0 0 0  1 1 1  1 1 0  1 1 1  1 1 0  1 1 1
1 1 1  1 1 1
;
PROC LOGISTIC DATA=LOGISTIC DESCENDING;
   TITLE "Predicting Accidents Using Logistic Regression";
   MODEL ACCIDENT = DRINK PREVIOUS /
      SELECTION = FORWARD
      RISKLIMITS;
RUN;
QUIT;
```

The logistic regression equation is:

$$\text{LOG(odds of accident)} = -1.9207 + 1.9559\ (\text{DRINK})$$
$$+ 1.7770\ (\text{PREVIOUS}).$$

The odds and probability of an accident for person 1 (no drinking history, no previous accidents) are .1465 and .1278 respectively. For person 2 (history of a drinking problem but no previous accident history), they are 1.0358 and .5088 respectively. The odds ratio is 1.0358.1465 = 7.07, which agrees with the PROC LOGISTIC output.

CHAPTER 10

10.1

```
DATA QUEST;
   INPUT ID 1-3 AGE 4-5 GENDER $ 6 RACE $ 7 MARITAL $ 8
      EDUC $ 9 PRES 10 ARMS 11 CITIES 12;
DATALINES;
```

```
001091113232
002452222422
003351324442
004271111121
005682132333
006651243425
;
PROC FACTOR DATA=QUEST
            ROTATE=VARIMAX
            SCREE
            NFACTORS=2
            OUT=FACT;
   TITLE "Example of Factor Analysis";
   VAR PRES ARMS CITIES;
   PRIORS SMC;
RUN;
```

10.3

```
DATA SCORE;
   ARRAY ANS[5] $ 1 ANS1-ANS5;
   ARRAY KEY[5] $ 1 KEY1-KEY5;
   ARRAY S[5] 3 S1-S5; ***Score array 1 = right,0 = wrong;
   RETAIN KEY1-KEY5;
   IF _N_ = 1 THEN INPUT (KEY1-KEY5)($1.);
   INPUT @1 ID 1-9
         @11 (ANS1-ANS5)($1.);
   DO I = 1 TO 5;
      S[I] = KEY[I] EQ ANS[I];
   END;
   DROP I;
DATALINES;
ABCDE
123456789 ABCDE
035469871 BBBBB
111222333 ABCBE
212121212 CCCDE
867564733 ABCDA
876543211 DADDE
987876765 ABEEE
;
PROC FACTOR DATA=SCORE
            OUT=FACTDATA
            NFACTORS=1;
```

[Continued]

```
    TITLE "Factor Analysis of Test Data";
    VAR S1-S5;
    PRIORS SMC;
RUN;
PROC PRINT DATA=FACTDATA;
    TITLE "Listing of Data Set FACTDATA";
RUN;
```

CHAPTER 11

11.1

```
*----------------------------------------------------------------*
| Program to score a five item multiple choice exam.             |
| Data: The first line is the answer key, remaining lines        |
| contain the student responses.                                 |
*----------------------------------------------------------------*;
DATA SCORE;
    ARRAY ANS[5] $ 1 ANS1-ANS5; ***Student answers;
    ARRAY KEY[5] $ 1 KEY1-KEY5; ***Answer key;
    ARRAY S[5] 3 S1-S5; ***Score array 1 = right,0 = wrong;
    RETAIN KEY1-KEY5;
    ***Read the answer key;
    IF _N_ = 1 THEN INPUT (KEY1-KEY5)($1.);
    ***Read student responses;
    INPUT @1 SS 1-9
          @11 (ANS1-ANS5) ($1.);
    ***Score the test;
    DO I = 1 TO 5;
       S[I] = KEY[I] EQ ANS[I];
    END;
    ***Compute Raw and Percentage scores;
    RAW = SUM (OF S1-S5);
    PERCENT = 100*RAW / 5;
    KEEP SS RAW PERCENT S1-S5; ***S1-S5 needed for 11-3;
    LABEL SS = 'Social Security Number'
       RAW = 'Raw Score'
       PERCENT = 'Percent Score';
DATALINES;
BCDAA
123456789 BCDAA
001445559 ABCDE
012121234 BCCAB
135632837 CBDAA
005009999 ECECE
789787878 BCDAA
;
```

```
PROC SORT DATA=SCORE;
   BY SS;
RUN;
PROC PRINT DATA=SCORE LABEL;
   TITLE "Listing of Student Scores in SS Order";
   ID SS;
   VAR RAW PERCENT;
   FORMAT SS SSN11.;
RUN;
PROC SORT DATA=SCORE;
   BY DESCENDING RAW;
RUN;
PROC PRINT DATA=SCORE LABEL;
   TITLE "Listing of Student Scores in Decreasing Order";
   ID SS;
   VAR RAW PERCENT;
   FORMAT SS SSN11.;
RUN;
```

11.3

```
PROC CORR DATA=SCORE ALPHA NOSIMPLE;
   TITLE "Computing KR-20";
   VAR S1-S5;
RUN;
PROC CORR DATA=SCORE NOSIMPLE;
   TITLE "Point-biserial Correlations";
   VAR S1-S5;
   WITH RAW; ***Same results if you use PERCENT;
RUN;
```

11.5

```
DATA KAPPA;
   LENGTH RATER_1 RATER_2 $ 1;
   INPUT RATER_1 RATER_2  @@;
DATALINES;
```

[Continued]

```
C C   X X   X X   C X   X C   X X   X X
C X   C C   X X   C C   C C   X X   C C
;
PROC FREQ DATA=KAPPA;
    TITLE "Inter-rater Reliability: Coefficient Kappa";
    TABLES RATER_1 * RATER_2 / AGREE NOCUM NOPERCENT;
RUN;
```

CHAPTER 12

12.1 (a)

```
DATA PROB12_1;
    INPUT GROUP $ SCORE;
DATALINES;
P   77
P   76
P   74
P   72
P   78
D   80
D   84
D   88
D   87
D   90
;
```

(b)

```
DATA PROB12_1;
    INPUT GROUP $ SCORE  @@;
DATALINES;
P 77 P 76 P 74 P 72 P 78
D 80 D 84 D 88 D 87 D 90
;
```

(c)

```
DATA PROB12_1;
    DO GROUP = 'P','D';
        DO I = 1 TO 5;
            INPUT SCORE  @@;
```

```
        OUTPUT;
    END;
END;
DROP I;
DATALINES;
77 76 74 72 78
80 84 88 87 90
;
```

(d)

```
DATA PROB12_1;
    DO GROUP = 'P','D';
        DO I = 1 TO 5;
            SUBJ+1;
            INPUT SCORE  @@;
            OUTPUT;
        END;
    END;
    DROP I;
DATALINES;
77 76 74 72 78
80 84 88 87 90
;
```

12.3

```
DATA PROB12_3;
    INFILE DATALINES DLM = ',';
    INPUT X1-X4;
DATALINES;
1,3,5,7
2,4,6,8
9,8,7,6
;
```

12.5

```
DATA PROB12_5;
    INFILE DATALINES DSD MISSOVER;
    LENGTH C $ 9;
    INPUT X Y C $ Z;
DATALINES;
```
                                                                    [Continued]

```
1,,"HELLO",7
2,4,TEXT,8
9,,,6
21,31,SHORT
100,200,"LAST LINE",999
;
```

12.7

```
DATA OFFICE;
    INFORMAT VISIT MMDDYY10. DX $10. COST DOLLAR8.;
    INFILE DATALINES MISSOVER;
    INPUT ID VISIT DX COST;
DATALINES;
1 10/01/1996   V075    $102.45
2   02/05/1997   X123456789 $3,123
3   07/07/1996   V4568
4   11/11/1996   A123    $777.
;
```

12.9

```
DATA PROB12_9;
    INPUT SUBJECT $ 1-3
          A1      $ 5
          X         7-8
          Y         9-10
          Z         11-12;
DATALINES;
A12 X 111213
A13 W 102030
;
```

12.11

```
DATA PROB12_11;
    INPUT @1 SUBJECT $3.
          @5 A1      $1.
          @7 (X Y Z) (2.);    /* OK to specify X, Y, and Z */
DATALINES;                    /* separately              */
```

```
A12 X 111213
A13 W 102030
;
```

12.13

```
DATA PROB12_13;
   INPUT @1   ID   3.
      @4    GENDER   $1.
      @10   (DOB VISIT DISCHRG)   (MMDDYY6.)
      @30   (SBP1-SBP3)   (3. + 5)
      @33   (DBP1-DBP3)   (3. + 5)
      @36   (HR1-HR3)   (2. + 6);
   FORMAT DOB VISIT DISCHRG MMDDYY8.;
DATALINES;
123M   102146111196111396   130 8668134 8872136 8870
456F   010150122596020597   220110822101028424012084
;
```

12.15

```
DATA PROB12_15;
   INPUT #1 @1   ID   2.
            @4   X   2.
            @6   Y   2.
         #2 @3   A1 $3.
            @6   A2 $1.;
DATALINES;
01 2345
   AAAX
02 9876
   BBBY
;
```

12.17

```
DATA PROB12_17;
   INPUT X Y  @@;
DATALINES;
```

[Continued]

```
1 2  3 4  5 6   7 8
11 12   13 14
21 22  23 24  25 26  27 28
;
```

12.19

```
DATA SURVEY;
    INPUT @12 TEST 1. @;
    IF TEST = 1 THEN
    INPUT @1  ID    $3.
          @4  HEIGHT 2.
          @6  WEIGHT 3.;
       ELSE IF TEST = 2 THEN
       INPUT @1 ID    $3.
             @4 AGE    2.
             @6 HEIGHT 2.
             @8 WEIGHT 3.;
    DROP TEST;
DATALINES;
00168155   1
00272201   1
0034570170 2
0045562 90 2
;
```

CHAPTER 13

13.1

```
LIBNAME A 'A:\';
DATA A.BILBO;
    INFILE 'A:FRODO' PAD; *(Don't forget the PAD);
    INPUT ID 1-3 AGE 5-6 HR 8-10 SBP 12-14 DBP 16-18;
    AVEBP = 2*DBP/3 + SBP/3;
RUN;
DATA A.HIBP;
    SET A.BILBO;
    WHERE AVEBP GE 100;
RUN;
/************************************************************
Alternative solutions using a WHERE data set option:
************************************************************/
DATA A.HIBP;
    SET A.BILBO(WHERE=(AVEBP GE 100));
RUN;
```

13.3

```
LIBNAME INDATA 'C:\SASDATA';
OPTIONS FMTSEARCH = (INDATA);
***Alternative is to use the default library name LIBRARY;
PROC FREQ DATA=INDATA.SURVEY ORDER=FREQ;
   TITLE "Frequencies for ICD_9 codes from the 1990 Survey";
   TABLES ICD_9;
RUN;
PROC MEANS DATA=INDATA.SURVEY N MEAN
     STD STDERR MIN MAX MAXDEC=2;
   TITLE "Descriptive Statistics for the Survey";
   VAR AGE;
RUN;
```

13.5

```
LIBNAME C 'C:\MYDATA;
DATA C.DEM_9697;
   IF END96 NE 1 THEN INFILE 'A:DEM_1996' END=END96;
   ELSE INFILE 'A:DEM_1997';
   INPUT @1   ID       $3.
         @4   AGE       2.
         @6   JOB_CODE $1.
         @7   SALARY    6.;
RUN;
```

13.7

```
DATA PROB13_4;
   INFILE 'B:SAMPLE.DTA' LRECL=320 MISSOVER;
   INPUT X1-X100;
RUN;
```

13.9

```
***DATA step to create MILTON;
DATA MILTON;
   INPUT X Y A B C Z;
DATALINES;
```

[Continued]

```
1 2 3 4 5 6
11 22 33 44 55 66
;
DATA _NULL_; ***No need to create a SAS data set;
    SET MILTON;
    FILE 'C:\MYDATA\OUTDATA';
    PUT @1 (A B C) (3.);
RUN;
```

CHAPTER 14
14.1 (a)

```
DATA GYM;
    LENGTH GENDER $ 1;
    INPUT ID GENDER AGE VAULT FLOOR P_BAR;
    ***GENDER is already declared a character variable by the LENGTH
        statement so a $ is not needed in the INPUT statement;
DATALINES;
3   M    8   7.5   7.2   6.5
5   F   14   7.9   8.2   6.8
2   F   10   5.6   5.7   5.8
7   M    9   5.4   5.9   6.1
6   F   15   8.2   8.2   7.9
;
```

(b)

```
DATA MALE_GYM;
    SET GYM;
    WHERE GENDER = 'M';
RUN;
```

or

```
DATA MALE_GYM;
    SET GYM(WHERE = (GENDER = 'M'));
RUN;
```

(c)

```
DATA OLDER_F;
   SET GYM;
   WHERE GENDER = 'F' AND AGE GE 10; ***IF statement OK;
RUN;
```

or

```
DATA OLDER_F;
   SET GYM(WHERE=(GENDER = 'F' AND AGE GE 10));
RUN;
```

14.3

```
DATA YEAR1996;
   INPUT ID HEIGHT WEIGHT;
DATALINES;
2 68 155
1 63 102
4 61 111
;
DATA YEAR1997;
   INPUT ID HEIGHT WEIGHT;
DATALINES;
7 72 202
5 78 220
3 66 105
;
DATA BOTH;
   SET YEAR1996 YEAR1997;
RUN;
```

14.5

```
DATA MONEY;
   INPUT ID INCOME : $1. L_NAME : $10.;
DATALINES;
```
[Continued]

```
3 A Klein
7 B Cesar
8 A Solanchick
1 B Warlock
5 A Cassidy
2 B Volick
;
PROC SORT DATA=GYM;
   BY ID;
RUN;
PROC SORT DATA=MONEY;
   BY ID;
RUN;
DATA GYMMONEY;
   MERGE GYM(IN=IN_GYM) MONEY;
   BY ID;
   IF IN_GYM;
RUN;
PROC PRINT DATA=GYMMONEY;
   ***Note: GYMMONEY already in ID order;
   TITLE "Listing of Gym and Financial Data";
   ID ID;
   VAR L_NAME GENDER AGE;
RUN;
```

14.7

```
PROC SORT DATA=BOTH;
   BY ID;
RUN;
DATA FREDDY;
   MERGE GYMMONEY(IN=ONE)
         BOTH(IN=TWO);
   BY ID;
   IF ONE AND TWO;
RUN;
PROC PRINT DATA=FREDDY NOOBS;
   TITLE "Listing of Data Set FREDDY";
RUN;
```

14.9

```
DATA FINANCE;
   LENGTH INCOME GENDER PLAN $ 1;
   INPUT INCOME GENDER PLAN  @@;
DATALINES;
```

```
A M W A F X B M Y B F Z
;
PROC SORT DATA=FINANCE;
   BY GENDER INCOME;
RUN;
PROC SORT DATA=GYMMONEY;
   BY GENDER INCOME;
RUN;
DATA FINAL;
   MERGE FINANCE GYMMONEY;
   BY GENDER INCOME;
RUN;
PROC PRINT DATA=FINAL NOOBS;
   TITLE "Listing of Data Set FINAL";
RUN;
```

14.11

```
DATA NEW;
   INFILE DATALINES MISSOVER;
   ***MISSOVER needed because of short lines;
   INPUT ID GENDER : $1. AGE VAULT P_BAR;
DATALINES;
3 . . .   6.7
5 . 15 8.1 7.2
7 F
;
PROC SORT DATA=NEW;
   BY ID;
RUN;
PROC SORT DATA=GYM;
   BY ID;
RUN;
DATA GYM_2;
   UPDATE GYM NEW;
   BY ID;
RUN;
PROC PRINT DATA=GYM_2 NOOBS;
   TITLE "Listing of Data Set GYM_2";
RUN;
```

An alternative way to create the update (NEW) data set is:

```
DATA NEW;
    LENGTH GENDER $ 1;
    INPUT ID= GENDER= $ AGE= VAULT= P_BAR= ;
DATALINES;
ID= 3 P_BAR= 6.7
ID= 5 AGED= 15 VAULT= 8.1 P_BAR= 7.2
ID= 7 GENDER= F
;
```

This is called NAMED input. You can learn more from the Online Doc™ or other SAS manuals;

CHAPTER 15

15.1

```
DATA PROB15_1;
    INPUT (HT1-HT5)(2.) (WT1-WT5)(3.);
    ARRAY HT[*] HT1-HT5;
    ARRAY WT[*] WT1-WT5;
    ARRAY DENS[*] DENS1-DENS5;
    DO I = 1 TO 5;
        DENS[I] = WT[I] / HT[I]**2;
    END;
    DROP I;
DATALINES;
6862727074150090208230240
64  68  70140   150    170
;
```

15.3

```
DATA OLDMISS;
    INPUT A B C X1-X3 Y1-Y3;
    ARRAY NINE[*] A B C X1-X3;
    ARRAY SEVEN[*] Y1-Y3;
    DO I = 1 TO 6;
        IF NINE[I] = 999 THEN NINE[I] = .;
    END;
```

```
   DO I = 1 TO 3;
      IF SEVEN[I] = 777 THEN SEVEN[I] = .;
   END;
   DROP I;
DATALINES;
1 2 3 4 5 6 7 8 9
999 4 999 999 5 999 777 7 7
;
```

alternative:

```
DATA OLDMISS;
INPUT A B C X1-X3 Y1-Y3;
ARRAY NINE[*] A B C X1-X3;
   ARRAY SEVEN[*] Y1-Y3;
   DO I = 1 TO 6;
      IF NINE[I] = 999 THEN NINE[I] = .;
      IF I LE 3 AND SEVEN[I] = 777 THEN SEVEN[I] = .;
   END;
   DROP I;
DATALINES;
1 2 3 4 5 6 7 8 9
999 4 999 999 5 999 777 7 7
;
```

15.5

```
DATA SPEED;
   INPUT X1-X5 Y1-Y3;
DATALINES;
1 2 3 4 5 6 7 8
11 22 33 44 55 66 77 88
;
DATA SPEED2;
   SET SPEED;
   ARRAY X[5] X1-X5;
   ARRAY Y[3] Y1-Y3;
   ARRAY LX[5] LX1-LX5;
   ARRAY SY[3] SY1-SY3;
```
                                                    [Continued]

```
    DO I =  1 TO 5;
        LX[I] = LOG(X[I]);
        IF I LE 3 THEN SY[I] = SQRT(Y[I]);
    END;
    DROP I;
RUN;
```

15.7

```
DATA PROB15_7;
    LENGTH C1-C5 $ 2;
    INPUT C1-C5 $ X1-X5 Y1-Y5;
    ARRAY C[5] $ C1-C5;
    ARRAY X[5] X1-X5;
    ARRAY Y[5] Y1-Y5;
    DO I = 1 TO 5;
        IF C[I] = 'NA' THEN C[I] = ' ';
        IF X[I] = 999 OR Y[I] = 999 THEN DO;
            X[I] = .;
            Y[I] = .;
        END;
    END;
    DROP I;
DATALINES;
AA BB CC DD EE 1 2 3 4 5 6 7 8 9 10
NA XX NA YY NA 999 2 3 4 999 999 4 5 6 7
;
```

CHAPTER 16

16.1

```
DATA FROG;
    INPUT ID X1-X5 Y1-Y5;
DATALINES;
01   4   5   4   7   3   1   7   3   6   8
02   8   7   8   6   7   5   4   3   5   6
;
DATA TOAD;
    SET FROG;
    ARRAY XX[5] X1-X5;
    ARRAY YY[5] Y1-Y5;
    DO TIME = 1 TO 5;
```

```
      X = XX[TIME];
      Y = YY[TIME];
      OUTPUT;
   END;
   DROP X1-X5 Y1-Y5;
RUN;
```

16.3

```
DATA STATE;
   INFORMAT STATE1-STATE5 $2.;
   INPUT ID STATE1-STATE5;
DATALINES;
1   NY   NJ   PA   TX   GA
2   NJ   NY   CA   XX   XX
3   PA   XX   XX   XX   XX
;
DATA NEWSTATE;
   SET STATE;
   ARRAY XSTATE[*] $ STATE1-STATE5;
   DO I = 1 TO 5;
      IF XSTATE[I] = 'XX' THEN XSTATE[I] =   ' ';
      STATE = XSTATE[I];
      OUTPUT;
   END;
   DROP I;
RUN;
PROC FREQ DATA=NEWSTATE ORDER=FREQ;
   TITLE "Frequencies on States Visited";
   TABLES STATE;
RUN;
```

16.5

```
DATA NEW;
   SET BLAH;
   ARRAY JUNK[*] X1-X5 Y1-Y5 Z1-Z5;
   DO J = 1 TO DIM(JUNK);
      IF JUNK[J] = 999 THEN JUNK[J] = .;
   END;
   DROP J;
RUN;
```

CHAPTER 17

17.1

```
DATA HOSP;
    INFORMAT ID $3. GENDER $1. DOB DOS MMDDYY8.;
    INPUT ID GENDER DOB DOS LOS SBP DBP HP;
    FORMAT DOB DOS MMDDYY10.;
DATALINES;
1 M 10/21/46 3/17/97 3 130 90 68
2 F 11/1/55 3/1/97 5 120 70 72
3 M 6/6/90 1/1/97 100 102 64 88
4 F 12/21/20 2/12/97 10 180 110 86
;
DATA NEW_HOSP;
    SET HOSP;
    LOG_LOS = LOG10(LOS); ***Part A;
    AGE_LAST = INT(YRDIF(DOB,DOS,'ACTUAL')); ***Part B;
    ***Alternative (YRDIF is preferred)
    AGE_LAST = INT((DOS-DOB)/365.25);
    X = ROUND(SQRT(MEAN(OF SBP DBP)),.1); ***Part C;
RUN;
```

17.3

```
DATA MANY;
    INPUT X1-X5 Y1-Y5;
```

(a)

```
MEAN_X = MEAN(OF X1-X5);
MEAN_Y = MEAN(OF Y1-Y5);
```

(b)

```
MIN_X = MIN(OF X1-X5);
MIN_Y = MIN(OF Y1-Y5);
```

(c)

```
CRAZY = MAX(OF X1-X5) * MIN_Y * (N(OF X1-X5) + NMISS(OF Y1-Y5));
```

(d)

```
IF N(OF X1-X5) GE 3 AND N(OF Y1-Y5) GE 4 THEN
   MEAN_X_Y = MEAN(OF X1-X5 Y1-Y5);
DATALINES;
1 2 3 4 5   6 7 8 9 10
3 . 5 . 7   5 . . . 15
9 8 . . .   4 4 4 4 1
;
```

17.5

```
DATA UNI;
   DO I = 1 TO 1000;
      N = INT(RANUNI(0)*5 + 1);
      OUTPUT;
   END;
   DROP I;
RUN;
PROC FREQ DATA=UNI;
   TABLES N / NOCUM TESTP=(.2 .2 .2 .2 .2);
RUN;
```

17.7

```
PROC FORMAT;
   VALUE DAYFMT 1 = 'SUN' 2 = 'MON' 3 = 'TUE' 4 = 'WED' 5 = 'THU'
      6 = 'FRI' 7 = 'SAT';
RUN;
DATA DATES;
   SET HOSP; ***From 17-1;
   DAY = WEEKDAY(DOS);
   MONTH = MONTH(DOS);
   FORMAT DAY DAYFMT.;
RUN;
PROC CHART DATA=DATES;
***Or GCHART;
   VBAR DAY MONTH / DISCRETE;
RUN;
```

17.9

```
DATA MIXED;
   INPUT X Y A $ B $;
DATALINES;
1 2 3 4
5 6 7 8
;
DATA NUMS;
   SET MIXED;
   A_NUM = INPUT(A,8.);
   B_NUM = INPUT(B,8.);
   DROP A B; ***Don't forget this;
RUN;
```

17.11

```
PROC FORMAT;
   VALUE AGEGRP LOW - < 20   = '1'
      20-40   = '2' /*OK SINCE INTEGERS*/
      41-HIGH   = '3';
RUN;
DATA NEWER;
   SET NEW_HOSP; ***From 17-1;
   AGEGROUP = PUT(AGE_LAST,AGEGRP.);
RUN;
```

CHAPTER 18

18.1

```
    DATA CHAR1;
    INPUT @1   STRING1 $1.
         @2   STRING2 $5.
         @3   STRING3 $8.
         @11 (C1-C5)($1.);
DATALINES;
XABCDE12345678YNYNY
YBBBBB12V56876yn YY
ZCCKCC123-/. ,WYNYN
;
DATA ERROR;
    SET CHAR1;
    LENGTH DUMMY $ 5;
    DUMMY = C1 || C2 || C3 || C4 || C5;
```

```
    IF VERIFY(STRING1,'XYZ') NE 0    OR
       VERIFY(STRING2,'ABCDE') NE 0   OR
       VERIFY(UPCASE(DUMMY),'NY') NE 0 THEN OUTPUT;
    DROP DUMMY;
RUN;
```

18.3

```
DATA PROB18_3;
    SET CHAR1; ***From 18-1;
    NEW3 = TRANSLATE(COMPRESS(STRING3,' -/.,'),'ABCDEFGH','12345678');
    IF VERIFY(COMPRESS(NEW3,'ABCDEFGH')) NE 0 THEN NEW3 =  ' ';
RUN;
```

18.5

```
DATA PROB18_5;
    SET CHAR1; ***From 18-1;
    ARRAY C[5] $ 1 C1-C5; ***Create a character array;
    DO I = 1 TO 5;
       C[I] = UPCASE(C[I]);
       IF VERIFY(C[I],'NY ') NE 0 THEN C[I] = ' ';
    END;
    DROP I;
RUN;
```

18.7

```
DATA PHONE;
    INPUT CHAR_NUM $20.;
    NUMBER = INPUT(COMPRESS(CHAR_NUM,' ()-/'),10.);
DATALINES;
(908)235-4490
(800) 555 - 1 2 1 2
203/222-4444
;
```

18.9

```
DATA EXPER;
    INPUT ID       $ 1-5
          GROUP    $ 7
          DOSE     $ 9-12;
    LENGTH SUB_ID $ 2 GRP_DOSE $ 6;
    SUB_ID = SUBSTR(ID,2,2);
    GRP_DOSE = GROUP || '-' || DOSE;
DATALINES;
1NY23 A HIGH
3NJ99 B HIGH
2NY89 A LOW
5NJ23 B LOW
;
```

18.11

```
DATA PROB18_11;
    SET EXPER; ***From problem 18.9;
    LENGTH ID2 $ 6;
    ID2 = ID;
    IF INPUT(SUBSTR(ID,4,1),1.) GE 5 THEN SUBSTR(ID2,6,1) = '*';
RUN;
```

18.13

```
DATA ONE;
    INPUT @1   GENDER     $1.
          @2   DOB    MMDDYY8.
          @10  NAME       $11.
          @21  STATUS     $1.;
    FORMAT DOB MMDDYY8.;
DATALINES;
M10/21/46CADY     A
F11/11/50CLINE    B
M11/11/52SMITH    A
F10/10/80OPPENHEIMERB
M04/04/60JOSE     A
;
```

```
DATA TWO;
   INPUT @1   GENDER        $1.
         @2   DOB      MMDDYY8.
         @10  NAME          $11.
         @21  WEIGHT        3.;
   FORMAT DOB MMDDYY8.;
DATALINES;
M10/21/46CODY     160
F11/11/50CLEIN    102
F11/11/52SMITH    101
F10/10/80OPPENHAIMER120
M02/07/60JOSA     220
;
DATA ONE_TMP;
   SET ONE;
   S_NAME = SOUNDEX(NAME);
RUN;
DATA TWO_TMP;
   SET TWO;
   S_NAME = SOUNDEX(NAME);
RUN;
PROC SORT DATA=ONE_TMP;
   BY GENDER DOB S_NAME;
RUN;
PROC SORT DATA=TWO_TMP;
   BY GENDER DOB S_NAME;
RUN;
DATA COMBINED;
   MERGE ONE_TMP(IN=INONE) TWO_TMP(IN=INTWO);
   BY GENDER DOB S_NAME;
   IF INONE AND INTWO;
RUN;
```

NOTE: There are no problems for chapters 19 and 20.

Index

Symbols